THE MANAGEMENT OF WORLD-CLASS MANUFACTURING ENTERPRISES

W. D. COMPTON

Prentice Hall
Upper Saddle River, New Jersey 07458

Library of Congress Cataloging-in-Publication Data
Compton, W. Dale
  Engineering management: creating and managing world-class operations/
  W.D. Compton
    p. cm.
  Includes bibliographical references and index.
  ISBN 0–02–324121-7
  1. Engineering--Management.   I.   Title
TA190.C65   1997   96–43445
658--dc20   CIP

Editor-in-chief: Marcia Horton
Production editor: Ann Marie Longobardo
Copyediting: Barbara Zeiders
Page composition: Ann Marie Longobardo, Joan Eurell
Director of production and manufacturing: David W. Riccardi
Art director: Jayne Conte
Managing editor: Bayani Mendoza DeLeon
Cover designer: Bruce Kenselaar
Manufacturing buyer: Donna Sullivan
Editorial assistant: Dolores Mars

© 1997 by Prentice-Hall, Inc.
Simon & Schuster / A Viacom Company
Upper Saddle River, NJ 07458

Printed in the United States of America.

10   9   8   7   6   5   4   3   2   1

ISBN 0-02-324121-7

Prentice Hall International (UK) Limited, *London*
Prentice Hall of Australia Pty. Limited, *Sydney*
Prentice Hall of Canada, Inc., *Toronto*
Prentice Hall Hispanoamericana, S.A., *Mexico*
Prentice Hall of India Private Limited, *New Delhi*
Prentice Hall of Japan, Inc., *Tokyo*
Simon & Schuster Asia Pte. Limited, *Singapore*
Editora Prentice Hall do Brasil, Ltda., *Rio de Janeiro*

PREFACE

The modern enterprise cannot be characterized by an organization chart, although organization is important to its success. It cannot be characterized by its technology, although technology is critical to its long-term success. It cannot simply be characterized by its present products or services, for it will survive only if it continues to offer products and services that satisfy the needs and wants of the customer. The successful modern enterprise will be characterized by a group of different characteristics, among them the responsiveness it displays to its customers, the attitudes that its employees have for the enterprise, and the leadership that its management provides. Achieving these characteristics in a world-marketplace environment is the challenge for modern management.

To be successful in the management of a modern enterprise requires that managers be leaders. They must be able to motivate people, build consensus, appraise the value of new ideas, and innovate. Success in accomplishing this often depends on the people skills of the manager. This, then, is a principal focus of this book–how to work with people, how to encourage them, and how to understand their reactions to various circumstances.

The managers' task does not end with people. Successful control of the physical system that generates the necessary added value, and in turn its sales and profits, is a critical element of managing a manufacturing system. There are many tools at the disposal of the manager for accomplishing this control. In this book we draw attention to only a selected number of these tools rather than attempting to survey the field. Modeling and simulation, statistical process control (SPC), MRP, MRPII, and JIT are examined.

This book focuses on approaches that are important in achieving a world- class manufacturing system–a system that can compete successfully in the world marketplace. The reader will recognize that the fundamental directions that must be followed can be applied to widely divergent situations, to the effective management of service activities, to the management of not-for-profit organizations, or to the management of government agencies. Although profits do not dominate the objectives of all of these organizations, it is clear that effectiveness, efficiency, and quality are of concern to all.

Management is not an exact science. Understanding the rationale for choosing a particular approach is best accomplished by discussion and debate. A frank and open exchange of ideas will be found to be an effective way to gain a better understanding of the importance of recognizing different points of view and to appreciate the concepts that are embedded in the foundations of good management practice. The questions at the end of the chapters are intended to encourage the examination of many of these concepts within contexts that are commonplace in the workplace. In considering these questions, students are encouraged to examine carefully the context in which a question is posed, identify possible reactions, and propose a solution that will be effective in assuring that the enterprise continues along the path of being world class. As is frequently the case, several solutions may be appropreate for some circumstances.

PROLOGUE

The issues that a manager regularly faces can be summarized as (i) managing the various elements of the enterprise, (ii) creating an atmosphere that ensures that the customer's wants/needs are considered of primary importance, (iii) controlling the physical system that is transforming material into a product or is generating a service, and (iv) assessing the future needs of the organization in terms of technology and new products. While the combination of talents needed to successfully accomplish this may appear daunting, an understanding of a limited number of fundamental concepts can provide the basis for becoming a successful manager. These fundamental concepts—ten in number—are called the Foundations of World Class Practice.[1] The emphasis on "foundations" conveys the idea that there are certain fundamental aspects of good management practice that is common to essentially all organizations. These are foundations. By world class is meant that some enterprise around the world is excelling in particular ways and, as such, deserves to be emulated, copied or, at the very least, understood. The Japanese describe the objective of achieving world class as dontotsu—translated to mean "being obviously number one" or "striving to be the best of the best." The view is taken in this book that in order to achieve "world class" the enterprise must embrace and practice the concepts contained in the following ten foundations.

The *Goals and Objectives* Foundation of World-Class Practice (p. 65)

"World-class manufacturers have established as an operating goal that they will be world class. They assess their performance....against their competition and against other world-class operational functions....They use this information to establish organizational goals and objectives, which they communicate to all members of the enterprise, and they continuously measure and assess the performance of the system against these objectives and regularly assess the appropriateness of the objectives to attaining world-class status."

The *Metrics* Foundation of World-Class Practice (p. 96)

"World-class manufacturers recognize the importance of metrics in helping to define the goals and performance expectations for the organization. They adopt or develop appropriate metrics to interpret and describe quantitatively the criteria used to measure the effectiveness of the manufacturing system and its many interrelated components."

The *Experimentation and Learning* Foundation of World-Class Practice (p. 96)

"World-class manufacturers recognize that stimulating and accommodating continuous change forces them to experiment and assess outcomes. They translate the knowledge acquired in this way into a framework, such as a model, that leads to improved operational decision making while incorporating the learning process into their fundamental operating philosophy."

---

[1] Heim, J. A. and W. D. Compton (eds.) 1992. *Manufacturing Systems: Foundations of World-Class Practice*, National Academy Press, Washington, DC:3-8 .

The *Employee* Foundation of World-Class Practice (p. 176)

"Employee involvement and empowerment are recognized by world-class manufacturers as critical to achieving continuous improvement in all elements of the manufacturing system. Management's opportunity to ensure the continuity of organizational development and renewal comes primarily through the involvement of the employee."

The *Management* Task Foundation of World-Class Practice (p. 197)

"Management is responsible for a manufacturing organization's becoming world class and for creating a corporate culture committed to the customer, to employee involvement and empowerment, and to the objective of achieving continuous improvement. A personal commitment and involvement by management is critical to success."

The *Organization* Foundation of World-Class Practice (p. 225)

"A world-class manufacturer integrates all elements of the manufacturing system to satisfy the needs and wants of its customers in a timely and effective manner. It eliminates organizational barriers to permit improved communication and to provide high-quality products and services."

The *Supplier or Vendor* Foundation of World-Class Practice (p. 225)

"A world-class manufacturer encourages and motivates its suppliers and vendors to become coequals with the other elements of the manufacturing system. This demands a commitment and an expenditure of effort by all elements of the system to ensure their proper integration."

The *Customer* Foundation of World-Class Practice (p.253)

"World-class manufacturers instill and constantly reinforce within the organization the principle that the system and everyone in it must know their customers and must seek to satisfy the needs and wants of customers and other stakeholders."

The *Describing and Understanding* Foundation of World-Class Practice (p.370)

"World-class manufacturers seek to describe and understand the interdependency of the many elements of the manufacturing system, to discover new relationships, to explore the consequences of alternative decisions, and to communicate unambiguously within the manufacturing organization and with its customers and suppliers. Models are an important tool to accomplish this goal."

The *Technology* Foundation of World-Class Practice (p. 375)

"World-class manufacturers view technology as a strategic tool for acheiving world-class competitiveness by all elements of the manufacturing organization. High priority is placed on the discovery, development, and timely implementation of the most relevant technology and the identification and support of people who can communicate and implement the results of research."

The importance and relevance of each of the Foundations will be treated in detail. You are urged to focus on them and to integrate them into your understanding of the needs that you, as a manager, will confront.

Contents

4. Operating Goals and Objectives   **71**

Part III   The Pathway to Current Management Practice   **105**

5. Early Management Focus—Historical Highlights   **107**

6. Current Management Focus–Participative Management   **132**

Part I Achieving an Effective Organization   **156**

7. The Employees   **157**

A Description of the Manufacturing System/Sector

PART 1

Part 1 of this book deals with the manufacturing system described so elegantly by Hanson (1992).

> The successful manufacturer will need to view itself from a new perspective. It will need to view itself in the broader context of a manufacturing enterprise and to understand that the factors that contribute to its manufacturing effort go far beyond the traditional production cycle. These factors encompass the entire range of activities that begin with market demand and end with customer satisfaction. Thus, the manufacturer must adopt a world-view in providing the utility that solves customers' problems. Developing this worldview begins by recognizing that even though all of the company's internal organizations–sales, marketing, engineering, and manufacturing–operate independently, they have a common focus and are committed to delivering customer satisfaction. But it cannot stop there. This entity must be expanded to include outside organizations such as suppliers, consultants, research centers, competitors, and the customers themselves. Each of these groups has resources and knowledge vital to the ultimate solutions. They all must be integrated into a cohesive "enterprise" working toward shared objectives. It is this *Integrated Enterprise* that allows both the manufacturer and customer to be successful.

In Chapters 1 and 2, some important factors will be examined that have led to the evolution of the modern manufacturing system and to the marketplace within which the system now must operate. In Chapter 1, the systems concept for manufacturing, including the interactions that the manufacturing operations have, in the classical sense, with other critical activities of the modern company is introduced. In Chapter 2, the importance of the environment within which the manufacturing activity operates, the definition of manufacturing, and measures of performance of the national manufacturing activity are discussed. We will also deal with the characteristics of the manufacturing enterprise, both historically and in terms of current trends, and the marketplace within which the current manufacturing enterprise operates.

1

The System Perspective of the Manufacturing Enterprise

System thinking finds its greatest benefits in helping us distinguish high- from low-leverage changes in highly complex situations. In effect, the art of systems thinking lies in seeing *through* complexity to the underlying structures generating change. Systems thinking does not mean ignoring complexity. Rather, it means organizing complexity into a coherent story that illuminates the causes of problems and how they can be remedied in enduring ways. (Senge, 1990)

The traditional view of manufacturing is that it deals with the transformation of materials–it is the combination of operations that receives the design of a product from *some*one and translates that design into a product that *some*one sells to *some* customer. The emphasis on "some" is meant to suggest that the manufacturing organization generally viewed the plant walls as circumscribing its responsibilities and opportunities. Thus, the manufacturing organization did not view itself as being responsible for understanding the demands of the customer. Neither did it perceive a role for itself in product design. It viewed its role to be much more limited–the making of the product at the lowest possible cost at some nominal level of quality.

Lest it be thought that the manufacturing organization was unique in taking such a limited view of its interest and responsibility, it is important to realize that most other parts of the organization acted similarly. The product development groups often had little concern for how their designs could be manufactured. They were concerned with performance. It was up to the manufacturing group to make the product. The marketing groups felt that it was their prerogative to determine the product attributes that the customer needed or wanted. The service group considered their domain of responsibility to be that of responding to the service needs of customers. Furthermore, management and workers did not trust each other and considered each other to have interests and intentions that were antithetical.

The result of this compartmentalization of responsibility within the organization created a situation in which barriers easily developed between groups. Competition for resources and for credit was inevitable. Communication among groups was difficult, with each group tending to develop its own vocabulary. The company became a collection of groups, frequently with incompatible goals and objectives, and frequently in competition.

> "The compartmentalization of knowledge" creates a false sense of confidence. For example, the traditional disciplines that influence management-such disciplines as economics, accounting, marketing, and psychology-divide the world into neat subdivisions within which one can often say, "This is the problem and here is its solution." But the boundaries that make the subdivisions are fundamentally arbitrary-as any manager finds out who attempts to treat an important problem as if it is purely "an economic problem," or "an accounting problem," or "a personnel problem." Life comes to us as a whole. It is only the analytic lens we impose that makes it seem as if problems can be isolated and solved. When we forget that it is "only a lens, we lose the spirit of openness. (Senge, 1990:283)

Poor communication, a lack of common organizational goals, and compartmentalization can have a debilitating effect on an organization. Product *redesigns* were often found to be necessary when it was discovered that manufacturing could not produce the first design within the cost objectives of the product. Changes in design to modify an attribute were often found to require changes in other features in order to be compatible. Engineering change orders (ECRs) were frequent during the launch of a new product. The control of costs was difficult. Manufacturing often found it difficult to provide a product with the desired level of quality at the desired cost objective. The time to bring a new product to market was often excessive.

This description of the environment prevailed throughout much of U.S. industry in the decades of the 1950s, 1960s, and 1970s. It would probably have persisted through the 1980s had U.S. industry not been challenged by more efficient manufacturers mostly in Japan–who had developed an approach that had almost none of the undesirable features noted above. In Japanese industry the design of a product was done jointly by groups with members drawn from marketing, engineering, and manufacturing. Everyone was expected to know and understand the wants and needs of the customer. Labor and management worked together as a team. Once production began, changes were few and usually relatively minor. The result of these arrangements was the creation of organizations that could bring a new product to the marketplace in a fraction of the time required by its U.S. competitor, at a lower cost, and at higher quality. With the evolving world marketplace, the Japanese manufacturer clearly had a significant competitive advantage.

The challenge presented by this entry of highly competitive products into the marketplace proved to be excruciatingly difficult for U.S. manufacturers, many of whom lost large segments of markets that they had previously dominated (e.g., in automobiles) or in some cases, entire markets, as in television sets and VCRs. The message was clear-change or perish. Although we will have occasion to identify many changes that U.S. firms made to cope with these challenges and to discuss the consequences of these changes, it is important to recognize that acceptance of these changes is not universal, even today. Not all companies or enterprises have embraced this need for change, nor have they put in place the procedures that are needed to ensure that further improvement will take place. Even in those companies that have attempted to revolutionize the approach they take to organization and employee participation, it is common to find a reluctance on the part of many employees to embrace this new way of working together.

> Although decisions to change management styles usually come from the top, the burden of carrying out these decisions falls upon middle managers. In most such situations, middle managers typically are asked to give up some of their power and status to win higher commitment from those who report to them. The fears of middle managers extend farther than a loss of the symbols of power and status. A flatter organizational chart means fewer supervisors and managers. Many middle managers feel that "participative" management styles mean that they will participate themselves right out of their job. (Yankelovich and Immerwahr, 1983)

It is a central thesis of this book that this new perspective is essential if an enterprise is to remain competitive and flourish in the evolving world marketplace.

1.1   THE STRATEGIC IMPORTANCE OF MANUFACTURING

Hayes and Wheelwright (1984) observe that

> If the manufacturing function assembles and aligns its resources through a cohesive strategy, it can play a major role in helping a business attain a desired competitive advantage. The notion that manufacturing can be a competitive weapon, rather than just a collection of rather ponderous resources and constraints, is not new, although its practice not very widespread.... most manufacturing-based businesses simply seek to minimize the negative impact that manufacturing can have. Even in many well-managed firms, manufacturing plays an essentially *neutral* role, reflecting the view that marketing, sales, and R&D provide better bases for achieving a competitive advantage.

The past decade has seen some significant movement away from this parochial attitude toward manufacturing. The more competitive companies have recognized the strategic importance of the manufacturing activity to their long-term health and well-being. This elevation of the manufacturing enterprise to a par with the other activities (e.g., design, engineering, development, sales, marketing, etc.) has been stimulated by the realization that many successful foreign companies view manufacturing as a competitive tool. Treating the manufacturing activity as one of several critical activities for achieving company objectives demands that a more unified and integrated view be taken of the entire enterprise. The systems view, as described in subsequent paragraphs, is consonant with this integrated approach.

1.2 THE MANUFACTURING SYSTEM

The manufacturing system is not simply an activity that operates within the boundaries of a plant, leaving it to other entities to decide what should be produced, how the product is to marketed, and what the strategy will be for servicing the product in the field. The manufacturing system receives inputs from the outside world in terms of orders, materials, and energy; it utilizes a set of resources to respond to those inputs; it transforms materials or components into a form that is needed or desired by a customer; and it operates within a set of constraints that are determined by the physical, financial, human, and political limitations of the environment.

Figure 1.1 presents a view of the many elements that play key roles in establishing the characteristics of the manufacturing system (Heim and Compton, 1992).

Operations are placed at the center of the enterprise, overlapping and interacting with the administration and management, the product and process engineering activities, the applied sciences, and the marketing, sales, and service activities. Overlapping and interacting with all of these functions are the customers for the products or services; the vendors and suppliers that provide materials, components and services to the enterprise; the community in which the enterprise exists; and the government that establishes regulations, rules and opportunities for the enterprise.

Figure 1.1 is intended to illustrate the interrelationships that exist in the manufacturing system. Although particular technologies are not identified in this description of the manufacturing system, it must be understood that they establish or enhance many of the capabilities of the various functions contained in the system....The applied sciences (shown as knowledge on Figure 1.1) provide the technical base for many of the areas. The material transformation processes, sometimes referred to as unit

Figure 1.1 Various elements of the manufacturing system included in the integrated enterprise. (From Heim and Compton, 1992:17)

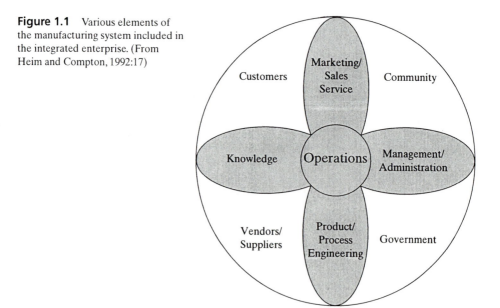

processes, are the means used to convert materials into components and subsystems. Computer-based systems provide the tools to enhance the capability and performance of the design, planning, scheduling, control, and sales of the products. The product and process engineering, the unit manufacturing operations, the marketing, sales, and services, the vendors and suppliers, and the management and administration each benefit from these systems and their capabilities to describe system performance. Describing the total enterprise in this way draws attention to the fact that no single unit operation or function can exist in isolation from all other components of the system....

An important conclusion to be drawn from this diagram is that if a manufacturing enterprise is to succeed, there can be no basic difference in viewpoints, values, and goals among its constituent groups. It is clear that the areas of responsibility are not neatly separated from one another but overlap to an important degree; financial and accounting systems, for example, have a major impact on operations and engineering. Perceived or artificially created boundaries between organizational units, such as those between marketing and engineering, production and purchasing, production control and marketing or employees and management, both restrict and complicate communication and cooperation. The performance of the system suffers....

As Krupka (1992) observes:

It is necessary to recognize that manufacturing operations–the activities that take place within the walls of a factory–can no longer be treated as the system to be optimized. Instead, in considering a manufacturer, we must think of several systems, of which manufacturing is but one....Customers' orders for products are conveyed by an ordering system to the manufacturing system, whose output then flows through a distribution system to the customer. Rapid and flexible response requires that materials and parts flow quickly into manufacturing; that requires a short and predictable interval for the material provisioning system. In addition to ensuring high performance for these systems, a successful firm must be capable of rapidly translating its designs into manufactured products. Hence, we need a well-crafted and rapid product introduction system.

In looking toward the future, Drucker (1990) offers the following predictions:

Today's factory is a battleship. The plant of 1999 will be a "flotilla", consisting of modules centered either around a stage in the production process or around a number of closely related operations. Though overall command and control will still exist, each module will have its own command and control.... This organization.... will allow rapid changes in design and product, rapid response to market demands, and low-cost production of "options" or "specials" in fairly small batches.

This organization requires...above all, different communication and information. In the traditional plant, each sector and department reports separately upstairs. And it reports what upstairs has asked for. In the factory of 1999, sectors and departments will have to think through what information they owe to whom and what information they need from whom. A good deal of this information will flow sideways and across department lines, not upstairs. The factory of 1999 will be an information network.

Consequently, all the managers in a plant will have to know and understand the entire process....In the factory of 1999, managers will have to think and act as team members, mindful of the performance of the whole. Above all, they will have to ask: What do the people running the other modules need to know about the characteristics, the capacity, the plans, and the performance of my unit? And what, in turn, do we in my module need to know about theirs?

1.3 INTERACTIONS AMONG THE ELEMENTS OF THE SYSTEM

The interactions that are possible among a set of activities, such as those depicted in Figure 1.1, are large in number and complex in form. It is these interactions that make the assessment, control, and management of the manufacturing system so complex and difficult. While managment objectives are frequently stated in terms that relate to optimizing certain measures of performance of the system (e.g., profit, quality, response to customer needs, etc.) the impact of changes in one area of the firm may not be appreciated properly in other segments. These interactions can often generate unforseen consequences, as General Motors discovered in a 1986 incident. In 1986 the car dealerships in General Motors were suffering sales declines and large inventories of unsold cars. In an effort to stimulate sales, GM offered customers an exceedingly low interest rate of 2.9 percent on the purchase of new cars. The result of this offering was a large demand for new products, as was hoped. What was not anticipated was that General Motors Acceptance Corp., the part of the GM organization that financed customer new car sales, did not have the financial capability to handle the large increase of sales at the very low interest rates. The result was that GM was forced to underwrite the low interest rates, even though this was normally the responsibility of GMAC. Actions by one part of the organization had unexpected and severe consequences on another part of the organization (*Wall Street Journal*, 1986).

In thinking about this in a mathematical sense, it is straightforward to show that the optimum of a combination of *independent* elements is the sum of the independently optimized parts. In other words, if there is no dependence among the elements that comprise the system, each can be treated as independent and each optimized independently. The combination–the system–will also be optimized for these conditions.

If the parts are not independent, this may also be true, but *only under very special conditions*. These conditions can be stated as follows: A system of interacting elements will be optimized by optimizing the individual elements if the system is describable by a function that is separable in terms of the parameters that are used to describe the system. This is discussed more fully in Appendix A1.1.

Unfortunately, the manufacturing system is seldom, if ever, simple enough to satisfy these conditions. No single element of the system exists in isolation from all other elements in the system. A change in one part of the system can produce a variety of results. For example, a change may have an important unexpected impact on the operation of another part of the system, or the operating conditions of the plant may prevent it from having any consequence. In this sense, the elements cannot be described by functions that are separable into independent functions. One cannot, in these circumstances, be assured that the optimization of each element of the system, taken independently, will lead to the optimization of the total. In fact, that is almost never the case.

The following simple example of a machining line illustrates these interactions. The amount of time that a part must wait in a queue before it can be machined depends on the availability of the machine to undertake the machining immediately. When the average time between successive arrivals is long (i.e., the arrival rate of parts at the machine is small) compared to the average time required to machine a part, the average waiting time will be small. As the arrival rate increases, thus reducing the average time between successive arrivals, the waiting time will increase and ultimately become very long as the machine utilization approaches, 100%. While, in principal, the arrival rate can be adjusted upward until the most heavily used machine has reached a utilization rate of 100%, the arrival rate cannot be increased further without producing a nonstable situation. Furthermore, as the utilization of one or more of the machines approaches 100%, the number of parts waiting to be serviced by these machines will grow increasingly larger. Thus, for a small increase in productivity, the plant may be forced to handle a large number of parts awaiting the availability of particular machines. Although it would be reasonable to state that the optimum that is desired for this line is a high level of productivity and a low work in process, it is seen that these two objectives are mutually incompatible. At high productivity the work in process is also high and at low productivity the work in process is low. It is obvious that there is no single point at which these two conditions are satisfied simultaneously.

As illustrated in this example, the result may be quite different from that which is desired. Since the increased arrival rate of parts into the system increase without saturating the system, this increase created more work in process and increased the amount of inventory being maintained in the plant. Unless the interactions among all elements of the system are known and understood, it may be impossible to forecast the impact that a given change will have on the system, even though its impact on one element of the system may easily be identified.

The optimization of a manufacturing system is not a simple problem. The usual approach to solving this problem is, first, to divide the system into individual parts, study the behavior of each part, and attempt to optimize each part as if its performance were independent of the performance of other parts. These initial solutions are then recombined, taking into account the interactions that exist among the parts. These interactions will, undoubtedly, affect the performance of the individual segments of the system.

The process is then repeated, starting with the understanding of how the interactions affected the earlier solutions. The system is again broken into parts, perhaps the same as before or perhaps different. Each is again assumed to be independent and a determination is made of a new optimum condition. The parts are then recombined along with the interactions among the parts and the total system examined again. These steps are repeated until a self-consistent answer is obtained. In this way it is hoped that a self-consistent answer is achieved that represents a solution to the total system problem. What becomes obvious is that an understanding of the nature and extent of the interactions becomes critical to this description. If a key interaction is omitted or not treated properly, an incorrect answer may result. Further, it is often difficult to ensure that the solution represents an optimal solution for all values of the variables. It may turn out that the solution that has been found is locally optimal but not globally optimal.

In the real world, it is necessary to emphasize on a continuing basis the importance of optimizing the total system and not allowing a preoccupation to develop with only seeking to improve the performance of each element. This demands that interactions among the various elements of the system be properly considered. Although this may appear to be completely obvious, the harsh reality is that many activities in an enterprise never explore the impact of their actions on other parts of the organization. For example, marketing may project that an increase in the opportunities that customers have to select product features–these are called product options–will expand the appeal and increase the share of the market that the product can achieve. Thus, they may appeal for increased option content. From the system view, the question must be asked what the impact of this increased complexity will have on other activities (e.g., manufacturing and servicing). Without this knowledge it is difficult to determine whether the value of adding the additional product will offset the increased cost to the total system.

This situation occurs in many of the high-option-content consumer products, of which automobiles provide perhaps the best example. U.S. manufacturers have traditionally followed a marketing strategy that offered a large number of options to the customer. An analysis of the frequency with which many of those options are actually chosen by customers reveals that a large fraction of the sales is contained within a small number of options. This is leading companies to concentrate more on packages of options rather than allowing the customer to have a nearly complete choice from a long list of options. Companies have found that they can offer fewer combinations at lower cost, thus benefiting both them and their customers. This, one should note, is the pattern that the Japanese companies have traditionally followed.

Whitney et al. (1988) offer an excellent example of the problems that can arise in not assessing the total impact of decisions on the combined system.

> An example will illustrate the type of problems that may arise when one unit is unaware of the needs imposed by other units on the design of a product. Consider a particular military product that depends critically on an infrared detector. The purchasing department switched to a lower-price vendor without determining the reproducibility of the detectors that would be provided. Although subtle differences between detectors can significantly affect performance, these cannot be found until the product is partially assembled with optics, power supplies, and so on. To increase ruggedness and reduce cost, the unit is glued together, making disassembly to replace the detectors very expensive. Naturally, the product could be redesigned with threaded joints to facilitate detector replacement as well as field repair. But the product in question is a single-use weapon. It must work the first time, its operating lifetime is only a few seconds, and its shelf life must be several years. Repair is simply not "in character" for this kind of an item. Thus, a decision to reduce costs by substituting one component for another can have consequences that can be appreciated only if the life history of the product is completely understood.

Why, then, do people not regularly consider the entire system when they are making decisions about individual elements of the operation? Many answers exist to this question. Unfortunately, it takes time to determine the system impact of a proposal. It is often necessary to discuss proposed actions with people in other segments of the company. It takes patience to explain the rationale for proposed actions and to enlist the support of other people. It takes courage to expose one's early plans to others, particularly when they may be in competition for resources to carry out their programs. Furthermore, if management has established a reward system that recognize the performance of the individual elements of the enterprise and not the total system, this will always encourage a tendency to optimize local performance even if it is to the detriment of the performance of the total system. Strong leadership is required to keep a constant focus on system issues.

The consequence of encouraging everyone to take actions that will optimize the system rather than just their area of responsibility has profound implications for an organization. Nothing can be more fundamental to the successful management and control of an enterprise than the appreciation of the many interactions that are possible and the need for understanding and controlling them. Describing the total enterprise in this way draws attention to the fact that the challenge to management is to find ways to take advantage of the strengths of the various elements and functional groups while discouraging any tendencies to work at cross purposes or toward conflicting goals. Achieving true involvement among the various activities requires, of course, more than simply reducing the barriers between groups. It requires an orientation away from the mentality of the "local" to the mentality of the "system."

QUESTIONS

1.1 What are some of the roles that the community and government will play in the system view of manufacturing as indicated in Figure 1.1?

1.2 What are some of the roles that vendors or suppliers and customers will play in the system view of manufacturing as indicated in Figure 1.1?

1.3 How are the policies and actions that are taken in achieving a Manufacturing System influenced by company size?

1.4 The optimization of a system is discussed in Appendix A1.1. In a practical sense, it is usually necessary to divide the total system into subsystems and to attempt to optimize each of these prior to recombining them into the total system. What are some rules that should be followed in dividing the system into subelements?

1.5 What are some of the effects of treating the manufacturing activity as having strategic importance to an organization?

1.6 Since the early 1950s, most U.S. corporations adopted a strategy that focused on achieving a competitive advantage in the marketplace through aggressive product design and unique product features. In recent years, Japanese industry has developed a strategy that combines product design and product features with an aggressive strategy focused on manufacturing. U.S. industry has found this to be a formidable challenge. Identify the circumstances under which one can compete successfully in the marketplace with a product while not being a leader in the processes that are used to manufacture that product. In contrast to this situation, what are the circumstances under which one must lead in the processes for manufacturing if one is to achieve market leadership with the product? Give examples of products and processes characteristic of each of these two circumstances.

1.7 You are manager of a research and development group in a large U.S. corporation. It is your task to support the worldwide product and manufacturing activity of the company. Each year you undertake to identify a number of opportunities that your activity will undertake during the coming year. What are the principal sources of information for identifying these opportunities? With whom do you discuss this, and what kinds of questions do you ask them as you attempt to identify opportunities? How will you choose among the many opportunities that are presented?

1.8 As the vice president for manufacturing, you have generally viewed your responsibility as ensuring that the manufacturing group has the capability to manufacture the necessary products in a cost-effective manner and to ensure a desired level of quality. You have always awaited receiving the future product plans before you decide what facility changes will be needed in the future. Upon returning from a company-wide seminar in which it was stressed that it will now be necessary to adopt a more "system" view of the company, you are contemplating how this will affect your activities. What are some of the things that will change in other parts of the company as a result of this new approach? How will you need to change to accommodate these new approaches? How is it likely to affect your operations?

READINGS

DRUCKER, P. F. 1990. The Emerging Theory of Manufacturing. *Harvard Business Review*, May-June, 1990:98-100.

HANSON, W. A. 1992. The Integrated Enterprise, *Manufacturing Systems: Foundationsof World-Class Practice,* J.A. Heim and W.D. Compton(eds.) National Academy Press, Washington, DC: 158-165.

REFERENCES

DRUCKER, P. F.1990. The Emerging Theory of Manufacturing. *Harvard Business Review*, May-June, 1990:98-100.

HAYES, R. H. AND D. C. WHEELWRIGHT. 1984. *Restoring Our Competitive Edge: Competing Through Manufacturing*, John Wiley & Sons, New York:35.

HEIM, J. A. AND W. D. COMPTON (eds.) 1992. *Manufacturing Systems: Foundations of World Class Practice*, National Academy Press, Washington, DC:15-18 .

KRUPKA, D. C. 1992. Time as a Primary System Metric in *Manufacturing Systems Foundations of World-Class Practice*, J. A. Heim and W.D. Compton (eds.), National Academy Press, Washington, DC:166-167.

SENGE, P. M. 1990. *The Fifth Discipline: The Art and Practice of the Learning Organization*, Doubleday Currency, New York:128.

Wall Street Journal. 1986. Buyers Respond to New GM Incentives: Questions Remain on Company Strategy, Aug. 29:2.

WHITNEY, D. E., J. L. NEVINS, T. L. DE FAZIO, R. E. GUSTAVSON, R. W. METZINGER, J. M. ROURKE, AND D. S. SELTZER. 1988. The Strategic Approach to Product Design in *Design and Analysis of Integrated Manufacturing Systems,* W. D. Compton (ed.), National Academy Press, Washington, DC 206.

YANKELOVICH, D. AND J. IMMERWAHR. 1983. *Putting the Work Ethic to Work: A Public Agenda Report on Restoring America's Competitive Vitality*, Public Agenda Foundation, New York: 37.

APPENDIX A1.1 SOME VIEWS CONCERNING SYSTEMS

The following discussion gives the views of early and later contributors to the concepts of how systems are to be considered. Churchman was one of the first to recognize the importance of the environment in the treating of the total system. It is interesting to note that he credits the philosopher Kant with concepts that form the basis for the modern methodology. O'Sullivan quotes Hegel as providing the bases on which General System Theory was developed.

As expressed by Churchman (1979):

> Industrial managers show a natural and strong concern about costs; to them the true costs of doing something (making a product, conducting a service, investing capital) are as crucial as the true measurement of a physical constant is to a physicist. Now industrial managers have lots of help in estimating costs; accountants, industrial engineers, and economists are the primary aides in this effort. But we found a peculiar aspect of the concept of cost relative to decision making. One of the first studies...was an attempt to determine the correct inventory of parts which a manufacturing firm should carry. The question can be stated in terms of costs: How much does it cost to carry an item in inventory, and how much does it cost not to carry it? These two costs can be broken down into subcosts, one of which ...is the cost of holding dollars in inventory. If an item lies on a shelf for sixty days, and I've spent ten dollars in making it, what does the idle time of the item cost me? If you think about this question from the point of view of the economics of the firm, the proper response seems to be that the firm has forgone the use of ten dollars for sixty days, because the part that sits provides it with no economic benefit.... If we accept the economic world view, then the problem of measuring the cost of keeping an item on the shelf depends on our ability to say how the inactive dollars should be spent if they had been available: that is, what is the best alternative use of the resource that has been captured by inventory? ...to be able to measure the cost of holding an item in inventory we must know how money should be spent by the firm: that is, we need to know how the financial system of opportunities should work....
>
> The cost data and other data in the systems approach are in effect "linkages" between systems; the cost of holding the dollars in inventory is a linkage between the inventory system and the financial system. Needless to say, the story of linkages doesn't end with just two systems; the financial cash flow system is linked to the marketing system, to the public relations system, to the personnel system, and so on....
>
> The systematic implication of the inventory models is that one cannot adequately determine inventory policy without determining financial and marketing policies; these latter policies will also require personnel, investment, and political policies for the firm. Thus, all so-called subsystems, such as inventory or service units, are really not "sub" at all and are strongly nonseparable from the whole system.
>
> Another, and perhaps more dramatic, way of putting this conclusion is to say that a necessary condition for solving the inventory or the waiting line problem is that one solve the firm's overall problem: if you really know how to control inventory and service, you know how to run the firm. The point is that is some general sense it does not matter where we begin in our search for ways of improving the human condition, because every specific problem unfolds into the general problem....

As expressed by O'Sullivan (1994):

> **Holism.** An open system is holistic, that is, it is an inseparable entity. A manufacturing system, for example, is comprised of many subsystems. To analyze any of these subsystems it is necessary to examine to some extent the whole. Dissecting the system and analyzing each subsystem individually causes loss of detail essential to the understanding of the system as a whole. Friedrich Hegel, the philosopher (1770 -1831) once wrote:
>
> 1. The whole is more than the sum of the parts.
> 2. The whole determines the nature of the parts.
> 3. The parts cannot be understood if considered in isolation from the whole.
> 4. The parts are dynamically interrelated and interdependent.
>
> His work is often cited as defining the origin of General System Theory (GST).

REFERENCES

CHURCHMAN, C. W. 1979. *The Systems Approach and Its Enemies*, Basic Books, Inc.: 16-17 and 46

O'SULLIVAN. 1994. *Manufacturing System Redesign: Creating the Integrated Manufacturing Environment*, Prentice Hall, Upper Saddle River, NJ:7

2

The Manufacturing Sector

> The physical and mechanical power which has enabled a community of woodcutters and farmers to become, in about 100 years, the greatest nation in the world, is the aggregate of the strong arms of men and women, aided by horsepower, machinery, and steam power, applied to the useful arts and sciences of everday life. (Mulhall, as quoted in Dodge, 1906)

Manufacturing refers to the many processes by which materials are transformed from one form into an alternative, with the alternative having a value to the customer that is inherently greater than that originally possessed *by the material*. The processes that are used may be simple or complex. They may require one or many steps. They may use a variety of materials. They may produce a product that is an assemblage of many sub-components and use a variety of materials. They may demand that materials and sub-components be handled many times and stored in many locations. They may require many people, some with highly perfected skills and some with lesser skills, to handle and work with the product. They may use testing to guarantee that certain levels of performance of the final product have been met. The emphasis that a given enterprise places on any of these steps depends on the industry in which it is located. In its totality, the manufacturing sector encompasses enterprises [*] that deal with all the elements by which materials are transformed into the many products that are offered to customers.

As a prelude to discussions concerning the management of manufacturing enterprises, it will be helpful to understand something about the impact of the sector on the U.S. economy, of the magnitude of the flow of goods among various countries that have been generated by manufacturing, and some of the trends that are taking place in manufacturing. We begin with a brief discussion of historical events that were critical in shaping the manufacturing sector.

---

[*] As used in this book, *sector* refers to the combination of all activities that are engaged in manufacturing. *Enterprise* is used synonymously with *company*.

2.1 DEVELOPMENT OF THE MANUFACTURING SECTOR

In surveying the evolution of the manufacturing sector in the United States, one can only marvel at the changes that have taken place in the past three centuries. The early economy of this country was dominated by the need to provide food and shelter to the population. In 1820, approximately 75% of the total population of 9.6 million were rural residents and were involved with the production of food for humans and the animals that served them. The nation was an agrarian society with manufacturing occupying the attention of only about 350,000 people. The classic rural manufacturer was the local blacksmith, who built and repaired many of the tools that the farmers used in tilling their fields. By 1880, the fraction of the population that were employed in agriculture had dropped to 50%, with a population that now totaled just over 50 million. The number employed in manufacturing had risen to over 3 million.

The industrial revolution in the United States followed on the heels of the revolution that began in England. Technology was being used to create new machine tools, which were, in turn, capable of making all new products. Carlsson (1984) notes that Watt's steam engine could probably not have been developed successfully had it not been for the earlier development of Wilkinson's new boring machine. The steam engine (James Watt, 1765), the water frame, which automated textile manufacture (Richard Arkwright, 1769), and the power loom (Edmund Cartwright, 1785) were early developments that enhanced the productivity of the textile industry, reduced the cost of goods, and made possible greatly expanded trade with the English. In the United States, the development of new machine tools appears to have been concentrated in specific industries, in particular the manufacture of small arms. Of the many important events that Carlsson identifies in his description of the machine tool industry, one is particularly interesting but was probably viewed as rather innocuous at the time. In 1895, the electric motor began to be used as a drive for individual machines, thus freeing the manufacturing operation from the complexity and rigidity that existed with the belt-and-pulley drive systems that operated from a central power source (Carlsson, 1984). With this development, the modern factory became possible.

Although the industrial revolution was in full force by the end of the eighteenth century, the proportion of the population that had become involved in manufacturing was only about 10%. The contribution of such a small fraction of the population to manufacturing is particularly impressive when one notes the estimates that the United States had become the sixth-largest industrial power of the developed world by 1830 (Kennedy, 1976). By 1870, manufacturing in the United States had reached a level that accounted for 23% of the *world* industrial production (Bairsch, 1982; Robertson, 1973).

The progress that has characterized the manufacturing sector has been substantial. Porter (1984) has described this rather vividly:

> Beginning with their origins in the federal armories of the early nineteenth century, Hounshell traces those interrelated technologies through a number of industries–the sewing machine, the woodworking industries, the reaper, the bicycle, and the automobile. The techniques were spread and improved by a close-knit network of key mechanics, who moved out in concentric circles from the armories to the machine tool companies, the sewing machine manufacturers, and the rest, all the way to the auto-

mobile makers. Mechanics who had learned parts of the growing body of ideas about how to create highly productive factories passed those ideas on to others, who, in turn, expanded and spread them. The extent of the linkages between the critical firms and individuals was astonishing. Over time, these men steadily widened the range of methods for manufacturing items of wood and metal. Factory organization, specialized machines, precision manufacture, interchangeable parts, carefully coordinated work sequences and materials flows, and new methods for stamping and welding metal eventually became standard items in the repertoire of America's production engineers.

Hounshell (1984) has described the progress made in manufacturing in the United States as follows:

> Ford's River Rouge plant....was the result of more than a century and a quarter of development since the United States Ordinance Department first committed the nation's resources to achieving uniformity in muskets and other small arms. The road from the Springfield and Harpers Ferry armories to Detroit was neither short nor direct. Though the advantages of the "uniformity principle" seemed obvious to the Enlightened, French-influenced officers of the Ordinance Department, and though they could draw extensively upon the public treasury, the actual attainment of uniform or interchangeable parts manufacture eluded almost two generations of the nation's most talented mechanics.
>
> The production of interchangeable parts entailed far more than its early proponents visualized, and it demanded more of them than anyone first expected. Not only did the armory mechanics have to figure out how their goal could be realized in theory but also in practice. This achievement required developing a system of jigs, fixtures, and gauges that were based on a "perfect" model and establishing procedures to maintain a constant vigil over the accuracy of these special devices and the machine tools with which they were used. Armory mechanics also pursued the development of machine tools. Their efforts paralleled those of other American inventors and mechanics in that the focus of much of their work was in the construction of special- or single-purpose machine tools.
>
> It was to these developments at Springfield that latter-day historians alluded when they wrote of the American system of manufactures. But to the mechanics of the nineteenth century, the developments at Springfield and other armories of special-purpose machines and tools; a model-based jig, fixture, and gauging system; and even the system of shop management were known simply as "armory practice". The mechanic who had been schooled in armory practice in the antebellum period became a fundamental agent in the diffusion of this technology to other manufacturing industries. Wherever armory practice was adopted, it was invariably the work of a mechanic or group of mechanics who had worked in the federal or the prominent private armories.
>
> In less than a decade, the Ford Motor Company brought together the best of armory practice with the rapidly developing techniques of pressed steel work. To these approaches, Ford production men added fresh thinking about work sequences, tool design, and controlling the pace of work in the Ford factory. When challenged to produce more of the standardized Model T's at lower cost and when allowed virtually free rein to experiment with new methods, the Ford team brought about a revolution in manufacture: mass production. The development of the moving assembly line was a key element in this revolution. But in addition, one must add Ford's ability to manufacture tremendous quantities of interchangeable parts, the installation of unprecedented numbers of highly specialized machine tools, and conveyor systems that were applied everywhere.

An abundance of natural resources and a growing market for goods that would assist in the expansion to the West created an incentive for the development of manufactured goods that would ever increase. Throughout the succeeding years, including periods of economic boom, great depressions, and wars, the economy has continued to grow. Improving the production efficiency of existing products and introducing new products and processes provided constant growth and stimulus to the economy, which, in turn, stimulated further development of manufacturing practice.

Historians have identified many events that have contributed importantly to the industrial development of this country. We will mention a few isolated events that have had an important impact on our industrial progress. These are chosen to illustrate the importance of various events, often having relatively minor impact at the moment, but whose long-term impact on development has been profound.

The successful drilling of the first commercial oil well in the United States near Titusville, Pennsylvania, in 1859, is mentioned first. Although there are many recorded uses of naturally occurring petroleum products–the use of pitch and asphalt is recorded in the Bible and natural gas is known to have been used by the Chinese as early as 1000 B.C.–this first commercially successful well in the United States is credited with launching the petroleum industry. The availability of a natural resource that could be readily transported, whose chemical form could be rather easily modified, and whose energy density (measured in Btu/barrel) was high, encouraged the development of new forms of mobile transportation. The diesel train locomotive, the internal combustion engine for cars and trucks, and the petrol-powered airplane all depended on this highly versatile energy source.

The second event that profoundly shaped the character of modern manufacturing was Orville and Wilbur Wright's first successful sustained flight of a powered airplane in 1903, at Kill Devil Hill, N.C., followed by the flight of a fully practical airplane in 1905. This event launched the aircraft industry and led to the development of low cost, reliable, fast air transportation. This, in turn, freed the manufacturer from the geographic constraints of local markets and made it possible to move products and supplies to markets located throughout the world.

The third event, occurring in 1948 in the Bell Laboratories, was the invention of the transistor. With this invention by Bardeen, Brattain, and Shockley, and the subsequent invention in the late 1950s by Noyes and Kilby of the integrated circuit, the information revolution was born. These events provided the capability to create computers that could manipulate vast amounts of data, that could measure, analyze, and control complex systems, and that could be used to describe analytically the performance of complex physical phenomena that might be harnessed for human improvement. It would be hard, if not impossible, to find any activity that has not, in some way, been affected by this event. There can be no doubt that this triggering of the information revolution created the capability by which manufacturers, designers, suppliers, and consumers can work together effectively over large distances.

Attention drawn to these events, not to ignore the importance of many others that contributed importantly to the evolution of our current social/political/economic environment, but to emphasize the impact that separate and largely nonpredictable events can have on the environment within which the manufacturing enterprise operates. Each of these have affected profoundly the way the manufacturing enterprise operates. All were done in areas that were not connected to manufacturing. Collectively, they and others have changed forever the markets that the manufacturing enterprise serves, and they have altered the role that people must play in operating these enterprises successfully.

There is an important lesson to be learned from this very truncated view of history. The environment that the manufacturing enterprise operates within has changed and continues to change, just as it was changed by action in 1859, 1903, and 1948. Our manufacturing enterprises operate within the general environment that is set by the society. This environment is complex and its impact on operations can be profound, as demonstrated by the many regulations that have been promulgated regarding waste disposal, emissions, trade agreements, and fiscal policy. The capability to influence the directions that future changes will take is, at best, limited. No manufacturing enterprise could have significantly altered the path that was created by the events noted above. None can completely control the future environment within which it will operate. If a modern enterprise is to be successful, it must develop the capability to sense changes, to adapt to them, and to be prepared to modify its structure radically to meet these new challenges. The essence of good management is to anticipate, to learn, and to lead the organization to capitalize on the opportunities that the changing environment provide. As Peter Drucker has said, "Change [is] normal and indeed [is] healthy...the major task of society–and especially in the economy–[is] doing something different rather than doing better what is already being done" (Drucker, 1985).

2.2 CATEGORIZING AND CHARACTERIZING THE MANUFACTURING ENTERPRISE

Many aspects of manufacturing refer to steps, operations, and procedures that are typically confined within the walls of the plant. Whereas the definition of manufacturing was expanded in Chapter 1 to include the many elements that interact and influence the efficiency of the entire manufacturing operation, the concerns in this section are limited to the actions and events taking place *within the walls of the factory*. Following is a classification of the many processes that are used in manufacturing:

1. Processes for changing physical properties
2. Processes for changing the shape of materials
3. Processes for machining parts to a fixed dimension
   a. Traditional chip removal processes
   b. Non-traditional machining processes
4. Processes for obtaining desired surface finishes
5. Processes for joining parts or materials

These processes are continually being modified, refined, improved, and increased in type and form. As new products become available or as new materials create additional product opportunities, it is often necessary to develop all new manufacturing processes. This can be seen for the wide range of composite materials-a physical mixture of two or more materials that are designed to provide a material with properties superior to that of either individual product-that are stimulating the development of a wide range of new processes designed to provide products with optimal characteristics[*]. A second example of new processes can be found in the semiconductor industry, where new processes are continually being developed that are tailored to the particular demands of the ever-smaller-scale structures that are characteristic of new integrated circuits.

While the key characteristics of manufacturing enterprises vary from one industry to another, a broad distinction is often made among the continuous process, batch process, and discrete parts industries. These descriptors are often used to describe roughly the petrochemical industry, the steel industry, and the automotive industry, respectively. In *continuous process industries*, as the name implies, processing of the material takes place in a continuous manner. A chemical process plant that converts a feed stock, such as the catalytic cracking of petroleum in a fluidized bed reactor, is an excellent example of a continuous process plant. Other examples of continuous flow processes can be found in the manufacture of sheet glass by the float glass process and the refining of metallic ores. While the processes are closely monitored and controlled at each step of the operation, there is little intervention into the process based on the attributes of the product itself. The attributes of the product are provided by control on the principal processes parameters (e.g., temperature, pressure, mixing time, flow rates, feedstock characteristics, catalyst characteristics, etc.). Furthermore, the operation of continuous processes is generally intended to be static over time; that is, the operations are expected to be in steady state.

In *batch process industries*, the product is treated in batch sizes. The characteristics of the batch may be repeated frequently, as occurs in the production of steel in the basic oxygen furnace, or they may be changed after each batch, as may occur in the formulation of batches of paints of unique characteristics. Another example of batch processing is heat treatment of a batch of parts in a furnace. An important characteristic of batch processes is that they operate as non-steady-state systems: that is, the transformation of the materials takes place over time, reaches some level of completion, and the final product is removed from the reactor and replaced by a new charge of original materials that then undergoes a similar transformation.

In *discrete parts industries*, the product is frequently composed of many components or subunits that are combined to provide the final products. Motors, automobiles, lamp bulbs, furniture, and clothing are examples of products that would be classified as belonging to the discrete parts industry. Assembly is usually a principal ingredient of a discrete parts manufacturing enterprise. Although it is far from a requirement, many of the assembly operations involve humans. An important characteristic of discrete processes is the variation in the times that occurs in moving from one station to the next station in the sequence and variabilities in the processing times at each station. While the sequence of events that are required will usually be specified, the stochastic nature of the processes introduces a major complication in terms of predicting the behavior of the system.

Law and Kelton (1982) define continuous and discrete systems in the following way:

---

[*] Composite materials are certainly not new. Wood and concrete are perhaps the most common examples of composites. Examples of more recent developments in composite materials include fiber-reinforced plastics, (e.g., glass fiber reinforced plastics) and fiber-and particle-reinforced metals, e.g., (aluminum containing fibers and particulates of Al_2O_3 or SiC).

A *discrete system* is one for which the state variables change only at a countable (or finite) number of points in time. A bank is an example of a discrete system, since its state variables, e.g. the number of customers in the bank, change only when a customer arrives or when a customer finishes being served and departs. A *continuous system* is one for which the state variables change continuously with respect to time. An airplane moving through the air is an example of a continuous system, since such state variables as position or velocity change continuously with respect to time.... a model of traffic flow on a freeway would be discrete if the characteristics and movement of individual cars were important. If the cars can be treated in the "aggregate," the flow of traffic can be described by differential equations in a continuous model.

As with most taxonomies, the classification into discrete and continuous is neither unique nor unambiguous. For example, an enterprise that makes bottles or cans for soft drinks, although creating discrete elements, generally operates in a fashion that approximates a continuous process. Similarly, although the manufacture of cars and trucks is usually considered to be a discrete process, the steel that is used will be made in batches, the body sheets in a rolling mill by processes that are continuous, and the engines by a series of discrete processes. As any chemical engineer will attest, the distinction between batch and continuous processes cannot be separated neatly. Both may change over time, processes that are expected to be steady-state must be started at some point in time, and a batch reactor can be considered as operating in a near steady-state if the characteristics of the reactor and its contents are always near equilibrium.

Although many of the methods and approaches that are used to manage and operate a manufacturing facility can be applied equally well to any of the manufacturing operations identified above, the analytical tools that are used in each of these are somewhat different. In the continuous and batch process industries, it is common to use analytical techniques that are derived from such basic sciences as thermodynamics, heat transfer, statistical mechanics, and hydrodynamics. As Law and Kelton pointed out in the quotation above, this usually means describing the processes by differential equations. In the discrete parts industry, the events occur stochastically, that is, at time intervals that vary. The analytical tools that are used for these systems are frequently derived from statistical, combinatorial, and optimization procedures. The types of problems that can be addressed by these techniques are discussed in future chapters.

In this book, attention will focus on the discrete parts industry. The modern discrete parts manufacturing enterprise has evolved through a number of forms and organizations. Remnants of each of these earlier forms can be found in modern systems. Although many elements of the organization and management of the enterprise are common to all types of manufacturing, the existence of certain unique characteristics of the manufacturing operation makes it necessary to adjust the organization and management practices to meet these special demands. It will be informative, therefore, to review briefly some of the more important organizations that have occurred in the evolution of the modern discrete manufacturing system. As noted above, all may exist in various forms today in certain limited circumstances.

2.2.1 The Artisan

The artisan was the first manufacturer. A person with the ability to translate the needs of a customer into a product through a particular set of processes that were under his direct control provided the artisan with a unique ability. In the early days, drawings were not used. It was the "artistry" of the individual that made these operations successful. As Voelcker (1988) notes:

> In the beginning, a product was its design–i.e., there were no explicit, separable specifications that could be called "a design for a product"; further, the designer and builder (and sometimes the customer) were one and the same.... As products became more elaborate, a class of artisan-designers (for example, the master cathedral builders and master shipwrights) became distinguishable from the artisan-builders and customer's specifications took on a more formal structure that evolved toward contractual descriptions and performance specifications.

2.2.2   The Job Shop

From the single artisan the manufacturing system began to evolve into a more complex structure that involved numerous artisans with individual specialties. In today's vernacular we describe these as being skilled tradespeople. Voelcker (1988) describes this evolution as follows:

> Through most of history, multiple copies of a product were almost always built by craft methods; i.e., single artisans or teams built a whole product from start to finish. This mode of fabrication, whether done in parallel....or singly for one-of products....may be termed adaptive serial assembly (colloquially, "file-and-fit") because parts were manufactured to fit the evolving assembly.

The job shop remains an important element of today's manufacturing sector. Throughout the world, speciality manufacturing shops can be found that employ a few too many workers, whose expertise is the making of a single or a few products to the specification of the customer, and who frequently use a rather restricted set of operations. Shops can be found that specialize in particular materials (e.g, plastics), in unique methods of joining (e.g, welding,) in particular metal removal processes (e.g., milling, drilling, or turning operations,) or in special surface treatment (e.g., surface carburizing or vapor disposition). The list is nearly endless.

The principal characteristic of these enterprises is their willingness and capability to make one-of-a-kind or small batches of products at a reasonable cost in a reasonably short time. While the equipment contained in a job shop is usually general purpose, in that it can be used in a variety of ways to make a variety of products, it is frequently constrained to do a limited set of operations. The job shop usually specializes in a limited set of operations and/or materials with which it works. Within these limitations, flexibility and responsiveness to the customer are the hallmark of the job shop.

2.2.3   The Flow Shop

As the markets for manufactured goods expanded, the need for greater and greater numbers of copies of a given product became a limiting factor. It was natural for the manufacturing system to evolve a structure that provided this capability. As Voelcker (1988) has noted:

> A major change came in the mid-nineteenth century, when the doctrine of gauge-based interchangeability began to be adopted.... Versatile artisans were replaced with specialist builders of standard parts and specialist assemblers and "gauge artisans" appeared as the precursor of manufacturing planners....

With the capability to specify tolerances and the capability to provide tools that could produce a large number of parts to those tolerances, the basis existed to group machines together to perform special tasks and then to develop the flow shop. The first example of grouping of tools to create a transfer line was in 1888 in watchmaking (Carlsson, 1984). In the flow shop, parts move throughout the plant, with operations being performed in sequence, frequently by different workers. The equipment in the flow shop is usually designed for a special purpose–to allow operation in the fastest and most reliable fashion many times in succession. As a result of concentrating their efforts on individual machines or groups of machines, the workers in flow shops often became very specialized in the tasks that they perform. This specialization, however, may be limited to doing a specific set of operations on a specific set of machines, as contrast to the skilled tradespeople who had developed a broad set of skills that can be adapted to enable them to operate a machine to accomplish a variety of tasks.

It was a logical step to move from this early configuration of the flow shop to the creation of the assembly line. The first stationary assembly line was created in 1899 by Ransom E. Olds for cars (Carlsson, 1984). Henry Ford introduced the moving assembly line in 1913. With the creation of the moving assembly line, the individual workers were required to do the same task repeatedly in a time frame that was determined by the movement of the parts along the line. Although the technical specialization that was characteristic of workers in a job shop is generally lost in the assembly line, the concept of the assembly line has provided a low-cost means of producing large numbers of nearly identical products. The assembly line, in various forms, is used in many discrete parts industries. Motors, cars, trucks, consumer electronic products, and consumer home products, for example, are generally created in this way. It is important to note that the specialization in equipment and its arrangement for a specific flow has the disadvantage of being inflexible to product change without alteration of the total system. Thus, the reduced cost of each part that the flow shop can achieve in large production can be offset if frequent changes are needed to provide a flexible group of products. Flow shops can also be found in the service industry (e.g., the check-processing operation of a large bank is a form of a flow shop).

2.2.4   The Automated Facility

The use of machines in circumstances where tasks are either dangerous or do not require the mental dexterity of human beings has long been the objective of manufacturers. Accomplishing this has been difficult, for it requires the development of systems that can begin to approximate the physical dexterity of a human being and have the controls necessary to assure the required repeatability in performance. The adaptability of the human being to new circumstances and our powers to perceive and analyze an unfamiliar situation and to reach a reasonable conclusion concerning necessary actions to take is still beyond the capability of machines. Nonetheless, great progress has been made in providing electro-mechanical and hydromechanical systems that can perform a wide variety of tasks that were formerly possible only with human labor. The robot has become an integral part of many manufacturing operations.

In the case of controls, digital controls were first demonstrated successfully in operating a three-axis milling machine in early 1952 (Reintjes, 1991). The first commercial numerically controlled (NC) machine tool was introduced in the late 1960s. The task of creating a program for control of these machines was onerous in the early days of this technology. With the development of relatively inexpensive programmable microprocessors, it was a natural step to combine this technology with that of numerically controlled machines. This created computer-numerically-controlled (CNC) machines.

A logical extension of the soft automation concept was to allow computers to make decisions concerning when and how a part should be made. This requires that computers be capable of scheduling the entry of jobs, planning the order in which certain machines will do specific tasks, and providing detailed instructions to each machine as to the operations that it is to undertake. Groupings of CNC machines into centers with flexible material handling systems, adaptive scheduling systems, automated tool cribs, and automated inspection systems has led to the recent development of flexible manufacturing systems (FMSs).

In approaching the development of physical systems with some of the dexterity of the human being, efforts have concentrated on the development of robots. The first robots were very inflexible. While capable of performing repetitive tasks with precision, they had little capability to adapt to changing circumstances. Current robots are frequently programmable–meaning that their tasks can be changed by altering their control software. They are used in many applications, including painting, welding, insertion of components, and testing. For highly repetitive tasks where human fatique may become a problem, robots are often used to assure that defects or failures in the operations are reduced.

Some degree of automation will be found in nearly every modern manufacturing facility. An important distinction must be made between hard automation and soft automation. *Hard automation* refers to processes that are automated through the use of mechanical systems that are designed expressly for that purpose. These systems are usually very inflexible, as is evident in such examples as overhead material conveyor systems, machining transfer lines, automated welding fixtures, and fixed inspection stations. In contrast, *soft automation* refers to automation that can be changed through a modification of software or controls. Automated guided vehicles for material handling, programmable robots, and controls that make the operation adaptive to part selection and orientation in an assembly operation are examples of soft automation.

2.2.5   The Intermediate Supplier

Earlier discussion focused tacitly on those entities that provide a final product. The modern manufacturing sector, however, contains many enterprises that supply intermediate elements of the finished products to a final product assembler or manufacturer. They provide basic materials such as iron and copper. They may specialize in a single item, such as wire or fasteners. They may provide one of the subcomponents (e.g., metal stampings, motor bearings, castings, or electronic components). They may be a supplier of a major subsystem that will be incorporated into the final assembly by the final assembler of the product, including, for example, a complete diesel engine for a truck, a turbine engine for an airplane, a display tube for a personal computer, or the energy source for a microwave oven. Some of these suppliers may be very large and may have facilities that are as complex and as costly as those of the final product assembler. Some may possess flow facilities. Some may have fully automated facilities. It is interesting to note that many specialty suppliers remain virtually unknown to the average person since their identity may not be made known to the final purchaser of the product. In other cases, the identity of the subcomponent supplier may be treated as a marketing advantage by the manufacturer (e.g., the air frame manufacturer may advertise that the aircraft engine is supplied by GE, Pratt and Whitney, or Rolls-Royce). Some issues that must be considered in determining the relative merit of buying components from a supplier versus manufacturing them are discussed in Section 9.6.

Because the final product is affected so strongly by the costs of purchases, the efficiency of those who supply the company with materials, components, subassemblies, and services is of great importance in determining the final cost of the product. A management team concerned with improving the efficiency of operations of a company must constantly encourage their suppliers to reach for the same level of efficiency that they are striving to achieve in their own operations. As one examines the efforts being made to improve the efficiency of all operations that affect an enterprise, it is critical that students constantly remind themselves that the boundaries of concern must extend well beyond the walls of the factory, to include suppliers of materials, parts, and components.

2.3   The Marketplace Environment

To better understand the environment within which manufacturing operates currently and can be expected to operate in the future, it is important to recognize the many stages through which the current marketplace evolved. While each of these developed in response to forces that were largely outside the direct control of the manufacturing sector, the experience that was derived from operating under their influence offers some insight into how enterprises can react in the future. As emphasized in various topics throughout this book, there are many advantages to retaining or recapturing some of these earlier aspects in the current environment.

2.3.1   The Local Market Served by a Local Manufacturer

In the early history of manufacturing, nearly every product that was manufactured was sold locally. In a sense, this represented a closely knit community of consumers, workers, and manufacturers. The management of the enterprise, being a part of that community, had ready access to the consumer and could easily participate in and respond to any changes that were taking place in the interests and needs of that community. In this environment the manufacturer had direct feedback from the customer. This continues to be the case for a special segment of the manufacturing sector (e.g., the builder of custom homes, local tradesmen, and services that are provided to the local populace). The current examples are far fewer, however, than would have been found in earlier times when the local communities tended to be more self-sufficient.

2.3.2 The National Market Served by a National System

With the industrial revolution came the railroad and new communication systems: the wireless, the telephone, and the postal service. These systems made it possible to move goods and information in a way that was not possible previously. The mail-order catalog created the first large-scale nonlocal market for goods and made it possible for manufacturers to reach consumers that were distantly removed from the source of the manufacturing. The manufacturer generally dealt with an intermediary who merchandised the product. In the early days, the marketer was often Montgomery Ward and Sears and Roebuck. In the early days of mail order, the novelty of being able to order goods was so great that the need to focus on the needs of the customer could be rather easily satisfied. In later years, to be successful the manufacturer needed to anticipate the needs of a consuming public–with whom they had limited personal contact–or they had to depend on the marketer to specify the products that would be offered to the consumer.

These early experiences with the opportunities and advantages of a national market led industry to develop other unique approaches in reaching their consumers. Many created a distribution system by which they sold their products through retailers in locations scattered throughout a region or the nation. Others depended on regional or national retailers who offered products from many manufacturers, including direct competitors. Some manufacturers have continued to approach localized markets (e.g., bottlers of drinks, specialized furniture manufacturers, and specialty food and drink producers). Just as many manufacturers now serve national markets (e.g., automobiles, kitchen appliance manufacturers, electronic equipment, and personal products), many companies engaged in services have also become national. Construction firms, architectural design firms, banks, and airlines operate at a national level by competing for projects or clients anyplace in the United States. Franchising has also evolved as an alternative way of creating local ownership while retaining a nationally recognized name and product. Although there are many other alternatives and combinations, higher education is somewhat unique in that it frequently operates in a national market by bringing students throughout the United States to a single local campus. While being national, in this sense, many universities also operate in a regional mode by providing campuses in various locations within their state boundaries.

2.3.3 The World Marketplace

As the capability to produce products continued to grow, the markets for those products expanded beyond the borders of one country. The capability to use the expertise developed in one country to generate products for customers in another country was recognized by early industrialists. The strategy that has been used by manufacturers to market their products in other countries has evolved over the years.

Henry Ford was one of the first industrialists in this country to expand into overseas markets. His philosophy, and for many who followed him, was to create a wholly owned subsidiary in the other country that would be essentially self-contained. The subsidiary designed, developed, manufactured, and marketed a product for the local national market. To the consumer, the existence of ties to the base company in another country was scarcely, if ever, visible. The subsidiary was expected to behave as a "good citizen" of the country in which it was located and to act in ways that were very similar to that of its competitors. The fact is that the ties were usually very loose, except for the repatriation of a portion of the profits to the parent firm. It is for these reasons that many British and Australians used to consider Ford Motor Company to be a company of British or Australian origin rather than an American company.

Such early multinational companies contributed some of the technology that was used, some of the management of the facilities, and the financial resources needed to make an overseas unit possible. The overseas unit often operated virtually independent of the parent. The value to the parent was the financial returns that it received from the overseas unit. These companies with their overseas subsidiaries were the first multinational companies. Although this pattern still exists to some degree, the modern multinational company is frequently more integrated than was the earlier multinational. It is not uncommon to find a company designing, developing, and manufacturing a product in one country and marketing that product in many locations throughout the world. This is a pattern that is being followed by many Japanese companies (e.g., the Japanese automotive and consumer electronics companies). Any unique feature that is needed for a particular market is provided by the manufacturer in Japan, (e.g., left- or right-hand-drive vehicles). This strategy is a modern phenomenon. Only in recent times has the technology for transportation become sufficiently dependable and been reduced to a low-enough cost that one could envisage shipping completed products around the world. An inexpensive transportation system coupled with highly reliable real-time communication systems has made this a realizable strategy for many manufacturers around the world. This capability is truly a product of modern technology.

There are both advantages and disadvantages to this strategy. Economically, the benefits to the home nation can be substantial. Not only does it benefit through the repatriation of profits from the sale of its products in the foreign country, but the localization of the manufacture in the home country increases the employment opportunities for its citizens. For a manufacturer, this strategy can eliminate some of the redundancies that would exist in the more dispersed organization as occurs if engineering and development groups are duplicated in each market. A single design group and a single development group should be more efficient than individual groups spread throughout the world. Furthermore, it should be easier to ensure that the latest technological developments are incorporated in the products that are marketed throughout the world. It should also be possible to benefit from economies of scale in manufacturing, since it is often possible to reduce costs by utilizing facilities more fully through the supply of multiple markets.

No strategy, unfortunately, has only positive benefits. On the negative side, it is usually more difficult to satisfy the various real or imagined needs of a particular market by a group of people who are removed from that market by large distances and separated by cultural backgrounds. It is easy for these barriers to become so dominant that the manufacturer loses contact with the customer in the more remote markets. Furthermore, it is often more difficult to develop an adequate service activity for the product when all of the other corporate activities are maintained oversees. Finally, there are potential political problems that can arise when a company retains all the manufacturing within their home country or moves it all to a market having low labor costs, thus depriving the workforce of the consumer nation access to jobs. It is interesting to note that Japanese industry is evolving somewhat toward the development of more complete subsidies in the foreign markets that it serves. Japanese automotive companies have established assembly plants overseas, thus providing some employment to people in the overseas market. It should be noted, however, that in many instances the high value-added components for the final assembly are still made in Japan. To become more sensitive to the customer, many Japanese companies are now locating design activities in their major overseas markets. They have not, however, taken the step of creating independent stand-alone companies in their overseas markets.

Irrespective of the strategy that is followed by a particular company, it is clear that the marketplace has assumed a character that can only be described as being a *world market*. Manufacturers can no longer expect that their products will be immune to competition from overseas. Simply because a local producer chooses not to export a product does not ensure that someone will not choose to enter the local market. The consequences of being forced to compete in a world marketplace are profound. The change in markets can be rapid because of the larger number of participants in the market. Products evolve through actions by distant manufacturers. New materials appear that make new products possible or allow improvements in existing products. New processes evolve, either to reduce cost, improve quality, or to accommodate new products or new materials. As competition increases with more and more manufacturers competing for markets, large homogeneous markets will probably fragment into smaller units, each with a specialization in particular items or features. Although specialization may offer some form of protection from competitors, it also carries the risk of becoming obsolete as the market moves away from a particular specialty product. Established markets may simply disappear when a new product or specialty enters the market. The current impact of the world marketplace on the United States is illustrated by the estimate that 70% of all goods manufactured in the United States face a competitive product that is manufactured overseas. Furthermore, this is not a situation that is unique to the United States. Essentially every developed country is experiencing a similar phenomenon.

Figure 2.1 is a dramatic demonstration of the effect of imports on the U.S. economy. The United States became a net importer of manufacturered goods in 1981. Between 1986 and the mid-1990s the excess of imports over exports varied between $100 billion and $150 billion annually. Figure 2.2 shows the share of the world export market that was held by the United States, Germany, and Japan in 1992 for a selected group of industry sectors. For those shown, the United States dominated in only the aircraft sector.

In addition to the consequences mentioned above, it is not uncommon to find countries taking actions to protect their indigenous industry from competition. This may take many forms, including:

- The establishment of import tariffs, thereby increasing the cost of selling a product in the host country and thus reducing its competitiveness relative to locally made products.

- The imposition of restrictions on the quantity of goods that can be imported into the country.

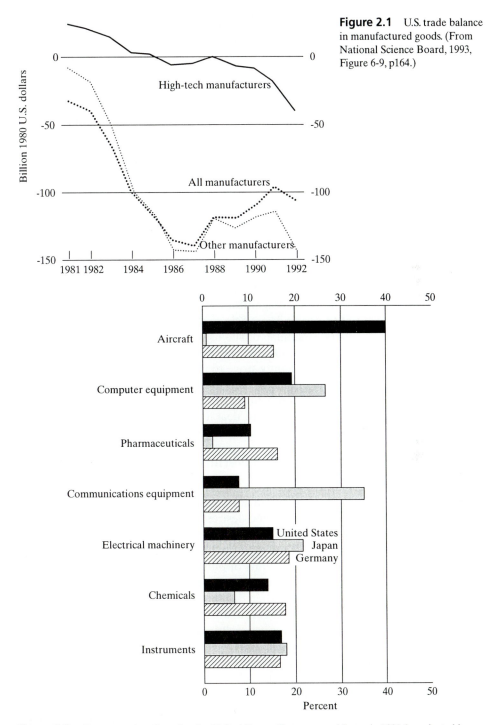

Figure 2.1 U.S. trade balance in manufactured goods. (From National Science Board, 1993, Figure 6-9, p164.)

Figure 2.2 Export market share for the United States, Germany, and Japan in 1992 for selected industry sectors. (From National Science Board, 1993, Figure 6-8, p163.)

- The imposition of local content requirements which state that the firm desiring to import into a country must manufacture a certain fraction of the components used in those imports within the country.
- The establishment of requirements for performance of the product that impose an additional cost on the company that is exporting the product.

Various international and bilateral agreements have or are being negotiated to reduce or eliminate barriers. This is, in part, the rationale for the creation of the various trading blocks. Goods are expected to flow between the United States and Canada freely as a result of the establishment of the U.S.-Canada free trade zone. The same is used to justify the U.S.-Mexico free trade zone. For the same reasons, the creation of the European Union is expected to enhance the uninhibited flow of goods among the member nations. The World Trade Organization (WTO) is intended to reduce tariffs on goods that are exported throughout the world by reducing or eliminating tariffs, local subsidies, and marketing restrictions.

The challenges and opportunities that the world marketplace poses to the manufacturing enterprise are enormous. While the world marketplace greatly expands the potential number of customers for a product, the challenges to operating successfully in this environment are very significant. If an enterprise is to survive and prosper, it must be aware of trends underway throughout the world. Information must be obtained and effectively assimilated. The organization must be prepared to encourage change and to welcome new ideas to take advantage of the changing opportunities that are identified. It must constantly strive to produce the highest-quality product, at the lowest cost, in the least amount of time, irrespective of who its current competitor is. It must be aware that a new competitor can appear at any time. These are some of the issues with which management must deal if they expect to remain competitive in the world marketplace. Meeting these challenges is the objective of this book.

2.3.4   The Philosophy of Government and Its Impact on the Marketplace

Various nations around the world have organized their economies in forms that they believe best support the needs and desires of their people. Some of these have a long tradition. Some are recent. Many are undergoing change. The United States has been the principal proponent throughout its history of the free market economy. While the political parties, the occupants of public office, economists, and political scientists often have detailed differences in their interpretation of this term, these differences are nearly insignificant when the differences between the philosophic approach of the United States and those of many of the other nations of the world are compared. *Free market economy* is generally taken to mean that the marketplace, *as it is embodied in the collective judgment of consumers*, is best able to judge the product that is of the greatest value to the consumer. In this context, the consumer may be an individual, a public entity such as a school or a government, or a company that is buying a product from another company. The fundamental philosophic precept in the free market economy is that an inefficient manufacturer or an out-of-date or inadequate product will not be sustained in the marketplace and that a more efficient manufacturer or a better product will prevail. In other words, failure is allowed and expected of those who cannot provide an adequate product for the consumer.

This philosophy does not, of course, mean that industry operates without restrictions. The government has introduced a vast array of rules and regulations that control interstate commerce, safety in the workplace, product safety, antitrust behavior, and antidiscrimination in the hiring and promotion of employees, to mention only a few. This "intervention" of the government into the marketplace is generally justified on the basis that a single manufacturer cannot be assured that a decision made for the general good of the public will optimize its performance and that if it does not, the same decision will be made simultaneously by all manufacturers, thus eliminating it as a competitive issue. As discussed more fully in a later section, this is an example of a situation in which optimizing the performance of a subunit of a system (i.e., the profit of an individual company, Appendix A2.1), does not necessarily result in optimization of the total system for the overall public good. Regulations are often imposed, therefore, to assure that all manufacturers, more or less simultaneously, adhere to some standard that is established for the benefit of all of the populace.

In principle, however, it is left to the marketplace to decide which company and which products perform best within the many legal guidelines that the legislative and legal system has established. This does not imply, of course, that the government remains unconcerned about the health of individual companies or of entire industries. The marketplace has often been affected by government actions to assist or protect certain industries or companies (e.g., the subsidies for farm goods, the guarantee of loans for Chrysler and Lockheed, the tariffs on certain textiles that are imported into the United States, etc.). It is an interesting consequence of the free market philosophy that industry generally decries the intervention of government when the industry is prospering, but argues forcibly for support and/or protection when it is doing badly. The challenge to government is to provide needed support without falling victim to the tendency to "support only the losers."

Perhaps the greatest contrast to the free market economy is that of centrally planned economies, exemplified by nations under Communist rule. In these political systems, the economy is planned centrally. Decisions concerning what, how much, who will make the products, the selling price of the products, and the manner in which they are distributed are decided centrally. The general philosophy is that through a form of central planning the general good can be served more efficiently than is possible in a free market economy. The clearest indication of the superior performance of the free market economy over the centrally planned economy is the failure of the centrally planned economies in Eastern Europe and the USSR. Although these nations may never fully embrace the form of free market economy as practiced in the United States, the current efforts to move toward more market-controlled economies are evident. We must await the future to determine how successful these efforts will be.

In the more socialist-oriented economies, the governments often choose to own those industries that are deemed to be of particular importance to the benefit of the public good. The public ownership often includes such industries as the railroads, the airlines, the telephone system, the television system, or the steel industry. Many other industries may be privately owned. Although the economies are not centrally planned and controlled as in the Communist nations, the mixture of public and private ownership of industry sectors may present challenging circumstances to industries operating in those countries. It is interesting to note that the competing political parties in many of these countries have chosen to adopt a philosophy of public or private ownership as one of the principal tenets of their philosophy, as, for example, between the Conservative and Labor parties of Great Britain. Thus, a change of governance in the country may lead to a radical change in the approach that the government takes toward the public ownership and operation of major industries.

Although the government in the United States does not own and operate large industries as is the case in some of the "planned" and socialist economies, it does offer some companies the right to operate as monopolies, in exchange for the right to regulate the services and profits of the monopolies. These may exist in some public utilities (e.g., electric power companies and the cable TV companies in many cities). Until recently, this was also the case with long-distance telephone service. Regulation may also occur in the setting of rates for certain services and the allocation of locations to be served. This was the case for the airlines prior to their deregulation in the mid-1980s and is currently the case with the electric utilities. The rate structure for the latter is controlled at the state level rather than at the federal.

From the "system" perspective, experience here and elsewhere demonstrates the difficulty of regulating the marketplace from a central control point. A nation's economy is composed of a vast array of elements that interact in complex and diverse ways. To presume that any individual or central agency understands sufficiently the many factors that determine the response of the system to changes in any one of them is almost beyond comprehension. Furthermore, a system can be controlled effectively only if it is possible to alter the stimuli or controls that are applied to it. Since many of the actions taken by government are frequently politically impossible to rescind at a later time (e.g., the difficulty that the federal government has experienced in attempting to reduce the level of farm subsidies significantly), it becomes a matter of great concern whether a systems view of the marketplace can be centrally taken.

Historically, government is most responsive when it is involved directly with its constituency. It is for this reason that local and state governments are frequently viewed as being more responsive to the needs of the public than is the federal government. It is also why experimentation and innovation appear to be most effective when carried out by state or local government. Mistakes in programs can be more easily corrected at the local or state level. New ideas can be tried on a small scale without the commitment of vast national resources. Learning can be accomplished more easily when bureaucratic involvement is least. This emphasizes the importance of state and local programs as a way of defining the extent of possible new federal programs.

2.4   MEASURES OF PERFORMANCE OF THE NATIONAL EFFORT

As society strives to improve the well-being of its populace, a constant concern deals with how well it is accomplishing these objectives. Various measures (metrics) are used to make this evaluation. Some are quite specific, such as the improvement in the average annual income of the citizenry. Others use per capita annual *gross domestic product* * as a measure. Still others use the average annual per capita income that is available for personal purposes, called disposable personal income. [†] In a sense, these and a variety of other metrics, offer a collective measure of what can be described as the "standard of living" of the populace. Table 2.1 offers a brief view of the trends in these statistics for the United States for the period 1929-1993. As is evident, the United States has been successful in achieving a tremendous improvement in its standard of living. The improvement in this measure of the "standard of living" of the average U.S. citizen was made possible by a rapid improvement in the average productivity of the workforce. Many factors, including availability of capital, abundant natural resources, the encouragement of entrepreneurial activities, and the availability of new technology, made this possible.

It is instructive to compare the more recent progress of the United States and several of the nations that are its principal trading partners. This is presented in Table 2.2 using a concept that is known as *purchasing power parity*. This calculation attempts to derive the implied purchasing power in various countries by determining the cost of a common collection of goods that are considered essential elements of existence. This process attempts to remove many of the factors that influence the exchange rates among countries (e.g., monetary policies) and disclose the cost in local currency that would be expected for the same quantity of commodities. By normalizing incomes using the relative magnitudes of the purchasing power parities, a more realistic comparison of the reported gross domestic product values can be made (Census, 1990:828).

The history of performance of the United States in improving the standard of living of its populace suggests that the United States has been less successful than has several of its major foreign competitors. Table. 2.3 presents the annual rate of change of productivity for various countries for the period 1980-1993. Although the absolute productivity of the U.S. populace remains high, the annual rate of improvement has been decreasing and is seen to have fallen below many of its competitors in recent times. The critical statistics on the rate of saving by individuals are also included in Table 2.3. The availability of capital for investing in new facilities and in productivity improving equipment is strongly influenced by the national savings rate, inasmuch as personal savings are a key source of funds for industry.

The message is clear. If the United States expects to retain the highest absolute standard of living it must continue to improve its overall productivity. This is the major challenge to the management of our enterprises, whether they be in manufacturing, agricultural, service, or government. Leading an enterprise to ever-higher levels of productivity is a challenge that all managers must accept if they aspire to excel.

Productivity is a useful measure of the efficiency with which goods are produced. A comparison of the gross domestic product that was produced by workers in the United States, Japan, Germany, and the other G-7 countries is presented in Figure 2.3. Productivity in Japan has been growing very rapidly. Statistics for the early 1990s suggest that the United States has achieved a higher rate of growth than Japan.

---

[*] "Gross Domestic Product (GDP) measures the output of production attributable to all labor and property located in a country. GDP relates to the physical location of the factors of production" (Census, 1990:421).

[†] "Personal income is the current income received by persons from all sources minus their personal contributions for social insurance" (Census, 1990:421). "Disposable personal income is personal income less personal tax and non-tax payments. It is the payments (net of refunds) by persons (except personal contributions for social insurance) that are not chargeable to business expense, and certain personal payments for general government that are treated like taxes" (Census, 1990:422).

TABLE **2.1**   **Per Capita Financial Statistics for the United States**

Year	Current Dollars			Constant (1987) Dollars	
	Gross domestic product	Personal income	Disposable personal income	Gross domestic product	Disposable personal income
1929	846	691	672	6,743	4,807
1930	734	611	594	6,079	4,402
1940	757	586	568	6,857	4,747
1945	1,523	1,215	1,068	11,453	6,367
1950	1,892	1,502	1,369	9,352	6,214
1955	2,446	1,903	1,693	10,699	6,842
1960	2,840	2,264	1,994	10,903	7,264
1965	3,616	2,845	2,527	12,712	8,508
1970	4,928	4,052	3,521	14,013	9,875
1975	7,343	6,053	5,329	15,036	10,906
1980	11,892	9,948	8,576	16,584	12,005
1985	16,933	14,170	12,339	17,944	13,258
1988	19,994	16,630	14,477	19,252	13,890
1990	22,189	18,699	16,205	19,593	14,101
1991	22,647	19,196	16,741	19,238	13,965
1992	23,637	20,139	17,615	19,518	14,219
1993	24,697	20,864	18,225	19,888	14,330

Source: Census, 1994. Adapted from Table 691, p. 451.

TABLE **2.2**   **Per Capita Gross Domestic Product for Selected Countries Using Purchasing Power Parities**

Per Capita Gross Domestic Product (dollars)						
Country	1970	1980	1985	1989	1990	1991
United States	4,933	11,891	16,786	20,920	21,866	22,204
Australia	3,468	8,597	12,326	15,261	15,887	16,085
Austria	3,004	8,664	12,156	15,401	16,596	17,280
Canada	3,571	10,082	14,632	18,629	19,049	19,178
France	3,499	9,247	12,785	16,310	17,301	18,227
Greece	1,442	4,247	5,797	7,176	7,403	7,775
Italy	2,962	8,317	11,585	15,046	16,012	16,896
Japan	2,798	7,871	12,035	16,130	17,645	19,107
Netherlands	3,534	8,847	11,949	14,821	15,951	16,530
Spain	2,239	5,814	7,953	10,880	11,738	12,719
Sweden	3,788	9,073	13,008	16,188	16,881	16,729
Turkey	597	1,595	2,340	3,017	3,363	3,456
United Kingdom	3,219	7,935	11,473	15,177	15,866	15,720
West Germany	3,678	9,660	13,535	17,020	18,307	19,500

Source: Census, 1994. Adapted from Table 1370, p.864.

Table 2.3 Selected International Economic Indicators

Country	1980	1985	1990	1991	1992	1993
GDP Growth Rates (Constant 1987 Prices)						
United States	-0.5	3.2	1.2	-0.7	2.6	3.0
Canada	1.5	4.8	-0.2	-1.7	0.7	2.4
France	1.6	1.9	2.5	0.8	1.2	-1.0
Italy	4.1	2.6	2.1	1.2	0.7	-0.7
Japan	3.6	5.0	4.8	4.3	1.1	0.1
Netherlands	0.9	2.6	4.1	2.1	1.4	0.2
United Kingdom	-2.2	3.8	0.4	-2.2	-0.6	1.9
West Germany	NA	NA	NA	1.0	2.1	-1.2
Percent of Disposable Personal Income Devoted to Savings						
United States	7.9	6.4	4.2	4.8	5.3	4.0
Canada	13.3	13.1	9.7	10.1	10.6	10.5
France	17.6	14.0	12.5	13.1	13.9	14.2
Italy	21.9	18.9	18.4	18.6	18.7	NA
Japan	17.9	15.6	14.1	14.9	16.0	NA
Netherlands	11.0	13.1	16.5	12.5	13.2	NA
United Kingdom	13.1	10.7	8.6	10.1	12.3	11.5
West Germany	NA	NA	NA	13.8	13.9	13.2

Source: Census, 1994. Adapted from Table 1367, p.863.

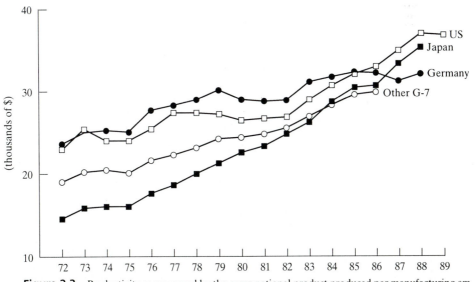

Figure 2.3   Productivity as measured by the gross national product produced per manufacturing employee (in constant 1980 dollars). (From Council on Competitiveness 1991and OECD National Accounts.)

2.5   TECHNOLOGY: AN OPPORTUNITY?

Finally, there is an important trend in the use of technology. As products and processes proliferate, so does the demand for the identification, development, and incorporation of new technologies into the manufacturing system. The enterprise that can accomplish this the most effectively will, undoubtedly, have a competitive advantage in the marketplace.

As noted above, a single technological achievement can have a profound impact on a sector of the economy. Consider the pervasive influence of the transistor. The impact that integrated circuits, derived from the technology of the single transistor, have on the electronics sector of the economy and the large contribution that the electronics industry makes to the U.S. gross domestic product suggest that the health of these industries–semiconductors and electronics–must be considered as critical to the United States. The electronics industry is currently the largest single employer in the United States. The comparison of U.S. performance relative to that of its most significant competitor signals the reason for much of the concern with our competitive status in these two areas. Figure 2.4 displays the large use of semiconductors in the electronics market. Figure 2.5 displays the trends in world market share for the semiconductor industry. It can only be viewed as ominous that the U.S. share of the semiconductor industry is continuing to drop while the use of semiconductors continues to grow. Figure 2.6 gives the level of exports by the United States, Germany, and Japan for high-tech products. The individual elements that are included in this definition of the high-tech segment of the market are displayed in Figure 2.7 for the U.S. global market share of these industries.

While technological improvements in the manufacturing enterprise are strongly influenced by electronics and computers, they are certainly not confined to these industries. Improvements in the unit processes, those individual operations that are found throughout the manufacturing process, has also been profound. New tools, entirely new processes, and new materials are among the innovations that have transformed the manufacturing enterprise. It is tempting to ask whether these technological developments have been the principal basis for the increase in productivity that we have experienced. Although the answer is yes, in part, it must be recognized that many nontechnical factors have also contributed to the improvements in productivity. A better appreciation of the factors that comprise good management practice, the training and nurturing of the workforce, and the importance of understanding the interactions of the subelements of the manufacturing enterprise are but a few of the topics that are critical to high productivity. These and many others are treated in detail in subsequent chapters.

2.6   CURRENT TRENDS AND RESPONSE

The flow shop led directly to mass production and the capability to produce large numbers of a product in a cost-effective manner. Although this approach to production was a key ingredient in the development of the United States as a world leader in manufacturing, it has become obvious that it has an enormous associated cost. Mass production implies that many customers will want the *same* product, that the volume of a single product–or class of products–that will be sold must be sufficiently large to justify the costs of the specialized facilities that are needed, and that the time interval between product changes will be long. In the world marketplace in which manufacturers must currently compete, these costs are becoming very difficult to justify.

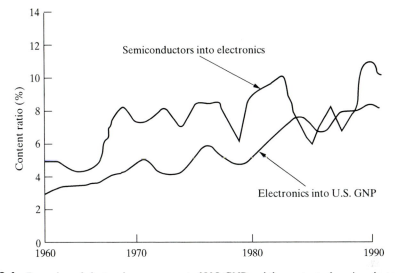

Figure 2.4   Pervasion of electronics as a percent of U.S. GNP and the content of semiconductor integrated circuits as a percent of the electronics market. (From U.S. Department of Commerce data.)

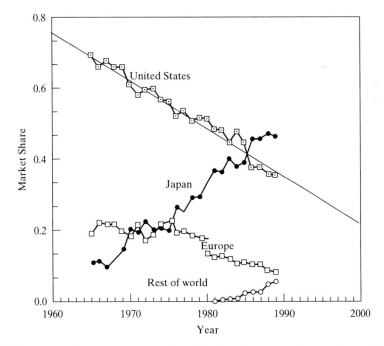

Figure 2.5   Semiconductor market share of world semiconductor merchant production. (From U.S. Department of Commerce data.)

Billions of constant 1980 dollars

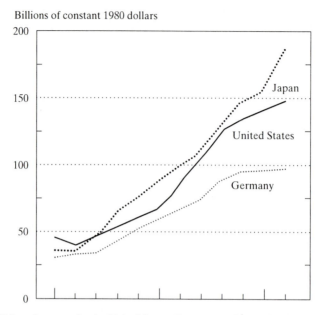

Figure 2.6   High-tech exports by the United States, Germany, and Japan in billions of constant 1980 dollars. (From National Science Board, 1993, Figure 6-7, p.163.)

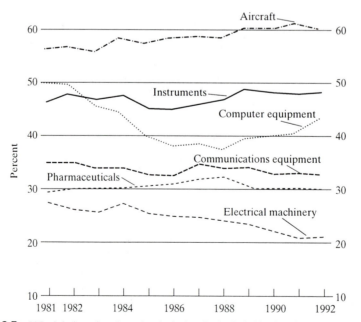

Figure 2.7   U.S. global market share for the industries included in the high-tech sector in Figure 2.6. (From National Science Board, 1993, Figure 6-5, p. 162.)

Product changes are appearing with startling rapidity as customers demand that their special needs be met by products with special attributes. Manufacturers who can bring a product to the marketplace in the shortest time find that they have a definite advantage, for they will be able to meet the changing demands of the customer and to anticipate future demands. The hallmark of the future will be responsiveness to a rapidly changing marketplace, sensitivity to the changing demands of the customer, capability to be profitable with production runs that are small, products that have very high quality, and facilities that have the flexibility to make a variety of products. These characteristics are variously described as *lean manufacturing* or *agile manufacturing*. The discrete parts manufacturer that has the greatest flexibility and responsiveness to change is likely to be the survivor in the future competitive marketplace. The competition that exists or that potentially can occur as a result of the entrance of a new competitor into the marketplace is a powerful force in driving improvement of the manufacturing system.

The implication of the development of a world market is also profound. As Drucker (1980) observed:

> Even small and purely local businesses may have to learn to think and to operate transnationally. And businesses already in the world economy will have to learn to think and to behave quite differently. The small business operating within a local or regional market may have to learn to organize its production transnationally and by stages of production rather than in one centralized "manufacturing" process; or it may have to learn to buy parts from all over and to assemble when today it buys finished products for resale.

An equally important trend is the changing role of the human being in the manufacturing environment. The increased use of robots and computers often introduces the need for highly specialized personnel who can program, maintain, and properly integrate this level of sophisticated technology into the manufacturing environment. Companies are moving to eliminate first-line supervisors in their manufacturing facilities, replacing them with self-directed work teams. The tendency is to provide these teams with technical specialists–usually engineers with advanced degrees–who can assist them in problem solving and in the analysis of unexpected situations.

Although technology for its own sake may be of limited value, it must be recognized that technology can be a powerful ally in achieving a superior level of competition, when the basic elements of good management practice are in place and are being utilized effectively at all levels of the operation. Once these are in place, the competitive advantage will go to those enterprises that can acquire, integrate, and most effectively use technology to their benefit.

For the management student, it is critical to realize that technology cannot be viewed as a single solution to many of the problems encountered in the manufacturing environment. For those who were looking for a quick answer to problems, an easy approach was often to develop and invest in technology because of its *inherent value*. Many instances can be found where an investment in automation was justified on the basis that it would reduce the requirements of expensive labor. When subsequent careful analysis often showed that the automation could be used only if the product was redesigned to enhance its capability to be produced and assembled in the automated facility, it came as a startling revelation to many people that the redesigned product required far less direct labor that did the old product, thus greatly reducing or eliminating the expected advantage of the automation. Great care must be used in attempting to use technology to solve problems that are basically nontechnical.

Unfortunately, the realization that technology is limited in its capabilities has not always prevailed when it comes to making decisions, many of which can be very costly.

This book deals primarily with the development of those management practices that are critical for achieving a highly competitive operation. It focuses on management of an enterprise, leaving the details of those technologies that are important to any manufacturing enterprise to another time and place. The intent here is to concentrate on the philosophy of management that needs to be followed, the actions that a good manager must take, and the types of tools that are available to managers to assist them in making better decisions.

It is the thesis of this book that a number of fundamental actions are needed to achieve and maintain a highly efficient and prosperous manufacturing enterprise. These fundamental actions are presented as *foundations*. These foundations recognize the complex combination of disciplines and technology that comprise the manufacturing environment. They recognize the importance of the attitudes of the management, the means by which the activities are organized, the influences of the environment, the importance of the customer, and the dominant contribution that can be made by a dedicated and committed workforce.

It is critical that students understand and be prepared to implement these foundations in their approach to problems they will face in the workplace. It is obvious that the rewards for good management can be large. It is equally true that the failures of poor management can be catastrophic. An implementation of these simple foundations can greatly enhance the likelihood of achieving a competitive manufacturing system.

REFERENCES

BAIRSCH, P. 1982. International Industrial Levels from 1750 to 1980. *Journal of European Economic History* 11(2):291.

CARLSSON, B. 1984. The Development and Use of Machine Tools in Historical Perspective, *Journal of Economic Behavior and Organization* 5:91-114.

CENSUS, U.S. BUREAU. 1990. *Statistical Abstract of the United States 1990*, 110th ed., U.S. Government Printing Office, Washington, DC

CENSUS, U.S. BUREAU. 1994. *Statistical Abstract of the United States 1994*, 114th ed., U.S. Government Printing Office, Washington, DC

DODGE, W.C. 1906. Invention as a Factor of American National Wealth in *The Making of America* R.M. LA FOLLETTE (ed..). p 267-277 The Making of America Co. Chicagp

DRUCKER, P. E. 1980. *Managing in Turbulent Times*, Harper and Row, New York:117.

DRUCKER, P. E. 1985. *Innovation and Entrepreneurship: Practice and Principles.* Harper & Row, NY :26.

HOUNSHELL, D. A. 1984. *From the American System to Mass Production, 1800-1932*, Johns Hopkins University Press, Baltimore: 327-329.

KENNEDY, P. M. 1976. *Rise and Fall of British Naval Mastery*, Allen Lane, London:189-90.

LAW, A. M. AND W. D. KELTON. 1982. *Simulation and Modeling*, McGraw Hill Book Co, NY:2-4

NATIONAL SCIENCE BOARD. 1993. *Science and Engineering Indicators,* U.S. Government Printing Office, Washington, D.C.

PORTER, G. 1984. In D.A. HOUNSHELL *From the American System to Mass Production: 1800-1932*, Johns Hopkins University Press, Baltimore:xv.

REINTJES, J. F. 1991. *Numerical Control: Making a New Technology*, Oxford University Press, New York.

ROBERTSON, R.M. 1973. *History of the Economy*. Harcourt Brace Jovanovich, New York:64.

VOELCKER.H. B. 1988. Modeling in the Design Process, *Design and Analysis of Integrated Manufacturing Systems*, W. D. Compton (ed.), National Academy Press, Washington, DC:191-92.

QUESTIONS

2.1 Three inventions and/or discoveries that have had an enormous influence on manufacturing are noted in Section 2.1. Identify three other important discoveries/inventions that have had a profound impact on manufacturing. Discuss these and explain their role in modern manufacturing.

2.2 How has the agriculture sector managed to reduce the fraction of the populace that is employed in generating food from 75% in 1820 to less than 10% in 1990? Is employment in manufacturing likely to follow the same pattern? If so, what are the probable consequences?

2.3 The U.S. government has recently adopted a new policy relative to the support of civilian-related industries. This new *industrial policy* focuses on assisting certain industries in achieving world competitive status. Give examples of the industries they are planning to support. Offer both pro and con arguments concerning the advisability of these actions.

2.4 The manufacturing sector was characterized in Section 2.2 according to whether it was batch or discrete, the type of activity that it engaged in, and the type of facilities that it employed. Identify examples in the service sector that have characteristics analogous to those of the manufacturing sector.

2.5 What elements of the manufacturing process differ significantly in the flow shop and in the job shop? What must be done to achieve high productivity in each?

2.6 You are president and principal shareholder of a small manufacturing company in Indianapolis, Indiana, that produces and services pumps used primarily for agricultural irrigation and oil pipeline systems. The former are marketed throughout the United States. You are currently considering a business arrangement that would require a large investment in a new plant. The feasiblity of such an investment depends on both a short-term expansion in the market for the agricultural systems and a long-term expansion into overseas markets, primarily in Eastern Europe and the Middle East. What issues must be addressed as you consider the advisability of making this investment?

2.7 You are the owner of a small company of fewer than 80 employees that manufactures and markets a product used by large manufacturing firms for measurement and verification of tolerances on large machined surfaces. Your product depends critically on a combination of sensors and software that you have created specifically for this application. The unique features of this product cannot be protected by patents. Recognizing that a demand for this type of system continues to develop rapidly throughout the world, therefore stimulating the development of new entries for the marketplace, how do you maintain an awareness of the status of the technology? How do you learn about developments in other companies of products that might compete with your product? How do you protect yourself against competition in your marketplace?

APPENDIX A2.1 A STATISTICAL DESCRIPTION OF THE MANUFACTURING SECTOR

To understand the role of the manufacturing sector, it is helpful to consider a variety of statistics that describe this important component of the economy. In exploring the ramifications of these statistics and extending them into areas that are of particular interest, you will soon be confronted with some disturbing facts. First, there is no single central source for the various statistics that may be of interest. Second, the means used to collect the statistics and the definition of the terms that are used have not been constant over time. This can make it very difficult to obtain a consistent long-term comparison of events. Third, various compilations of statistics will be found to use different bases for comparisons and will thus yield somewhat different values for seemingly the same variable. Despite these discrepancies and difficulties, it is possible to obtain a very significant amount of information and insight by looking and analyzing the data available. Although an exhaustive treatment of the available statistics is not presented here, this appendix will serve as an introduction to the sources of available data and as a framework for understanding the importance of the manufacturing sector to the overall economy.

The principal source of data on the economy of the United States is the federal government, including, but not limited to the Department of Commerce, the Census Bureau, the Office of Management and Budget of the Executive Office of the President, the Council of Economic Advisers of the President, the Office of Technology Assessment of the Congress, the National Science Foundation, the Bureau of Labor Statistics, and the Security Exchange Commission. Various private activities study the data provided by these and other sources and publish regular analyses. All of these can be useful in understanding the manufacturing sector.

The various sectors of the economy are identified through the Standard Industrial Classification (SIC) system established by the Executive Office of the President through the Office of Management and Budget (*Standard Industrial Classification Manual*, 1987).

> The Standard Industrial Classification (SIC) is the statistical classification standard underlying all establishment-based Federal economic statistics classified by industry. The SIC is used to promote the comparability of establishment data describing various facets of the U.S. economy. The classification covers the entire field of economic activities and defines industries in accordance with the composition and structure of the economy. It is revised periodically to reflect the economy's changing industrial organization....

Table A2.1 contains a list of the major divisions used for the classification of industry, along with a more detailed list of the major groups for the manufacturing sector. The individual industries contained in the codes for the sectors other than manufacturing division D can be found in the SIC *Manual*.

By way of illustration of the details that can be gleaned from these classifications, consider SIC Code 35–industrial machinery and equipment. There are 52 subcategories under this heading. These include: SIC Code 351–engines and turbines; 352–farm and garden machinery; and 353–construction and related machinery. SIC

Code 351 is further subdivided into SIC Code 3511–turbines and turbine generator sets and 3519–internal combustion engines. The total number of unique subdivisions, (i.e., unique SIC Codes) within the manufacturing division is in excess of 200.

Focusing now on the manufacturing sector, the employment by the various groupings within manufacturing are displayed in Table A2.2 The total number of establishments in each of these groupings and the number of establishments, grouped according to the number of employees, are also given.

> An establishment is a single physical location at which business is conducted or where services or industrial operations are performed. It is not necessarily identical with a company or enterprise, which may consist of one establishment or more(SIC *Manual*, 1987).

The data in Table A2.2 offer some interesting insights into the structure of the manufacturing sector. Note that 82.7% of the establishments involved in manufacturing employ fewer than 50 people. Table A2.3 presents a view of the distribution of the number of companies along with the value that they add to the products that they sell. This also illustrates the number of companies that possess a single establishment. While it is common to think about the large plants that dot the landscape as being characteristic of the manufacturing sector, only 1,598 establishments have 1000 or more employees. Manufacturing in the United States remains a collection of small establishments. A more complete analysis of all manufacturing activities would demonstrate the often quoted statement that more than half of the manufacturing in this country is accomplished in companies with fewer than 100 employees.

Figure A2.1 displays the relative share of the gross national product (GNP) contributed by manufacturing and the service sectors between 1950 and 1985. Figure A2.2 displays the percentage of total employment contained in these two sectors. While the percentage of GNP contributed by the manufacturing sector has remained relatively constant over time-varying only between about 20 and 23% the percentage of employment in manufacturing has dropped from about 28 in 1950 to a bit below 20% in 1985. This reflects the increased productivity of the manufacturing sector that has occurred during this period.

A comparison of the increased productivity of U.S. workers is compared with the productivity of our trading partners in Figure A2.3 for the period between 1960 and 1986. The United States is taken as the base against which Canada, Italy, West Germany, Japan, and the United Kingdom are compared. Purchasing-power-parity exchange rates have been used to determine the levels of productivity. Note that all of the trading partners of the United States are approaching the U.S. level of productivity. The rate at which they are approaching this level of productivity is of some concern to economists and policymakers. Figure A2.4 illustrates the average annual productivity growth in manufacturing in the United States and its principal trading partners in 1988. It is particularly striking that the rate of increase in productivity in Japan is greater than in the United States by a factor of 2 to 3.

TABLE **A2.1** **Standard Industrial Classification**

Division	SIC Codes
A	Agriculture, forestry and fishing (SIC Codes 01, 02, 07-09)
B	Mining (SIC Codes 10, 12-14)
C	Construction (SIC Codes 15-17)
D	Manufacturing Code
	20 Food and kindred products
	21 Tobacco products
	22 Textile mill products
	23 Apparel and other finished products made from fabrics and similar materials
	24 Lumber and wood products, except furniture
	25 Furniture and fixtures
	26 Paper and allied products
	27 Printing, publishing, and allied industries
	28 Chemicals and allied products
	29 Petroleum refining and related industries
	30 Rubber and miscellaneous plastics products
	31 Leather and leather products
	32 Stone, clay, glass, and concrete products
	33 Primary metal industries
	34 Fabricated metal products, except machinery and transportation equipment
	35 Industrial and commercial machinery and computer equipment
	36 Electronic and other electrical equipment and components, except computer equipment
	37 Transportation equipment
	38 Measuring, analyzing, and controlling instruments; photographic, medical and optical goods; watches and clocks
	39 Miscellaneous manufacturing industries
E	Transportation, communications, electric, gas, and sanitary services (SIC Codes 40-49)
F	Wholesale trade (SIC Codes 50 and 51)
G	Retail trade (SIC Codes 52- 59)
H	Finance, insurance, and real estate (SIC Codes 60-67)
I	Services (SIC Codes 70, 72, 73, 75, 76, 78, 79-84, and 86

Figure A2.1 Relative shares of GNP by servide and manufacturing sectors (constant 1982 dollars).

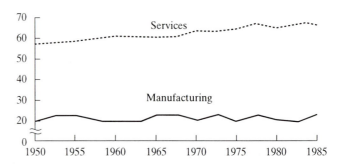

TABLE **A2.2**

Number of Employees and Establishments for Manufacturing by SIC Code 1991

Number of Establishments by Employment-Size Class

SIC Code	Number of Employees	Total Number Establish- ments	1 to 49	50 to 99	100 to 249	250 to 499	500 to 999	1000 or more
20	1,471,535	20,256	14,584	2,164	2,064	886	417	141
21	37,867	142	77	20	22	10	5	8
22	621,591	6,233	3,984	700	822	457	207	63
23	973,229	23,515	18,884	2,198	1,672	573	155	33
24	635,024	34,458	31,565	1,608	1,033	210	35	7
25	458,531	11,253	9,314	902	665	257	88	27
26	622,656	6,364	3,591	1,046	1,201	304	155	67
27	1,502,760	62,649	57,043	2,946	1,801	532	205	122
28	865,888	12,224	9,380	1,258	928	355	180	123
29	114,872	2,235	1,862	150	122	52	30	19
30	846,320	15,397	11,195	2,077	1,534	428	115	48
31	106,028	1,990	1,513	191	164	99	19	4
32	476,986	15,962	14,000	1,017	639	206	81	19
33	688,280	6,867	4,546	883	872	327	150	89
34	1,383,586	35,967	29,573	3,363	2,222	576	164	69
35	1,854,527	52,534	45,964	3,212	2,149	713	341	155
36	1,481,703	16,810	12,013	1,830	1,661	749	356	201
37	1643,264	10,571	7,893	938	885	394	210	251
38	926,047	10,310	7,852	905	849	341	226	137
39	368,948	16,912	15,393	780	517	160	47	15
Total	17,079,642	362,649	300,226	28,188	21,822	7,629	3,186	1,598

Statistics exclude administrative and auxiliary establishments
1994. US Bureau of the Census, County Business Patterns, 1991. US Government Printing Office, Washington, DC.

Figure A2.2   Manufacturing and private services as a percentage of total employment, 1920-1985. (Based on various U.S. government publications.)

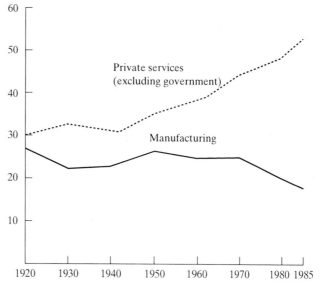

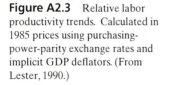

Figure A2.3 Relative labor productivity trends. Calculated in 1985 prices using purchasing-power-parity exchange rates and implicit GDP deflators. (From Lester, 1990.)

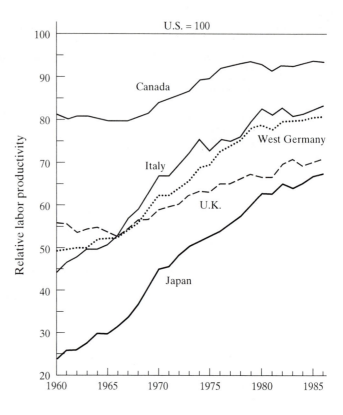

TABLE **A2.3**   Manufacturing Statistics by Type of Organization And Industry, (1987)

Companies With Single Establishments				
Type of organization	Total companies	Total number	With 19 or fewer employees	Value of shipments
All Industries	310,341	287,994	222,105	2.48×10^{12}
Motors & Generators (SiC 3621)	349	233	152	525×10^{6}
Telephone & Telegraph Apparatus (SiC 3661)	403	270	153	1.13×10^{9}
Printed Circuit Boards (SiC 3672)	950	819	432	1.91×10^{9}
Semiconductors & Related Devices (SiC 3674)	755	590	386	2.40×10^{9}
Motor Vehicle Parts and Accessories (SiC 3714)	2,306	1,926	1,381	3.98×109

Source: Census of Manufactures: 1987.

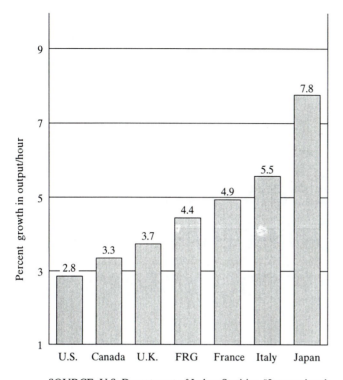

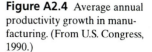
Figure A2.4 Average annual productivity growth in manufacturing. (From U.S. Congress, 1990.)

SOURCE: U.S. Department of Labor Statitics, "International Comparisons of Manufacturing Productivity of Labor Trends, 1988," June 1989 Table 1.

REFERENCES

CENSUS OF MANUFACTURERS. 1987. U.S. Government Printing Office, Washington, D. C.

COUNCIL ON COMPETITIVENESS. 1991. *Competitiveness Index 1991.* Washington, D.C.

Lester, R.K. 1990. Private communication.

STANDARD INDUSTRIAL CLASSIFICATION MANUAL, 1987. U.S. Government Printing Office, Washington, D.C.

PART 2

Goals and Objectives for the Enterprise

In the following two chapters, the focus is on the need to establish clear and concise goals for the corporation and for each unit within its boundaries. The processes by which these tasks are accomplished and the involvement of the employees in ways that encourage their "buy-in" of the goals is a critical part of good management. The establishment of long-term goals and objectives for the manufacturing system–the strategic goals and objectives–is introduced in Chapter 3. In Chapter 4, we discuss the establishment of shorter-term operating goals and objectives. We emphasize the importance of comparing your performance with that of your competitors–benchmarking–and of establishing and using the theoretical limits that exist for your processes in the setting of operating goals and objectives.

3

Strategic Goals and Objectives

We should recognize that there is no single perfect definition; the elegant crafting of abstractions is only one part of this endeavor. The most important aspect is the dialogue, the process, the give-and-take that comes from thinking seriously about what you are doing in the company of others who are engaged in the same enterprise (Keohane,1993).

If an organization is to be successful, its goals and objectives must be defined and communicated clearly to every member of the organization. This is not an idle admonition. Far too often, the goals and objectives of the organization, of principal divisions of the organization, or of individual operating units are established by the upper management of the activity but not communicated effectively to all the members of the activities affected. When an understanding and focus on common goals is missing, the system faces the possibility of having confused operating goals and mixed priorities. It is the obligation of management to ensure that every member of the organization understands and appreciates the importance of the common goals and objectives.

Without clearly defined goals and objectives, it is difficult to marshal resources, choose among alternatives, and take the necessary actions to ensure that the directions being followed are in the best interest of the enterprise. As someone observed, there is nothing worse than following a path that leads to the wrong destination in the shortest possible time.

An organization must be capable of establishing goals and objectives at a variety of levels. There are the broad goals that the company wishes to achieve. There are the goals that are set for the products being offered to the customer. There are the goals that each activity of the organization (e.g., the sales and marketing group, the manufacturing activities, the engineering activities, etc.) must achieve if the entire system is to be successful. Goals must be set for both the short-term operational goals and the long-term strategic goals. How the various goals are established, how progress is measured in regard to achieving them, and how one marshals the resources of the system to ensure a constancy of purpose are principal tasks of management.

3.1 CORPORATE GOALS AND OBJECTIVES

The goals and objectives for the corporation generally include a clear identification of the products or services that the enterprise will offer its customers, a statement of the markets in which it expects to be active, an enumeration of the customers that it plans to serve, a statement of its obligations to and aspirations for its stockholders and employees, and a statement concerning its obligations to comply with certain accepted standards of ethical conduct in the marketplace and in the community within which it operates. While the forms that these statements take are varied, their intent is the same: to declare to all concerned that the company has certain aspirations, that it plans to do certain things, and that it is willing to take a public stand on certain issues.

Figures 3.1 to 3.7 display recent statements of the goals and objectives of six prominent U.S. companies: Ford Motor Company, the Timken Company, AT&T, Saturn Corporation, UT Automotive, Intel, and a federal manufacturing laboratory, the Naval Air Warfare Center (formerly called the Naval Avionics Center). Despite the differences that exist in the products and services that each of these organizations provides, the similarities among the statements are striking. These statements are a combination of strategic and operating goals. The strategic element focuses on the long term; the operational focuses on the short term.

It is reasonable to inquire why statements such as these are important. After all, it may be asked, "doesn't everyone in the organization know these?" The answer, perhaps surprisingly, is that unless the goals and objectives of an organization are stated clearly, distributed widely, and repeated frequently it is easy for members of the organization to lose sight of what their company or their unit in the company is trying to achieve. The larger and more complex the organization, the greater the diversity of its products and services, and the greater the geographical separation of the units, the more important it is to have the statement of goals and objectives displayed prominently and discussed frequently. It is important that everyone be reminded that day-to-day decisions should be made consistent with these goals and objectives. Joseph F. Toot, Jr., president of the Timken Company (Toot, 1987) identified the importance of the mission statement as follows:

> The success of an organization such as The Timken Company is the result of the combined efforts of every individual employee working toward common objectives. Everyone must be pulling in the same direction. This unified direction has to be clearly understood by everyone and must reflect the organization's basic character and personality.
>
> Fundamental to our collective success is, first, that each employee understand what this document says, and second, that we all agree that these are the things we can and must do.
>
> This MISSION STATEMENT provides guidance and insight for each employee's day-to-day job responsibilities. It provides meaning to our work and gives us something greater to relate to than just the job itself.
>
> It presents a challenge–achieving the corporate objectives–and prescribes how to measure success. By performing your job with your best effort, you are contributing to the success of our Company.

COMPANY MISSION, VALUES, AND GUIDING PRINCIPLES

MISSION—Ford Motor Company is a worldwide leader in automotive and automotive-related products and services as well as in newer industries such as aerospace, communications, and financial services. Our mission is to improve continually our products and services to meet our customers' needs, allowing us to prosper as a business and to provide a reasonable return for our stockholders, the owners of our business.

VALUES—How we accomplish our mission is as important as the mission itself. Fundamental to success for the Company are these basic values:

• **People**-Our people are the source of our strength. They provide our corporate intelligence and determine our reputation and vitality. Involvement and teamwork are our core human values.

• **Products**-Our products are the end result of our efforts, and they should be the best in serving customers worldwide. As our products are viewed, so are we viewed.

• **Profits**-Profits are the ultimate measure of how efficiently we provide customers with the best products for their needs. Profits are required to survive and grow.

GUIDING PRINCIPLES

• **Quality comes first**-To achieve customer satisfaction, the quality of our products and services must be our number one priority.

• **Customers are the focus of everything we do**-Our work must be done with our customers in mind, providing better products and services than our competition.

• **Continuous improvement is essential to our success**-We must strive for excellence in everything we do: in our products, in their safety and value-and in our services, our human relations, our competitiveness, and our profitability.

• **Employee involvement is our way of life**-We are a team. We must treat each other with trust and respect.

• **Dealers and suppliers are our partners**-The Company must maintain mutually beneficial relationships with dealers, suppliers, and our other business associates.

• **Integrity is never compromised**-The conduct of our Company worldwide must be pursued in a manner that is socially responsible and commands respect for its integrity and for its positive contributions to society. Our doors are open to men and women alike without discrimination and without regard to ethnic origin or personal beliefs."

Figure 3.1 Company mission, values, and guiding principles of the Ford Motor Company. (From Ford 1984.)

THE TIMKEN COMPANY

MISSION STATEMENT

Mission

We are an independent organization with a leadership position in quality anti-friction bearing and alloy steel products. To maximize shareholder value and sustain our competitive position, we will capitalize on the relationships between our businesses and emphasize the application of technology to products and processes combined with unmatched customer service.

Objectives to Achieve Our Mission

- Achieve customer recognition as a company offering higher value in products and services than our competitors.
- Increase penetration in markets providing long-term profit opportunities.
- Achieve an inflation-adjusted return on assets of approximately 13 percent. This will provide our shareholders with growth in their investment and dividends. It will also provide the funds necessary for reinvestment in our businesses to maintain and advance our leadership position.
- Maintain excellence through investment in technology with a continuing commitment to timely implementation of the results.

How We Will Achieve Our Objectives

- We will take a long term view in our decision making by strategically managing our businesses.
- We will adopt an organizational style that encourages bigger changes at a faster pace to capitalize on opportunities in our ever-changing environment.
- We will be dedicated to the highest degree of ethics and integrity in the conduct of our business.
- We will attract and retain superior employees, provide for the full development of their capabilities, and foster a spirit of team effort.
- We will enhance our competitive advantage through each individual employee's superior performance in his or her job responsibilities.
- We will meet or exceed customer expectations through each employee's active participation in and dedication to customer service.
- We will achieve excellence by emphasizing the fundamental operating principles of quality, cost, investment usage, and timeliness.

Figure 3.2 Mission statement of the Timken company. (From Toot, 1987:8)

AT&T'S BUSINESS

AT&T's business is moving and managing information, domestically and globally. That includes providing long distance telecommunications services through the company's World-wide Intelligent Network, as well as systems, products and services that combine communications and computers.

Our mission is to apply the talents, knowledge and skills of our people to make the company the global leader in enabling customers to reap the benefits of information technology.

Our major customer markets are:

Business and government. For large and small businesses and federal and state governments, AT&T offers a range of voice and data transmission services and sophisticated communications, computer and data networking products and systems. AT&T develops customized data networking solutions that connect incompatible and widely dispersed computer systems into integrated networks.

Consumers. AT&T provides customers with long distance services and high-quality telephones and related products.

Telecommunications Industry. AT&T supplies switching systems, transmission equipment and operations support services to the telecommunications industry.

Electronic Equipment Manufacturers. AT&T designs and manufactures advanced electronic components for AT&T business units and for sale to other high-technology firms.

Figure 3.3   AT&T's business, (from AT&T 1988 annual report.)

NAVAL AVIONICS CENTER
CONTINUOUS IMPROVEMENT GOALS AND STRATEGIES
Preamble

Recognizing customer satisfaction as our highest priority, and consistent with our vision and values, the following goals and strategies are established to provide focus to our continuous improvement efforts.

GOAL I

NAC will provide world-class response to our customers.

Strategies

1. Continuously reduce the Center's cycle times.
2. Meet or exceed competitive schedule and quality standards in the delivery of products and service.
3. Strengthen the Center's focus on satisfying customers.

GOAL II

NAC will employ business practices that result in value added to our customers.

Strategies

1. Assure the timely and orderly integration of our information technology systems.
2. Support customer information requirements.
3. Improve resource and cost allocation processes to assure that our products and services provide the highest value.

GOAL III

People and environmental values will be an integral part of NAC's business practices.

Strategies

1. Pursue an innovative culture that capitalizes on diversity among people and treats everyone as a respected member of the organization.
2. Work toward eliminating hazardous materials and wastes at NAC.
3. Foster an innovative safety culture.

GOAL IV

NAC will pursue a product focus on avionic and electronic systems in support of its Warfare Center leadership areas and responsibilities.

Strategies

1. Clearly establish NAC's role in the front-end process of advanced and integrated avionics.
2. Seek work opportunities that maintain capabilities in design, manufacturing, manufacturing technology, acquisition and production support.
3. Develop corporate partnerships with Air Warfare Center members and other Warfare Center activities consistent with our leadership areas and our product goals.

Figure 3.4   Continuous improvement goals and strategies, Naval Air Warfare center. (From a statement issued October 1, 1991.)

The Mission of Saturn

• To market vehicles developed and manufactured in the United States that are world leaders in quality, cost, and customer satisfaction through the integration of people, technology, and business systems.

• To transfer knowledge, technology and experience throughout General Motors.

The Philosophy of Saturn

We, the Saturn Team, in concert with the UAW and General Motors, believe that meeting the needs of Saturn customers, members, suppliers, dealers, and neighbors is fundamental to fulfilling our mission.

• To meet our customers' needs, our products and services must be world leaders in value and satisfaction.

• To meet our members' needs, we will create a sense of belonging in an environment of mutual trust, respect and dignity. We believe that all people want to be involved in decisions that affect them. We will develop the tools, training, and education that each member needs. Creative, motivated, responsible team members who understand that change is critical to success are Saturn's most important asset.

• To meet our suppliers' and dealers' needs, we will create real partnerships with these organizations. We will strive for openness, fairness, trust, and respect with them, and we will work with them to help them feel ownership of Saturn's mission and philosophy.

• To meet our community neighbors' needs, we will be a good citizen and protect the environment, and we will seek to cooperate with government at all levels.
By continuously operating according to this philosophy, we will fulfill our mission.

The Core Values of Saturn

As a final complement of Saturn's mission and philosophy statements, the founding team enunciated five core values: commitment to customer enthusiasm, commitment to excel, teamwork, trust and respect for the individual, and continuous improvement.

Figure 3.5   Mission, philosophy, and core values of Saturn Corporation. (From Lefauve and Haux, 1992.)

UTA Vision

It is the vision of United Technologies Automotive to be the best supplier of components and systems to the worldwide automotive industry. We will achieve this as a team, with all UTA associates contributing. We will manage change in a global environment of broad-based leadership that results in the following:

• Customers who are absolutely certain of UTA's unsurpassed reliability in terms of both product and people.

• Product design, innovation, quality and value that inspire customer allegiance and competitor respect:

> - Electrical systems that achieve the most reliable performance in the world.
> - Interiors that achieve the highest aesthetic and functional standards.

• Recognition of UTA as an employer of choice by our own associates and by high quality candidates who aspire to become part of UTA's culture—an exciting, innovative environment that is founded on trust, openness, individual dignity and powerful teamwork.

• Communities that are proud to have UTA as a valued corporate citizen of unimpeachable ethics, and stockholders who recognize UTA as an outstanding growth investment opportunity.

Figure 3.6 UTA vision. (From UTA news, September, 1993: 2)

Our Mission

Do a great job for our customers, employees and stockholders by being the preeminent building block supplier to the computing industry.

Our Values:

Discipline

The complexity of our work and our tough business environment demand a high degree of discipline and cooperation. We strive to:

- properly plan, fund and staff projects
- pay attention to detail
- clearly communicate intentions and expectations
- make and meet commitments
- conduct business with uncompromising integrity and professionalism

Quality

Our business requires continuous improvement of our performance to our Mission and Values.

- set challenging and competitive goals
- do the right things right
- continuously learn,develop,and improve
- take pride in our work

Result Orientation

We are results oriented. We strive to:

- set challenging goals
- execute flawlessly
- focus on output
- assume responsibility
- confront and solve problems

Customer Orientation

Partnerships with our customers and suppliers are essential to our mutual success. We strive to:

- listen to our customers
- communicate mutual intentions and expectations
- deliver innovative and competitive products/service
- make it easy to work with us
- serve our customers through partnerships with our suppliers

Risk Taking

To maintain an innovative environment, we strive to:

- embrace change
- challenge the status quo
- listen to all ideas and viewpoints
- encourage and reward informed risk taking
- learn from our successes and mistakes

Great Place to Work

A productive and challenging work environment is the key to our success. We strive to:

- respect and trust each other
- be open and direct
- work as a team
- recognize and reward accomplishments
- be an asset to our community
- have fun

Figure 3.7   Intel mission and values.

3.2 STRATEGIC GOALS

The strategic goals for an organization are a statement of where the organization hopes to be at a future date. They establish the framework within which current and future decisions are to be made. For Ford, its strategic goals are contained in its mission statement: "to be a leader in automotive and automotive-related products and services." Ford states that how they accomplish their long–term goals is as important as the goals themselves. They identify people, products, and profits as critical to accomplishing the strategic goals. Timken states that its long-term strategic mission is to be "an independent organization with a leadership position in quality anti-friction bearing and alloy steel products." Prior to recent restructuring, AT&T stated that its strategic goal was "moving and managing information, domestically and globally." Each of these statements identifies the company's long term-strategic-goals and objectives.

These goals must be established and understood by everyone so that sensible decisions can be made concerning resource allocation, entry into new markets, acquisitions, technology developments, and so on. *The process* by which strategic goals are established can be nearly as important as the goals themselves. This is not to minimize the importance of the goals. It emphasizes, instead, that the discussions that must be undertaken to establish long-term goals and the involvement of everyone in an organization in their development are critical in building consensus concerning the long term strategic goals of the organization. The *process* offers an important opportunity to enhance the understanding of each employee and organizational unit in their responsibility *to the system*. This is not, of course, a static situation. Although the strategic goals of an organization are not expected to change frequently, they are subject to modification and their appropriateness requires regular discussion and assessment. This can be seen in Ford's statement of goals and objectives . Since this statement in 1984, Ford has chosen to divest itself of its aerospace, communications, and farm equipment business. The current statement of their mission would certainly be somewhat different from that contained in Figure 3.1. Similarly, the recent sale of the Naval Air Warefare Center will lead to a goals statement, that is characeristic of its new products. The implications of a change in mission are stressed in the following illustrations by Beckhard and Harris (1987).

> Traditionally, in a medical school, producing physicians is the core "reason to be." But in today's world, the main purpose may be to provide physicians who can function in the community in which the school is located; the real mission is to *provide services* to that community. Or it may be that the mission is to find the cure for cancer through the school's research activities. Which role a school sees for itself is defined by its stewards, and the decision controls the organizational structure, budget priorities, and so on.
>
> In a youth organization such as the Boy Scouts or Girl Scouts, the mission may change from providing educational and citizenship activities for young people to mobilizing them to take stands on social issues.

The emphasis on the importance of the process by which the strategic goals are developed raises an important issue. If the process is so important, you may ask, how is the process made "real" to employees who enter the company after the goals have been established. Or is this a one-time event? Many companies attempt to convey an understanding of its strategic goals and objectives during the indoctrination or the early training that is provided new employees. In these sessions the new employee will be introduced to the goals, a presentation or a videotape of a presentation may be made by the president emphasizing the goals, subsequent speakers will usually couch there remarks in the framework of the goals, and an opportunity will be given the new employees to raise questions regarding the goals. In these ways, the reality of the goals will be emphasized to the new employee. Although this indoctrination may be less meaningful than was experienced by employees who participate in the creation of goals, such steps are important in encouraging people to take them seriously and to understand their significance.

It is not sufficient for an organization to stop with the broad statements of strategic goals and objectives. Each unit in the organization must develop statements of the goals and objectives that are specific to its function while ensuring that they are compatible and consistent with those of higher levels of the organization and with the organization's broader goals and objectives. It is not surprising, therefore, that the research unit in a company will have a different statement of goals and objectives than will the manufacturing plant or sales group. This *cascading* of the goals and objectives through an organization is an important part of ensuring that each unit is operating consistently with the overall goals and objectives of the company and that its specific goals and objectives are in concert with those of the corporation. As is illustrated in Figure 3.8, the specificity of the statements increases as the process moves to lower and lower levels in the organization. By the time the goals and objectives have been completed for individuals, they may be quite specific and may approach the objectives that are used in the setting of individual work objectives (Chapter 7).

3.3   OPERATING GUIDELINES

There are other goals that have as much strategic importance to the long-term success of the enterprise as do the products and services provided and the markets served. These are the general principles that are used to guide the organization in its daily operations. These principles determine the response that people will make to new situations and to the approaches that they will take in solving problems. Several of these are often listed as operating goals (see, e.g., the Intel statement in Figure 3.7). The following represent fundamental operating principles that have been adopted by successful enterprises.

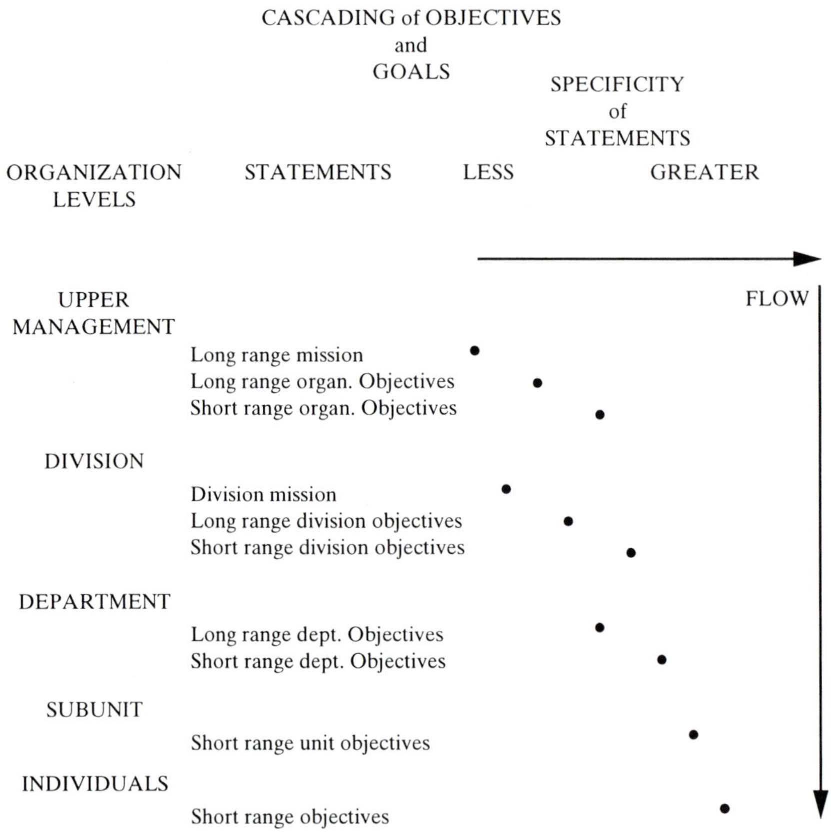

CASCADING of OBJECTIVES
and
GOALS

SPECIFICITY
of
STATEMENTS

ORGANIZATION LEVELS	STATEMENTS	LESS	GREATER

FLOW

UPPER MANAGEMENT
Long range mission
Long range organ. Objectives
Short range organ. Objectives

DIVISION
Division mission
Long range division objectives
Short range division objectives

DEPARTMENT
Long range dept. Objectives
Short range dept. Objectives

SUBUNIT
Short range unit objectives

INDIVIDUALS
Short range objectives

Figure 3.8 Cascading the goals and objectives to all levels of the organization..

- *Employees* are the *most important* asset to our organization.
- The *quality* of the product or service *must never be compromised*.
- The system must *continuously improve* all aspects of its operation.
- *Every action* that is taken *should add value* to the final product or service.
- *Timely response* of the system to changing external conditions is paramount to success.
- *Integrity* in all actions by all employees is paramount.

This list of operating principles can be greatly extended and broadened in detail. The important issue is not the exhaustive extent of the list but the understanding of everyone in the organization of the importance of each principle. Nothing can be more disruptive for an organization than for various parts to be operating as if they have different operating principles, or to have a set of principles that are stated but never used seriously when evaluating and encouraging the performance of groups and individuals.

3.3.1 Employees

- *Employees* are the *most important* asset to our organization.

Employees who are respected and highly motivated are usually happy and, in turn, very productive. World-class companies have learned that listening to employees, giving them responsibility to enhance operational efficiency, and including them in the decision-making process not only benefits the organization but enhances the content of the job for employees. These operating principles are variously described by terms such as participative management, employee involvement, and employee empowerment. These are discussed extensively in Chapters 6 to 8, but it is worth noting here a common misconception that revolves around these concepts. Employee involvement and employee empowerment do not translate into an abrogation of decision making by management. Some decisions are better left to the employees, who have detailed knowledge of the operations, with management respecting and accepting their judgment. The responsibility for other decisions must remain with management. Neither does participative management mean management by consensus. It means listening, inquiring, seeking advice and counsel, making the best decision for the organization and fully informing all who participated in the process of the rationale for the decision even though some may hold differing opinions.

3.3.2 Quality

- The *quality* of the product or service *must never be compromised.*

Perhaps no single strategic goal is more important to success than *quality*. In the strategic sense, this refers to quality in all aspects of system operation. It refers to quality of the product, quality of the manufacturing operation, quality of the service provided to customers, and the quality of the interactions among the various elements *within* the organization. It is this total commitment to quality in *all* aspects of the business that is contained in the term *total quality management.*

Asking your customers what they mean by the term *quality* will probably produce a set of definitions that are rather different from your perception of quality. This seeming ambiguity has led some to allege that quality is only in the eyes of the beholder. Does this mean that quality is not definable as a general attribute? Does this mean that quality cannot be given an operational definition? As Garvin (1984) observes, the various definitions of quality that are used "result [from] a host of competing perspectives, each based on a different analytical framework and each employing its own terminology." It is critical that a more complete understanding be obtained of what customers mean by quality if the producer is to respond to their admonitions that they want products of better quality.

Garvin (1984) has explored this topic thoroughly in a paper entitled "What Does 'Product Quality' Really Mean?" In this paper he identifies five distinct approaches that are commonly taken toward the concept of quality:

1. The transcendent approach, which is couched in philosophical terms.
2. The product-based approach, which is derived from economics.

3. The user-based approach, which combines economics, marketing, and operations management.
4. The manufacturing-based approach, which focuses on operations management.
5. The value-based approach, which derives from concepts in operations management.

Transcendent approach [Quality] is both absolute and universally recognizable, a mark of uncompromising standards and high achievement. Nevertheless, proponents of this view claim that quality cannot be defined precisely; rather, it is a simple, unanalyzable property that we learn to recognize only through experience.

Product-based approach Product-based definitions....view quality as a precise and measurable variable....Differences in quality reflect differences in the quantity of some ingredient or attribute possessed by a product....Because quality reflects the presence or absence of measurable product attributes, it can be assessed objectively, and is based on more that preferences alone.

User-based approach User-based definitions start from the premise that quality "lies in the eyes of the beholder." Individual consumers are assumed to have different wants or needs, and those goods that best satisfy their preferences are those that they regard as having the highest quality. This is an idiosyncratic and personal view of quality, and one that is highly subjective.

Manufacturing-based approach Manufacturing-based definitions focus on the supply side of the equation, and are primarily concerned with engineering and manufacturing practice. Virtually all manufacturing-based definitions identify quality as "conformance to requirements." Once a design or a specification has been established, any deviation implies a reduction in quality. Excellence is equated with meeting specifications, and with "making it right the first time".... Quality is defined in a manner that simplifies engineering and production control. On the design side, this has led to an emphasis on reliability engineering; and on the manufacturing side, to an emphasis on statistical quality control.

Value-based approach Value-based definitions....define quality in terms of costs and prices. According to this view, a quality product is one that provides performance at an acceptable price or conformance at an acceptable cost.... The difficulty in employing this approach lies in its blending of two related but distinct concepts. Quality, which is a measure of excellence, is being equated with value, which is a measure of worth. The result is a hybrid–"affordable excellence"–that lacks well-defined limits and is difficult to apply in practice.

If the marketing group is seeking to determine what the final consumer desires in a product, the meaning of quality is likely to be that represented by the product- or user-based approach. If a supplier who provides materials or components to a manufacturer is attempting to understand what customers want, the supplier is likely to hear quality expressed in terms of the manufacturing-based approach. In attempting to ascertain the customer's needs and wants, it is critical that the reference of that customer be clearly understood.

Ambiguity in the meaning of terms can be a source of confusion, even to the point of introducing conflict in the setting of objectives for an organization. Since the manufacturing activity will expect that improved quality in its operations will reduce errors, eliminate rejects, and improve productivity, it will probably assert that enhanced quality will reduce costs. Since the marketing activity will probably expect the customer to interpret quality in terms of enhanced product attributes, it may be prepared to argue that improved quality will add cost, or at least that a product with higher quality will command a higher price in the marketplace.

To avoid conflict and misunderstanding, the successful manager must insist that everyone arrive at an understanding of the terms that are being used and that goals, plans, and resource allocations be established only after this has been accomplished. Although this need not require that everyone accept and use a universal definition of terms such as quality, it clearly implies that the various meanings that common terms are given cannot be allowed to confuse the priorities that an organization establishes. Insisting that one understand the definition of the terms being used is a first step in this process.

The ambiguity in meaning that different viewers have of common terms is not limited to quality. It is equally apparent with such terms as *reliability, serviceability, dependability*, and *affordability*. Retaining an appreciation for the various meanings that can arise because of the natural biases introduced by one's environment or responsibilities is essential for a successful manager.

3.3.3 Continuous Improvement

- The system must *continuously improve* all aspects of its operation.

The Japanese word *kaizen* describes their focus on improving capability continuously through attention to incremental enhancement of all operations. All employees must seek to improve the way they carry out their responsibilities. The system must encourage and reward employees for identifying improved ways of operating. A philosophy of seeking always to improve must be instilled as a strategic objective of the organization.

U.S.-based companies have historically operated with the strategy that a new plant will be built or a new process installed that is the best available. Following installation, the system will be operated essentially without change until a radical change is required in the plant or process. The consequence of this attitude has been that progress has generally been made by large innovative "jumps" in performance. While the Japanese strategies did not deny the importance of innovations, they also emphasized the importance of improving existing processes or systems continuously through incremental improvements. The enterprise that is impatient with its performance, that continuously seeks to improve it, and that rewards employees who achieve this improvement will ultimately surpass the enterprise that depends on innovative improvements alone.

3.3.4   Increase Value at Every Step

- *Every action* that is taken *should add value* to the final product or service.

The value–added principle encourages people to eliminate all actions that do not add value (e.g., unnecessary movement of materials, unnecessary inspection, unnecessary inventory, etc.). In the spirit of continuous improvement, any step or procedure that is not adding value is a candidate for elimination, with a net increase in efficiency and an increase in net value to the system. If data are assembled but not analyzed, if reports are required but are not read, if material is moved repeatedly without being modified, if time is being spent waiting for others to act, the efficiency of the system will be improved by eliminating these unnecessary actions. The system must emphasize that its strategy is to identify and eliminate unnecessary steps. Insisting that each step add value is a good means of separating those that should be retained from those that can be eliminated.

3.3.5   Responsiveness

- *Timely response* of the system to changing external conditions is paramount to success.

Markets are changing constantly and the needs of the customer are being modified continually, either by changing conditions or through actions taken by competitors. The enterprise that has the shortest production time, that uses the least time to bring a new product to market, or that can modify a facility in the shortest time will have a distinct advantage in the marketplace. A strategic goal of the organization must be to respond quickly and effectively to the needs of customers and to capitalize on all opportunities quickly and efficiently. This requires that the organization encourage cooperation among all of the units and all employees, that it maintain its strategic directions so that the employees can seek continuously to achieve their goals, and that it organize itself to be effective in achieving a responsiveness to changing conditions.

3.3.6 Integrity

- *Integrity* in all actions by all employees is paramount.

Every organization and every employee will be confronted with circumstances that require that choices be made among alternatives, some of which may be more legal, more ethical, or more responsible than others. The credibility of an organization and an enterprise is a precious attribute. Customers, employees, suppliers, and the community must have confidence that the enterprise will behave in a legal and ethical manner. Each employee must operate in a manner that will never embarrass the organization or detract from the organization's reputation as a responsible member of the business or local community in which it is located and the customers it serves.

3.4 LEADERSHIP IN SETTING GOALS AND OBJECTIVES

As important as it is to establish realistic goals and objectives for each element of an organization, their existence will be of no avail unless they are communicated to everyone in the organization and unless every member of the organization understands their role in achieving those goals and objectives. Furthermore, the reward system must be consistent with the objectives established. Accomplishing this is the responsibility of management.

As noted earlier, the highest level of management has the responsibility of articulating clearly the organizations goals and objectives, not just once but in a recurring manner. This will be accomplished through a variety of means; the written media (e.g., memos and directives), the oral media (e.g., talks, meetings, and personal statements), and through the example that the management makes by its actions and responses to various situations.

Ensuring that all members of an organization fully understand the commitment the organization has to accomplishing its goals can be a daunting task in a large activity. Translating the overarching corporate goals into the specific objectives and goals of each subactivity is essential if employees are to view these as truly representing the priorities of the company. Consistency between rewards and measures of success is also an essential element of success.

3.5 THE GOAL OF WORLD-CLASS PERFORMANCE

To compete successfully in the global marketplace, a manufacturer's goal must be to become a *world-class manufacturer*, a term used to convey the sense of excelling–excelling in creating the highest-quality product, in achieving the lowest product cost, in being the most responsive to its customers, in having the most effective management system, and in achieving and operating the most efficient system of manufacturing. The first step in achieving this is the recognition that competition in the world marketplace can be brutal. In that marketplace, the corporations that will ultimately survive and prosper are those that have achieved world-class performance. Nothing else will guarantee success: not government protection, not monopoly control, not special markets. The Japanese describe this drive to excellence by the term *dontotsu*, the striving to be the "best of the best." This must be our corporate objective for survival in the world marketplace. This objective forms the first foundation of achieving world-class status.

3.6 THE *GOALS AND OBJECTIVES* FOUNDATION OF WORLD-CLASS PRACTICE

FOUNDATION: World-class manufacturers have established as an operating goal that they will be world class. They assess their performance....against their competition and against other world-class operational functions....They use this information to establish organizational goals and objectives, which they communicate to all members of the enterprise, and they continuously measure and assess the performance of the system against these objectives and regularly assess the appropriateness of the objectives to attaining world-class status. (Heim and Compton, 1992)

REFERENCES

AT&T. 1988, *Annual Report*. New York, NY.

BECKHARD, R. AND R. T. HARRIS. 1987. *Organization Transitions: Managing Complex Change*, Addison-Wesley Publishing,Company, Reading, Massachusetts:5.

FORD, H. 1984. *Our Corporate Culture: The Way I see it.* Ford Worldwide Management Meeting, Boca Raton, FL.

GARVIN, D. A. 1984. What Does "Product Quality" Really Mean? *Sloan Management Review, Fall*, 1984:25.

HEIM, J. A. AND W. D. COMPTON (eds.) 1992. *Manufacturing Systems: Foundations of World Class Practice*, National Academy Press, Washington, DC:4:28-30 .

KEOHANE, N. O. 1993. The Mission of the University, *Proceedings of the American Academy of Arts and Sciences,* 122 (4): 101.

LEFAUVE, R. G. AND A. C. HAX. 1992. Managerial and Technological Innovations at Saturn Corporation, *MIT Management*, Massachusetts Institute of Technology, Cambridge, MA.:8-20.

TOOT, J. F. 1987. *Timken* June:10.

QUESTIONS

3.1   Under what circumstances should the goals and objectives for a company be modified? Who is responsible for changing them? What procedures should be used in accomplishing the new goals and objectives?

3.2   Is it as critical to have a statement of goals and objectives for a company of 50 employees as for a company with 5000 employees? What influence will the size of the company have on the statement of goals and objectives for the company?

3.3   In the sense of the five different definitions of quality that are presented by Garvin, how did U.S. manufacturers view product quality in the past? How did it affect their share of the market? How did it affect productivity?

3.4   You have conceived of a new product and have developed a unique manufacturing process for its manufacture. Being convinced that this is a excellent opportunity to enter into the business of making and marketing this product, you have established ABC, Inc. As the leader of this new startup company, when should you become concerned about creating a goals and objectives statement for the company? Draft a goals and objectives statement for the company. How might this differ from the statement for an established company?

3.5   You have just been hired as general manager of the specialty materials division of a large diversified company. Your business unit has maintained good profitability by focusing on supplying specialized plastic components to kitchen appliance manufacturers. During the first week on the job, the head of your product planning activity comes to you with a proposal to enter the marketplace with a new plastic product that will be marketed directly to those who buy kitchen appliances. Many, but not all, of the processes needed to make this new product are currently available within the division. What questions must you answer as you explore this request? How will you obtain answers to these questions? Who will you involve in the process of making this decision?

3.6   Create a goals and objectives statement for a retail service company that is currently a seller of specialty coffee and related items in a retail mall but aspires to become a regional or perhaps national chain of coffeehouses.

3.7   You are the president of a well-established company that designs, manufacturers, and sells a specialty line of shoes for men. You have just been exposed to a lecture on quality management that emphasizes the importance of having a goals and objectives statement. How do you go about creating a statement for your company? How do you handle the conflict that develops between two key employees in your sales area who have very different visions of what the future of the company should be–one arguing that the company must develop its own set of retail stores and the other contending that all sales should be through prestigious national department stores?

3.8   Using Garvin's eight dimensions of quality (see Appendix A3.1), examine the strategies that Ford seems to have taken in its offering of two very different products, the Lincoln Town Car and the Escort. How would the content of the advertising be likely to differ for each of these products if the quality of each is to be used as an important basis on which the customer should choose these products over those of similar offerings by competitors?

3.9   Develop a matrix of the critical elements that constitute a high-quality college education. Use Garvin's eight dimensions of quality (see Appendix A3.1) as a basis for creating this matrix. If you were the president of a major university, how would you use this to enhance the quality of the education that is offered at your institution?

READINGS

GARVIN, D. A. 1984. What Does "Product Quality" Really Mean? *Sloan Management Review, Fall* 1984:25.

APPENDIX A3.1
THE DIMENSIONS OF QUALITY

Garvin (1984) identifies eight dimensions of quality that can serve effectively as a framework for examining the basic elements of product quality. In examining the quality of a product, he identifies and explains these dimensions in the following manner.

1. *Performance*. Performance refers to the primary operating characteristics of the product (e.g., fuel economy for an automobile, ease of programming for a VCR, or clarity for a television set). This dimension combines elements of the product and user-based approach, in that quantification of the products attributes are used. Since the connection between performance and quality depends somewhat on individual preference, this dimension is not a unique measure of quality. "The performance of a product would correspond to its objective characteristics, while the relationship between performance and quality would reflect individual reactions "(Garvin, 1984:30).

2. *Features*. Features are the secondary characteristics that each product possesses. Although they are not critical to the basic performance of the product, they enhance its performance. Examples of features include the redial function on the telephone, coin changers on vending machines, and buzzers that signal that the lights have been not been turned off when you are leaving your car. "Features, like product performance, involve objective and measurable attributes; their translation into quality differences is equally affected by individual preferences. The distinction between the two is primarily one of centrality or degree of importance to the user"(Garvin, 1984:30).

3. *Reliability*. Reliability is a measure of the probability of failure of a product over a period of time. Direct measures of reliability are mean time between failures (MTBF) and for a class of products the number of failures per 100,000 products within one or three years.

4. *Conformance*. Conformance is closely related to the success of the product in fulfilling the design and operating characteristics that were intended for the product. This applies to both the manufacturing operations as measured by the number of defects that occur during manufacture and lead to in-plant repairs and failure of the product to meet design specifications in the field as measured by the number of recalls by the manufacturer or the failure to meet certain declared objectives of the design (e.g., fuel economy objectives, energy efficiency of an appliance, or distortion in an audio amplifier). Improvements in the conformance of the product are generally translated into improved quality.

5. *Durability*. Durability is a measure of the usable life for a product. A product that has a long useful life is considered durable. If occasional and not excessively costly, repairs may be used to extend the usable life of the product. In this regard, there is a close connection between reliability and durability.

6.   *Serviceability.* A product is judged to have good serviceability if repairs can be made rapidly and reliably. Design for serviceability has come to mean that designers must be aware of the needs of service personnel and must provide designs that minimize the time required to service a product. A quantitative measure of servicability is the mean time to repair a given failure (MTTR).

7.   *Aesthetics.* Aesthetics is closely related to the user-based approach. It reflects personal judgment and individual preferences (e.g., Gucci handbags, Rolex watches and Mikasa crystal).

8.   *Perceived quality.* This is an indirect measure of quality and depends less on the objective measures of the product and more on the perceived characteristics of the manufacturer. This can be found in loyalty to particular manufacturers (e.g., Chanel for fine perfume, Waterford for excellent crystal, and Tiffany for elegant jewelry). This loyalty may even arise from a favorable association of a particular person with one company's products (e.g., Lee Iaccoca with Chrysler).

> A recognition of these eight dimensions is also important for strategic purposes. A firm that chooses to compete on the basis of quality can do so in several ways; it need not pursue all eight at once. Instead, a segmentation strategy can be followed, with a few dimensions singled out for special attention....This....suggests the importance of carefully targeting one's quality niche. The selection of a defensible niche, however, is only a first step. Operational requirements must also be met, for each dimension of quality imposes its own demands on the firm. High performance requires careful attention to design and a strong design staff; superior durability requires the use of long-lived or "derated" component and close cooperation between the engineering and purchasing departments; superior conformance requires attention to written specifications and precision in assembly; and exceptional serviceability requires a strong customer service department and active field representatives. In each case, a different function enjoys the lead role, and different tasks are required for success. The managerial implications of this analysis should be obvious: after selecting the dimensions of quality on which it hopes to compete, a firm must tailor its organization and operations to meet these specific needs. Otherwise, the wrong departments may be elevated in status, or the wrong tasks pursued. (Garvin, 1984:33)

REFERENCE

GARVIN, D. A. 1984. What Does "Product Quality" Really Mean? *Sloan Management Review*, Fall, 1984:25-39.

4
Operating Goals and Objectives

> When you can measure what you are speaking about and express it in numbers, you know something about it; but when you cannot express it in numbers, your knowledge is of a meager and unsatisfactory kind. Lord Kelvin

Operational goals and objectives relate to the general operating objectives that an enterprise has established, as well as the specific goals that an enterprise has for its products, services, and processes. The operational goals are usually short range. They may emphasize profits for the current and next few quarters, the desired production volumes for each product, and even levels of overtime or inventory that are expected for the next period of operation. The validity of the goals that are set for the products or services that are to be offered by the enterprise or the plant must be tested continuously against the demands of the marketplace.

Questions such as the following must be answered in the process of establishing the operational goals and objectives. What attributes must the product or process have for it to be competitive in the marketplace? What are the changes that must be planned for succeeding generations of products? When should a new technology be introduced? In the manufacturing plant, what are the objectives for productivity, inventory, cycle times, quality, yield, and so on? What improvements are needed over time, and how much investment can be afforded in achieving these objectives? On the financial side of the business, the goals and objectives must be concerned with the desired profit levels, the rate of growth of the enterprise, and the return on investment.

Each unit of an organization must establish short-term goals. Having done so, it is the task of management to understand the interdependencies of the goals of the various units and to ensure that the organization can work collectively to achieve them. The marketing organization cannot meet the goal of increased market share if the products are not competitive in the marketplace. The profit goals cannot be reached if the manufacturing operation is inefficient. The manufacturing operation will have difficulty in meeting a goal of producing high-quality products if the product engineering activity is constantly changing the product.

In managing manufacturing operations, a variety of operational goals will need to be set. It is necessary to be concerned with minimizing the defects produced during manufacture, the amount of inventory (both work-in-process inventory and finished goods inventory) that is being carried, and the number of pieces being produced per unit of time. Even more detailed goals will be needed concerning the performance of the unit processes that are employed in the manufacturing. These may involve the amount of time the machines are idle, the amount of waste that occurs at each step of the process, or the length of time it takes to change a die or to implement a new machine setting.

It is not uncommon practice to establish many of these goals based on a percentage of change from the preceding year or preceding reporting period. Common practice has been to require an operation to improve by a few percent each year. Often, the basis for the choice of the level of improvement was historical. Since we achieved $x\%$ during the preceding period, we should be able to do the same again. This became the basis for establishing the goals for continuous improvement. When the product changed or a new process was introduced, the entire procedure began anew.

There are, however, alternative ways to establish operating objectives. As an example, a variety of demands that must be met to operate a just-in-time manufacturing system are discussed in Chapter 12. For the current discussion, only one from the group known as the seven zeros of JIT will be used for illustrative purposes.

Consider the objective of achieving zero defects in the manufacturing activity. How realistic is it to use this as an objective? Inasmuch as one can never achieve better than zero defects, this objective truly represents the theoretical limit for the operations. If there are theoretical limits, will people take them seriously given that these objectives will probably never be reached?

Contrasting the differences in approach that a Japanese and a U.S. company frequently take in attempting to accomplish a reduction in defects is instructive. Assume that the creation of defective parts is occurring at a frequency of 20 in 100,000. A normal approach to defect reduction in a U.S. company would be to establish an objective of reducing defects during the coming year by some amount, perhaps an annual improvement of 5%. This translates into reducing the frequency of defects from 20 in 100,000 to 19 in 100,000.

In the setting of local objectives for the organization, the following questions will arise. If the production activity is large and complex, how does one determine what action will lead to the reduction of one defect for a run of 100,000 products? Who is given the task to accomplish this? Furthermore, what arguments will ensue if the objective was set at a 10% reduction in defects instead of 5%? How many reasons could be offered by the operating people that this objective is not possible, or that it is not cost-effective, or that it is simply unreasonable?

Contrast the difference in philosophy that an organization will develop if everyone in the organization is committed to reducing the number of defects to zero. No time and effort will be lost in establishing the goal. No time and effort will be spent in finding the one problem that can be solved to reduce the defects by two instead of one. No one will be concerned about the assignment of responsibility for improving, since all employees accept that they share the responsibility of meeting the goal of reduced defects. If the organization is committed to this as an objective, everyone will find *all* the ways to improve and will continue to look for those ways until the theoretical limit is approached.

The challenge to management is to determine what goals should be established. Goals that are set unrealistically high and cannot be met can result in a disillusioned workforce and a lack of incentive to improve. Goals that are set too low will be met but will not achieve improved performance and the highest possible return on the investment in facilities and people. There are two important tools that should be used regularly in setting short-term goals and objectives. Use of these tools can help assure management that the enterprise's goals are realistic and are appropriate for the achievement of world class status. These tools are known as benchmarking and theoretical limits. Benchmarking is particularly useful in setting short-term objectives, whereas the identification of theoretical limits is important for defining both short- and long-term goals.

4.1 BENCHMARKING: LEARNING FROM OTHERS

Searching for and identifying the practices that represent the best that are being achieved by *any* manufacturer, irrespective of the industry that they are in or the product that they produce, is known as *benchmarking* (Camp, 1989; Compton, 1992). To accomplish this, a quantitative comparison of performance is required. A simple ranking is not sufficient. Choosing the companies or the activities against which to measure oneself is critical. Since a leader in one part of the business will not necessarily be a leader in all parts, it will be necessary to benchmark against more than one competitor, all of whom are chosen because they are leaders in specific areas. Furthermore, since some activities are not specific to a given industry, it will be appropriate to benchmark performance against the best company, whether in your industry sector or in another . As an example, while certain manufacturing technologies may be unique to one industry, the various approaches taken to managing the manufacturing system will probably not be industry specific. As noted by Compton (1992):

> The metrics that are useful for benchmarking are numerous and varied. In some industries it is common to emphasize a few operating characteristics, while in others the focus will be on a wide range of parameters. In some cases the parameters that describe system performance will be highlighted. In others, it may be more common to examine the performance of subelements of the system. Some metrics may be reasonably easy to obtain. Others may need to be estimated based on limited available data. However it is accomplished, benchmarking is a critical foundation for the successful operation of a manufacturing enterprise.

Consider the problems that arose in the late 1970s for the U.S. automotive industry. During the period between the mid 1940s and early 1970s, domestic manufacturers were almost completely preoccupied with competition from other domestic manufacturers. They paid little attention to the level of performance that was being–or could be–achieved by manufacturers in other countries. Their efforts were focused on producing vehicles comparable in performance, quality, and cost to those of other U.S. manufacturers. Although this could be viewed as a valid approach during the early part of this period, since the worldwide leaders were U.S.-based companies, it certainly was not appropriate for the late 1970s.

The relative performance of the U.S. and Japanese manufacturers in and following the late 1970s attests to the consequences of this lack of concern of what others could accomplish. Although the Japanese were not major factors in the U.S. marketplace prior to the 1970s, the German manufacturers–with their high quality–were present in the U.S. marketplace but were not considered as serious competitors, since they generally concentrated their product offerings in the high-price segment of the market. Thus U.S. manufacturers did not view them as principal competition and did not attempt to learn seriously from them. Some of the same situations existed, in varying degrees, for the U.S. consumer product and semiconductor industries.

The foregoing illustrates two very critical aspects of benchmarking. First, benchmarking against the wrong competitor can lull an enterprise into a sense of false security that can be disastrous. Second, as new competitors emerge, as the marketplace changes, or as the market is restructured because of events occurring beyond the enterprise's national borders, the focus for benchmarking must shift. It becomes important, therefore, for a firm to review continuously the appropriateness of its choices for comparison.

In initiating an effort at benchmarking, the following questions must be asked:

- What activity should be evaluated? For example, will the focus be on the competitiveness of the product, the manufacturing activity, or the field service activity?
- What specific metrics can be identified that measure the effectiveness of the activities that are to be examined?
- What company or companies will be chosen for comparison? Who are the world leaders for this activity?
- Will attention be confined to companies in our industry sector, or does the performance of companies in other industries offer important comparisons?

The metrics that are useful for benchmarking are numerous and varied. In some industries it is common to emphasize a few operating characteristics, while in others the focus will be on a wide range of parameters. In some cases the parameters that describe system performance will be highlighted. In others it may be more common to examine the performance of subelements of the system.

The operational parameters that should be considered for evaluation in a comprehensive comparison of the performance of operations with that of a competitor include:

1. Financial metrics
2. Product performance metrics
3. Unit operation metrics
4. System operational metrics
5. Aggregated measures of performance
6. Measures of long-term competitive capability
   a. Technological capability

    b. Technical personnel

7. Interactions with other activities

It must be emphasized that the appropriate measure of performance relative to a competitor can be obtained only when the appropriate data are available. This emphasis on data–sound data that are reflective of the parameters that determine performance–is a continuing theme of this book. There is no substitute for facts and facts are found in data that are obtained in an unbiased measurement of phenomena. Putting in place a system that encourages careful and systematic collection of such data can be expensive and time consuming. Without it, however, the information that is normally available may be limited and conclusions based on it may be erroneous. Benchmarking is the process by which data are obtained regarding the performance of the leaders in a field. It is these data that you can use to compare the performance of the firm to the world's best and to determine what operating objectives must be achieved to remain competitive and to achieve world-class status. Some of the important elements that are included in each of the foregoing metrics are indicated below.

4.1.1   Financial Metrics

Establishing the proper metrics to characterize the operation of your system and to compare it with others is a critical first step. As William E. Butler, chairman and CEO of Eaton Corp. has noted (1993), the choice of the proper metrics can be critical to a successful business: "In my prior experience as a general manager and group president, I was never given productivity goals as part of my incentive plan. We focused on profitability, primarily. We never measured productivity. Now we are using a measurement, not unique to us, that includes *all costs*."

Traditionally financial metrics have been collected and used to assess performance. These include:

1. Return on investment (ROI), measured as a percent of total investment

2. Return on assets (ROA), measured as a percent of the value of total assets

3. Return on sales (ROS), measured as a percent of the total sales

4. Ratio of debt to equity (percent leveraged)

5. Number of turns of inventory each year

6. Number of units produced per employee

7. Number of units produced per hour

8. Sales per employee

9. Profit per unit of production

10. Volume at which break-even occurs

Each item above is a measure of the composite performance of the system.

With the exception of the last three, they can reasonably be compared from one year to the next to assess relative performance over time. The last three can also be used over time if proper adjustment is made for inflation and the value of money over time. All of the above are important system metrics.

4.1.2   Non-Financial Metrics

Although, traditionally, a variety of nonfinancial metrics have also been collected and analyzed for the manufacturing system ,it has become common practice only recently to examine the long-term trends of such metrics. The following examples are useful in assessing the performance of the system:

1. Average number of defects detected by customers in the first months of ownership
2. Average number of defects detected during manufacture and repaired
3. Hours lost to production from unscheduled maintenance
4. Work-in-process (WIP) in the plant
5. Number of machines per worker
6. Length of time to change a machine or introduce a new operation (e.g., time to change a die in a stamping machine)
7. Number of job classifications in the plant
8. Amount of material made obsolete by model changes
9. Average units of energy used per unit of product manufactured
10. Rate of absenteeism of the workforce

With competition in the world marketplace requiring more attention to the rate at which new products must appear, it has become necessary to attempt to identify a set of metrics that are effective in assessing this aspect of the performance of the system. These are a measure of the responsiveness of the system to changes that will occur in the customer demand for products. Some of these are:

1. Number of months required to introduce a new product model
2. Number of engineering change requests (ECR's) required during a new product program
3. Number of units produced prior to a model change

Stalk and Hout (1990), Krupka (1992), and Hammer and Champy (1993) argue that a number of the metrics above can be captured by paying attention to time, in their view the critical metric for determining competitiveness. Hammer and Champy illustrate this through an interview with Regis Filtz of Bell Atlantic, recounted in Appendix A4.2. Stalk and Hout (1990:39) describe the importance of time as a metric in the following way:

> Time is a fundamental business performance variable. Listen to the ways in which managers talk about what is important to the success of their companies: response *time*, lead *time*, up *time*, on *time*. Time may sometimes be a more important performance parameter than money. In fact, as a strategic weapon, time is the equivalent of money, productiv-

ity, quality, and even innovation....

   Today, time is on the cutting edge of competitive advantage. The ways leading companies manage time–in production, in sales and distribution, in new product development and introduction–are the most powerful new sources of competitive advantage.

At a higher level of aggregation, it is important to attempt to understand the management philosophy that a company is employing, the ways in which it is utilizing its employees, and its organization structure. This leads to the needs for information of the following types:

1. Management philosophy
2. Organizational structure
3. Emphasis on employee involvement and employee empowerment
4. Utilization of teams

   It is recognized, of course, that these nonfinancial metrics have a significant impact on the financial metrics. While these various metrics are not fully independent, they each provide the organization with a set of operating statistics that can be used to determine performance relative to other successful enterprises. In this way, they can be used to identify problems that are generating unfavorable financial results.

   Stalk and Hout (1990:209) note that it should not be assumed that in areas in which information cannot be found, competitors are behaving just like you. They argue that the most useful thing that you can do is to assume that they are not and to imagine how they might be approaching a problem differently. They offer the following keys to good competitor benchmarking.

   • Look for unusual practices. Ask how they could make sense. View them hypothetically as missing links, not anomalies.
   • Talk to customers he sells but you can't.
   • Imagine him to be one-tenth your actual size. Ask how work then would be organized and managed.
   • Imagine his ideal information flow. Figure how many material flow problems would then disappear.
   • Study his distribution–frequency, composition of lots, channels, customer selection, and so forth–as clues to inside workings..

Values for the metrics above can be determined in a reasonably straightforward manner. It is not uncommon, however, to describe desirable characteristics of manufacturing systems by terms that are not so readily quantified. These include:

1. Flexibility
2. Productivity
3. Responsiveness
4. Capacity
5. Complexity

Use of these terms to describe the performance of a manufacturing system often results in confusion and ambiguity. This can be illustrated by considering the alternative definitions that can be found for *flexibility*.

> The flexibility of a manufacturing system is its ability to respond to change in a positive manner. (Choobineh, 1986)
>   The ability of a manufacturing system to accommodate undesirable or threatening contingencies that can occur in the short, medium or long term. (Graham and Rosenthal, 1986)
>   The ability of the system to accommodate changes in demand, deviations of incoming material from specifications, or arbitrary sized batches for each model. (Nevins et al.,1989)

In examining these for flexibility, it is difficult to design a proper metric. For example, how should the response to change be measured? How is the ability to accommodate undesirable contingencies to be measured? What is the proper measure of the ability to accommodate arbitrary sized batches? Flexibility in a manufacturing system can be critical in remaining competitive. If your competitors can accommodate a new model in 12 months and your system requires 24 months to respond, it is certain which system will survive. Although it is not so difficult to identify the desirability of low or high values for these metrics, it is extraordinarily difficult to offer a value for them that can be compared unambiguously with a value determined by someone else.

Productivity is often defined in terms of the number of items produced per hour of employment in the plant. As the level of automation in the plants has increased, this definition becomes somewhat less meaningful. Since productivity depends on both the level of capital that is employed and the number of employees that are used in the production, these two secondary metrics must be coupled in a proper manner before productivity can be used as a meaningful metric for benchmarking.

Complexity may refer to the range of operations that are used in the plant, the variety of the products that are made in the plant, or the variety of markets that are served by the plant. While any of these definitions can be meaningful, this term, as a descriptor of a manufacturing enterprise, must be defined carefully before it can be used as a metric for the purpose of benchmarking.

This illustrates the difficulties that can arise when there is no standard definition for terms. Attempting to develop precise values for ambiguous terms can present great difficulty in interpretation. Meaningful comparisons of performance among different enterprises is best done when the metrics are quite precise.

4.1.3   Product-Related Metrics

Since the metrics that relate to specific aspects of the manufacturing operations will be discussed in future chapters, the current discussion will be limited to illustrations of the type of questions that need to be addressed and to a mention of some of the metrics that might be used in benchmarking the performance of a product against a competitor's product (Compton, 1992:100–106).

> The performance of the product in the hands of the consumer is a key indicator of the capability of the development, production, and marketing system. The successful manufacturing enterprise does not, however, simply wait to see emerging trends in market share and then respond. A constant assessment of the relative merits of competing products

and the continuous incorporation of the appropriate response into one's products are a mark of the successful firm. The "appropriate response" must be determined, of course, within the context of the needs and wants of the customer. The following are some of the actions that successful enterprises have institutionalized to obtain this information:

- Evaluation of competing products under conditions similar to those used to test your products
- Tear down analysis of competing products
- Reverse engineering of competing products
- Customer surveys of competing products

From tests and evaluations of the type described above, information is obtained on the following important metrics:

- Part counts
- Material types used
- Material utilization in each component
- Processes used in production (e.g., assembly techniques)
- Estimate of product costs
- Service capability (e.g., field repair versus field replacement)
- FMEA (failure-mode effects analysis)

From customer surveys undertaken either directly or through industry-wide surveys carried out on behalf of the industry, it is possible to obtain information regarding:

- Quality of the product as experienced by the customer
- Long-term durability of the product
- Fraction of sales to repeat customers
- Responsiveness of the producer to service requests

Some industry, consumer, and government groups join together to undertake benchmarking studies of all products within a class. Standard tests are devised that are used for all products of a given class. Examples include the speed with which a class of computers can execute a program, the number of instruction per second that a computer can execute, the length of time that a computer vision system requires to locate and recognize a specific pattern in a scene, the fuel economy of new cars and trucks, the emissions from a vehicle that is driven over a fixed test cycle, the noise level of engines, and so on. These comparisons offer useful broad-based comparisons of products and should be examined regularly. The question always arises, however, whether the particular test that is used for the comparison is appropriate for all products of the class. For example, a general-purpose computer may not perform as well in a test as one that is designed specifically for the type of problem being run. It is important, therefore, to assess carefully the appropriateness of the test as well as the results derived by that test.

Determining the properties and performance characteristics for a competitor's product will be useful only if comparable information is available for your products. To use effectively information derived from benchmarking, it is necessary for a firm to establish a system that assesses the same quantities for their products regularly and systematically. This is just one more example of the importance of collecting good data and using the information that it provides for achieving further improvement.

The list above can be expanded and will need to be augmented for many products. The following are offered as examples of some of the questions that might arise that will demand the identification of additional metrics. Some products may have such unique features or attributes that they must be probed and examined in greater detail. For example, does the minimal use of material suggest that the designer had available a set of analytical tools that allowed the design to be optimized in ways that reduced its cost? Perhaps it is obvious from the design that the product can be used under much more severe conditions than are normally encountered with your product. Does this suggest that the manufacturer of this product has or is planning on entering another market that would lead to expanded production and perhaps lower unit costs, thus making their products more competitive in your marketplace? For products containing both a software and a hardware component, has the trade-off between these two been made to emphasize flexibility in controls or to minimize hardware costs?

A firm that regularly practices benchmarking should be one of the earliest purchasers of a competitor's product. Detailed examination of the product, analysis of the results of that examination, and an understanding of the trends that the analysis yields can offer important guidelines to any manufacturer striving to set short-term goals and objectives.

4.1.4   Relationship Among Metrics

The identification of important metrics and a quantification of them represents only the first step in understanding a system as complex as a manufacturing operation. If these important variables are to be useful in guiding actions that will enhance the competitiveness of the enterprise, it is necessary to understand the relationships that exist among them. What, for example, is the relationship between operational variables such as flexibility and responsiveness to the firm's share of the total market? What is the quantitative relationship between the time needed to introduce a new model or new product and the number of critical processes in the manufacturing system that are under tight control? What is the quantitative relationship between the financial and the nonfinancial metrics (e.g., between profits per unit of production and the various nonfinancial metrics listed in section 4.1.2)? Although there are some generally accepted rules-of-thumb concerning these relationships, studies have not been carried out that offer convincing evidence of their validity. A striking example of the importance of these interactions in achieving high productivity is provided in Appendix A4.2. The lack of quantitative understanding of the system represents a serious deficiency. It will be solved only through a careful research effort that develops the necessary background and data for a number of products in a variety of industries.

4.1.5   Choosing the Companies to Benchmark

The choice of the company against which benchmarking is to be done will be determined by the particular set of metrics that are to be explored. A comparison of product performance will focus attention on firms that produce similar or competitive products. Comparison of unit-operational performance may include a wide variety of companies outside the immediate competitive market of the firm, since similar processes may be used by a large number of firms. Financial comparisons are most often appropriate among firms within the same industry sector, since the level of capital investment, labor practices, and so on, tend to be more uniform within sectors than across sectors. Comparison of management philosophies and organizational structure can be made among companies in many industries, without close attention being given to the products that are produced.

General financial data are available for public companies from a variety of sources, including corporate annual reports and SEC 10-K filings. Information on privately held companies is much more limited. Product performance metrics can be determined for products that are offered for sale and are discussed below. Assistance in identifying companies with good performance in their unit operations can often be obtained from suppliers of production equipment that is being used by the companies. Determining the companies that are leaders in system and aggregate performance can often be assessed by examining total outputs from a plant or company and relating it to total employment, total investment, materials costs, inventory levels, and so on, much of which can be determined from public financial reports. Identifying companies who are leaders in the ability to compete over the longterm is more difficult and usually requires extensive study of trends in investment, product offerings, market share, and so on. Fisher (1992) offers some insight into the types of questions that must be asked in analyzing long-term trends.

4.1.6 Action Plans

The collection and analysis of benchmarking data and an understanding of the performance of your firm relative to that of world-class performers are merely the initial steps in the process of achieving world-class perfomance. To be effective, the information that has been developed must be used to create an action plan. The conclusions that are drawn from these studies must be disseminated throughout the organization. Goals for the organization must be adjusted to reflect the opportunities conveyed by the benchmarking studies. Each group must use these results to determine how its performance must be altered to allow it to achieve the levels of performers that the benchmarking study will have demonstrated are needed. This cascading of new objectives for the organization must follow the procedure outlined in Section 3.3. The benchmarking studies will have limited value if they are not used to change the thinking of the organization. The resulting action plans must create the opportunity for genuine change in the organization. The opportunity that these provide to make significant improvements must not be allowed to be lost.

4.1.7 Limitations of Benchmarking

Despite the important benefits of benchmarking, and there are many, it is important to recognize its limitations. First, benchmarking is unlikely to signal that an entirely new product or process will be forthcoming in the future. That is, it tends to offer information relevant to current practice. Second, it is unlikely to anticipate that a new competitor will enter the marketplace. It will, of course, allow one to evaluate the performance of new competitors when they appear. Finally, as will be seen in a number of examples in subsequent chapters, some of the most important information about a company's performance may be extraordinarily difficult to obtain. Thus it may be impossible to achieve the objective of obtaining significant indicators for all of the desirable metrics. In these cases, surrogate metrics may need to be used. For example, if unit costs of a product are desired but cannot be determined, it may be necessary to estimate them from a knowledge of the sales price and an estimate of the profits per unit. If the productivity for each of several products made within a plant is desired, it should be possible to estimate this based on knowledge of the average number of each product made by the plant in a year, adjusted for the complexity of each part, as related to the average annual employment and the investment level in the plant. Obviously, the more estimates that must be made, the more uncertainty there is in the validity of the final number. Despite these problems and the uncertainties that they may introduce, it is critically important for any company who desires to be a world-class leader to benchmark themselves continuously against the best performers in the world.

4.2   THEORETICAL LIMITS

In any field, technological improvement is eventually limited by the laws of nature. The ultimate strength of a fiber is limited by the strength of its intermolecular bonds. The number of transistors that can be placed on a silicon chip is limited by the crystal structure of the silicon material. The goal of technical management is to identify the limits of any given technology early, as a first step in determining what finally can be accomplished with it.

Industry, though, is usually far from these natural limits, and it is more likely to come up against practical, physical barriers that represent the current state of the art. The differences between the technical and the state-of-the-art limits determines the technology's potential for performance improvements; the greater the distance, the more the potential. This can have dramatic strategic implications.

When a major manufacturer of a mature product looked at its business from a context of technical limits, for example, the company was surprised. Each year its lab's goal had been to improve productivity over the past year. Using that amorphous target, productivity had increased 20% in the previous 10 years. But by defining the best that could be done, the company boosted productivity 40% in just two months.

Once a company becomes aware of what it can hope for by defining the limits of its technology, it can then determine how far it wants to go to approach these limits. It accomplishes this by applying research–knowledge acquisition–and development–knowledge application–to narrow the gap between the state of the art and the natural limit. (Foster, 1982:24–33)

Richard N. Foster has long been an advocate of the practice of determining the theoretical limits that are possible and using these to determine the goals and objectives to be established for the enterprise. He notes that it is not possible to determine *properly* the goals that should be established and the resources that need to be allocated unless one has determined how much improvement is still available. If the theoretical limits for the performance of a material are being approached, the benefit from achieving that last measure of performance–if it is small–is not likely to be worth the expenditure of large resources or a long period of development. Rather, the appropriate question becomes, is it better to look for a new product or a new concept that does not suffer from this limitation? If, on the other hand, the performance is far from the theoretical limit, an improvement that moves nearer to the theoretical limit may be very worthwhile.

Turnbull et. al., (1992) regularly use the concept of theoretical limits in their analyses of the strategic management of Alcoa operations.

> A sound strategic analysis requires the compilation of a rich set of data. Data bearing on the fundamentals of any manufacturing enterprise must be included: the factors driving overall demand for its products, customer satisfaction criteria and performance against those criteria, the nature and strengths of the competition, and the efficiency and effectiveness of the manufacturing system.
>
> The purpose of data in strategic analysis is to contribute to a better understanding of the processes for the business. Therefore, the data are assembled in the language of the processes themselves, for example, energy consumption in British thermal units per pound, productivity in pounds per hour, and delivery performance in days.
>
> Understanding theoretical limits provides both an outer bound for forecasts of potential process performance and a framework for clarifying the principles that govern the process. Theoretical limits are not goals or plans; they are numerical estimates of the ultimate conceivable level for a process variable, supported by the fundamental principles and reasoning that led to that numerical estimate.
>
> Theoretical limits often introduce stretch into considerations of future process potential. Sometimes, however, considering historical performance against the theoretical limit will indicate that there is little further opportunity for improvement in a particular process feature; this is also an important strategic consideration.

Theoretical limits can be determined for a variety of systems. Thermodynamics establishes the theoretical limits with which a device can convert thermal energy into mechanical energy. For example, the Otto cycle–the basis of the spark-ignited internal combustion engine–has a theoretical efficiency of 67% for a compression ratio of 9:1, assuming ideal gases. The efficiency with which an engine in a vehicle converts fuel energy to useful work–movement of the vehicle over a particular driving cycle–may be less than 20%. Thus the overall efficiency of the vehicle system has ample opportunities for improvement. Similar estimates offer insights into the strength of structures. As is well known, the strength of glass depends critically on the thickness of the material. A further example is shown in Figure 4.1. The strength of a glass fiber 1/10,000 of an inch in diameter (10^{-4} in.) is approximately 20 times that of a piece of bulk glass. Also, the strength of a rod of glass drops drastically if there are scratches on its surface. For the product design, the challenge is to attempt to conceive of ways in which the theoretical strength limits for thin fibers can be approached in systems that demand bulk material. This is a partial reason for the interest in systems consisting of composites of glass fibers and plastics. The strength associated with the fibers and the low cost and weight of the bulk plastics can be combined to give very attractive properties.

As a further example of the theoretical limit that is imposed on a process by the current physical system, consider a machining operation that can be described by two time constants. The first, tc, is the length of time that is required for the machine to execute the desired cut in the material (i.e., the metal removal process). The second time, ts, is the length of time that is required for the material handling system to remove a finished piece from the machine and insert a new piece for machining. Clearly, the total cycle time for the process, the time from beginning to ending of the machining of a piece, is simply $tc + ts$. A great deal of effort has been made to reduce the cycle time by developing new cutting

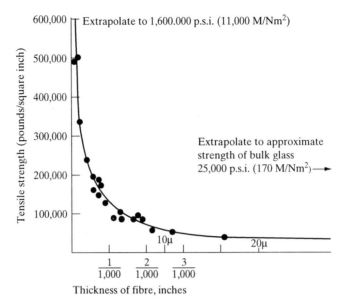

Figure 4.1 Strength of glass fibers as a function of thickness. (From Gordon, 1976. p.75.)

tools that allow faster and deeper cutting, thereby reducing tc. Without making changes to the material handling system, the theoretical limit for the cycle time is ts. In other words, even if tc can be brought to nearly zero, the total cycle time will be determined by ts. For this example, the theoretical limit for the cycle time is set by ts. This does not say that ts cannot be reduced by appropriate actions. It simply means that all contributions to the cycle time must be considered if efficiency is to be maximized.

These simple examples are intended to draw attention to the need to understand the entire process–the system–when setting priorities. This understanding, combined with an understanding of the theoretical limits, becomes a powerful tool for analyzing and establishing goals for the operations. Turnbull et al. (1992) illustrate the importance of understanding the level of current performance relative to that of the theoretical limits for the process in Figure 4.2. Also illustrated is the engineering limit, an estimate of the level that the processes could attain using known technologies. As the performance of the process approaches the theoretical limit, it can prove as wasteful of resources to attempt to "wringout" the last few percentage points of improvement from a process as it is to focus on the wrong priority in attempting to achieve an improvement. Turnbull points out that further improvement in the Hall process for aluminum production is unlikely, as developments have brought the parameter *current efficiency* to levels that approach those that are theoretically feasible.

The Japanese have demonstrated that using the theoretical limits for objectives encourages everyone to take actions that move the system closer and closer to these objectives. This was illustrated in the early paragraphs of this chapter with the example of zero defects for manufacturing. To be successful, employees must be prepared carefully, including discussion and commitment, if these objectives are to be introduced into a system that has operated under the more historical set of objectives. By building a consensus and bringing everyone together as a team, these objectives can be introduced and can become a powerful tool in helping the system to be improved continuously.

The use of theoretical objectives will be successful only if management has adopted the proper view of how individuals and sections of the organization are to be measured and judged in meeting their goals. If the management measures and rewards performance based on continuous improvement, these theoretical goals can be powerful stimulants to improvement. If, on the other hand, management penalizes the organization that does not reach a firm numerical goal, the use of theoretical limits for goals can be a serious detriment and even a barrier to progress.

4.3   THE IMPORTANCE OF LEARNING FROM PAST EXPERIENCE

An understanding of the manufacturing system can be achieved only when reliable quantitative data of system performance have been collected and analyzed. It is not sufficient to *imagine* that one understands the system or to depend on *rules of thumb* to determine how well a system is operating. Experience can be as important a teacher for an organization as for an individual. This requires that both successes and failures be analyzed and reported. It also requires that data be collected on performance over long periods of time and that the trends in those data be understood. Far too often, organizations engage in data collection sporadically, use different assumptions in the measurement process, and forego the opportunity to correlate performance over long periods. This pattern effectively prevents the organization from learning from its experiences and from using past performance to guide it in setting objectives for improvement or for recognizing that a change has occurred in a level of performance.

Traditionally, learning curves have been used to estimate the probable unit cost of a product in the future, based on experience accumulated in its earlier production. This is possible because a large number of studies suggest that the unit cost of many different products follows a pattern of dropping in a more or less predictable fashion with each doubling of production. What will be demonstrated in this section is that other important parameters of the manufacturing system also improve through learning and that experience can be extrapolated if the organization has engaged in the systematic collection, analysis, and dissemination of good performance data over a period of time. It is unfortunately true that few managers insist on the collection of the types of data that are essential for achieving the needed understanding. This arises for a number of reasons. Collecting the data is expensive and time consuming. Product and plant designs change, making it difficult to achieve an understanding of the historical

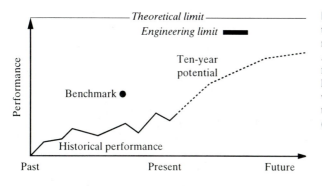

Figure 4.2   Representation of the historical and potential performance of a manufacturing process. Also shown is the theoretical limit for the process, the engineering limit, and the performance of a world class competitor as given by the benchmark information. (From Turnbull, 1992. p.227.)

significance of the data. Finally, and perhaps most important, it has not been tradi-tional to keep such data. Achieving world-class competitive status demands that en-terprises change their attitudes toward data collection and use.

4.3.1   The Learning/Experience Curve

The *learning/experience* curve is a means of representing long-term trends in perfor-mance. It has been shown that these curves can also be useful in projecting future levels in performance. The learning curve is a representation of a phenomenon that was first de-scribed in 1936 by T. P. Wright in connection with the manufacture of aircraft. Wright (1936) examined the cost of producing a single aircraft and discovered that the average labor costs decreased by 20 percent with each doubling of the number of aircraft produced. A power law is the proper mathematical formalization of this general relationship:

$$\overline{C_N} = C_1 N^{-\alpha} \tag{4.1}$$

where $\overline{C_n}$ is the average cost of all N units produced and C_1 is the cost of the first unit. For Wright's observation of a 20 percent decrease in average cost for each doubling of production, the exponent a would have the value 0.322. Often Eq. (4.1) is refor-mulated in terms of X_n, the cost of production of the nth unit, in which case

$$X_n = K N^{-\alpha} \tag{4.2}$$

and $K = C_1 (1\text{-}b)$. The differential form of this relationship is the familiar

$$\frac{dX}{X} = -\alpha \frac{dN}{N} \tag{4.3}$$

The graphic form for these equations is represented in Figure 4.3 and Figure 4.4. It should be emphasized that the unit cost values plotted on the ordinate of these curves must be calculated in constant dollars to ensure that inflation does not dis-tort the data over time.

The observation by the Boston Consulting Group that a large number of prod-ucts follow this general trend has lent great credence to it. Although there is no gener-ally accepted theoretical basis for predicting the slope of the curve in Figure 4.4, a vast majority of cases group around a value that corresponds to a reduction to 85% of its former value for each doubling of the production. This is the origin of the *85% learn-ing curve*. Several different examples of this are given in Appendix A4.1. Henderson and Levey (1965) and Argote and Epple (1990) have examined the validity of this in a wide variety of manufacturing environments. The studies have included a comparison of the performance of various plants making the same product as well as plants that make

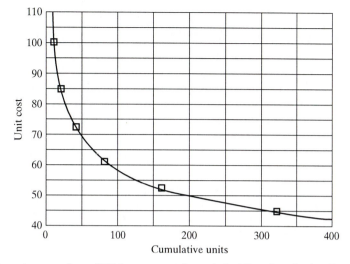

Figure 4.3   Learning curve for an 85% improvement for each doubling of production, plotted as a linear curve.

widely different parts. The latter studies confirm that the learning curve phenomenon is generally valid for the manufacturing environment (see Appendix A4.1).

   In the casual interpretation of this learning phenomena, it might appear that the improvement with increased production could be expected to occur simply as a result of building more units. This is certainly not a true interpretation of the observation. Improvement in performance occurs *through concerted action on the part of the management and the employees.* The improvement is a result of a collection of many actions that the organization must take to accomplish the desired result of continuously improving the process and the procedures that are being used to operate it. Aaker (1984) and Allen and Hammond (1975) have commented extensively on this issue. They have identified a large number of actions that must be taken if the continuous improvement demanded by the learning curve is to be sustained. Some of these are:

- An increase in production provides advantages of scale. Indirect costs, energy, R&D expenditures, and so on, are spread over increasing volumes, thereby reducing the average unit costs of each of these items of expense.

- Improved efficiency in the utilization of labor. Repeating a task leads to elimination of wasted effort and a reduction in errors.

- Introduction of improved methods and improved processes. With continuing production, improvements such as the reduction in cycle times, reduced material handling, improved tooling, or faster tool changes will improve efficiencies.

- Redesign of the product can reduce manufacturing costs and improve the yields of the manufacturing process.

- Standardization of the product enhances the capability of employees operating the process to optimize their actions and to improve their efficiencies.

- Substitution of less expensive materials or replacement of an existing process

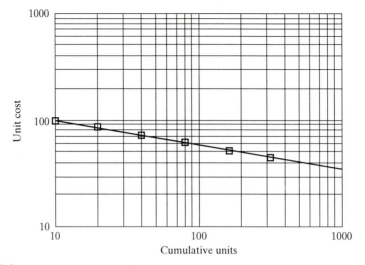

Figure 4.4 Learning curve for an 85% improvement for each doubling of production, plotted as a log-log curve.

with materials that are less expensive or more reliable will lead to improved efficiencies.

Each of the improvements noted above results from an action that is taken by management and/or employees. The learning-curve phenomenon is a consequence of all of these, many of which are occurring simultaneously. It is not possible, therefore, to unfold the learning curve and to determine the specific actions that were responsible for a particular improvement in a specific time frame.

4.3.2 The Experience Curve as an Aid in Developing Pricing Strategies

If it is not possible to relate the specific aspects of the learning curve to particular actions, if might be questioned whether the existence of a learning curve is helpful in understanding the operation of an activity. The answer is generally yes. First, if the progress that is being made in reducing unit operating costs can be sustained over a long period, the learning curve offers a basis on which *future* costs can be estimated. This can be exceedingly important in determining the strategy that will be followed in pricing the product.

Consider the implications of the following three strategies, as shown in Figure 4.5, of alternative pricing strategies for one particular cost curve. Note that the abscissa in this case should be, to the extent that it is possible to know, the total accumulated units that have been, or are likely to be, produced. The various pricing strategies that one might follow is represented by price curves A, B, and C. The cost curve, shown with the points, is the anticipated curve that can be expected as predicted by a learning curve.

Strategy A. Strategy A assumes that you are entering a market that is highly competitive and that your costs at the time you enter are comparable with that of your competitors. Because of the presence of competitors in the market, you are only able to realize a profit for each unit that is a fixed percentage above your costs. The price curve that you would expect to follow in this case is parallel to the cost curve and a fixed factor above it.

Strategy B. Strategy B assumes that you are introducing a new product into the marketplace that has no immediate competition and has an immediate appeal. The initial sales of the product are large and are reasonably insensitive to the price. The pricing strategy attempts to take advantage of this initial market appeal by setting the price at a high level and keeping it constant for as long as the initial market appeal and no competition persist. Ultimately, the initial euphoria will disappear or competitors will note that your profits are sufficiently large that they will be encouraged to enter the marketplace with a competing product. The strategy, then, is to drop the price to a level that allows a more normal rate of return. The price will continue to decline in proportion to the decline in costs that can be achieved. Presumably this will can be predicted by the learning curve.

Strategy C. Strategy C assumes that a market for the new product does not already exist and that you must encourage its development by an attractive pricing policy for the early products. This means that you must be willing to lose money on early sales in order to create a market. As the market begins to develop, you will expect to recover your initial loses. At this point strategy C becomes essentially equivalent to strategy B.

The principal issue with both strategies B and C is when to set the price to the level that reflects a more or less standard level of profit for each unit. The longer one delays, the more encouragement you offer to a competitor. The earlier that you drop the price, the more your total profit suffers. Furthermore, one cannot be assured that a new competitor in the marketplace, having seen your success, having examined your product, and having analyzed your probable costs, will not be able to enter the market with an improved product that can be manufactured at lower cost than that which you have already achieved through improvement in your product and processes.

Although the learning curve is an important contributor to understanding the

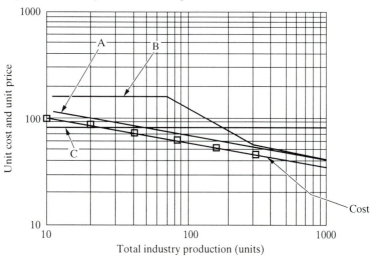

Figure 4.5   Relative relationship between price and cost.

likely improvement that will occur in your cost structure, it must be combined with strong intuition about the direction the market will move, the capabilities of your competitor, and the direction the industry will take.

4.3.3 Extension of the Learning Curve Concept

In considering the actions that management must take to achieve the improvements in cost that a learning curve represents, it seems reasonable that many of the same actions would be expected to lead to improvements in other characteristics of the system. It seems likely, for example, that many of these actions would lead to improvement in the quality of the product or process. Remembering the various definitions of quality that were suggested by Garvin (Chapter 3), it is necessary to recognize that the measure that might be used to reflect improved quality includes such metrics as reduced number of defects, increased yield from a process, or increased mean time to failure. What, then, is the situation relative to the existence of learning curves for quality, setup times, average processing times for operations in a service activity. The answer appears to be that this has been explored only in a preliminary manner. Compton et al. (1992) have examined the applicability of learning curves to quality improvement for various products. This study focused on a variety of manufacturing products involving many different processes. An example of the improvement in quality that results from the same type of a learning process is given in Figure 4.6. In this figure, the quality index is a measure of the operating-life failure rate.

Since the change that takes place in the quality index throughout a production cycle is often not large, the sensitivity of the plots to the form of the learning curve is often not clearly established. Compton et al. (1992) note that the data are often neither sufficiently accurate nor span a sufficiently large range of values to discern which of the laws the learning curve is following. If the quality index improves by less than a factor of 10, it is unlikely that the specific mathematical formulation of the law will be clear, whether the volume that is produced is as low as 10^3 or as high as 10^{10} units. To project performance of the quality index, it is important to recognize that the quality index, QI_n, of the nth part may follow any of the following laws:

$$QI_n = (QI)_1 N^{\pm m} \qquad \text{(power)} \qquad (4.3)$$

$$QI_n = (QI)^* + (QI)_0 e^{\pm N} \qquad \text{(exponential)} \qquad (4.4)$$

$$QI_n = (QI)_a \pm (QI)_b N \qquad \text{(linear)} \qquad (4.5)$$

Use of the $\pm$ sign is to signify that the quality index may decrease with increasing volume, as would occur if QI is a measure of rejects, or increase with cumulative volumes, as would occur if it is a measure of the fraction of good products that are produced.

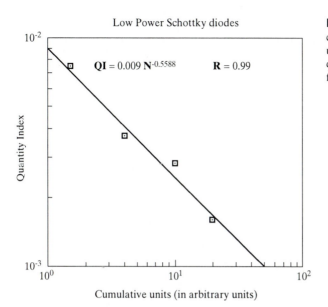

Figure 4.6 Quality index versus cumulative volume (in arbitrary units) for a Schottky low power diode. The quality index represents failure rates in operation.

Figure 4.7 is a replot of the data from Figure 4.6 using Eq. (4.4.) The closeness of the correlation values–$R=0.99$ and $R=0.94$–suggests that either Eq. (4.3) or (4.4) represent these data properly. To understand the trends of the processes, it makes little difference which of these laws is followed.

Studies have shown that when plotted over long periods of time, the quality index can show some interesting trends that deserve to be explored. In Figure 4.8, the quality index is presented as observed in the production of light bulbs. Each

Figure 4.7 Quality index versus cumulative volume (in arbitrary units) for a Schottky low-power diode. The quality index represents failure rates in operation. (Replotted from figure 4.6 using equation 4.4)

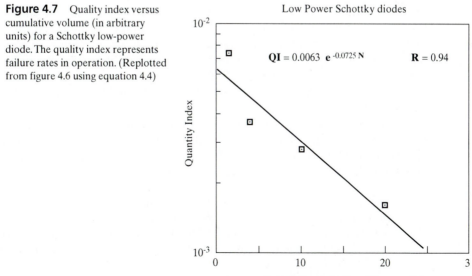

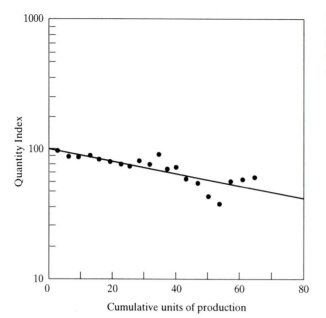

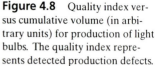

Figure 4.8   Quality index versus cumulative volume (in arbitrary units) for production of light bulbs. The quality index represents detected production defects.

point represents the average quality index for one calendar year of production. Data are presented for a 20-year period for a single plant. While the downward slope demonstrates a general improvement in the index, as would be expected for learning with increasing production, it is particularly interesting to note the rather abrupt improvement in quality that occurred and then vanished in the latter part of the period. Had a plot of the form shown in Figure 4.8 been available at the time that these deviations occurred, it is probable that an explanation for this behavior could have been determined. If the explanation is something other than a different way of measuring the quality index or something similarly artificial, it is regrettable that the company did not learn from this experience and take advantage of the improvement that was demonstrated for a brief period.

Schneiderman (1986) and Strata (1989) have examined quality improvement as a function of time. In Figure 4.9, Strata uses data collected at his firm–Analog Devices–to illustrate the importance of learning. Quality improvement is plotted on a semilogarithmic scale as a function of time. Two observations can be made using these graphs: first, improvement occurred dramatically with the introduction of the QIP (quality improvement process), and second, from plots of improvement against time, the *half-life* for improvement of a particular amount can be determined. As Strata (1989:69) observes:

> The slope of the learning curve is determined by how long it takes to identify and prioritize the causes of the problem and to eliminate those causes. The skills of the people and the level of resources do have an impact, but surprisingly the time required for each cycle of improvement is largely a function of the complexity and bureaucracy of the organization. Or, to put quality improvement in the larger context....the slope of the characteristic half-life curve is determined by the rate of organizational learning.

Although Strata chooses to draw a distinction between the concept of learning that he shows to be time dependent and the processes that have been shown to be dependent on total production volume, as described by the Boston Consulting Group, it is better to view all of these as representations of the same phenomena. Organiza-

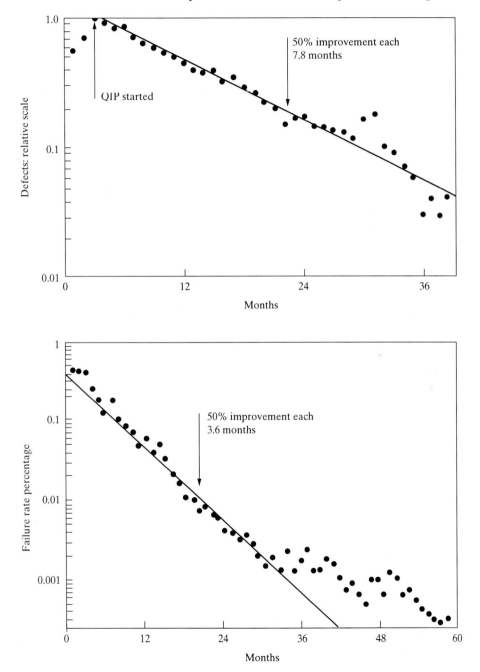

Figure 4.9 Examples of quality improvement versus time. (From Strata, 1989. p.69.)

tional learning must take place for any of these representations to exist. The important issue is the collection of the necessary data over a period of time that is sufficient to allow an understanding of the trends and to ensure that any changes that occur in the overall quality index be clearly understood and well documented. The importance of good data that allow a systematic examination of the consequences of actions is particularly important for anticipating the level of improvement that is possible for items such as the quality index. Studies of learning curves for related variables, many of which are critical to the operation of a manufacturing activity, would make a valuable contribution to an understanding of the system of manufacturing. With the exception of the work of Schneiderman (1986), Strata (1989), and Compton et al. (1992), such studies have not been reported in the literature.

4.4 CONTINUOUS IMPROVEMENT THROUGH LEARNING AND EXPERIMENTATION

The operating objective that is represented by its theoretical limits cannot be satisfied until the organization has found *all* ways to improve. There can never be disagreement about the need to improve. There can never be disagreement about the need for progress. There can never be disagreement about the commitment to become better. This is not to say that this philosophy is either easy or can be accomplished without problems. It requires an organizational commitment, a high level of employee involvement, strong employee empowerment, and an enlightened management.

Continuous improvement in all elements of the system must be the objective of a world-class enterprise. This improvement must occur at all levels and it must be encouraged and fostered. As Heim and Compton (1992) observe:

> Individual learning experiences are not automatically converted to organizational memory and made available for all members to draw and build upon. The people with access to new data and information, authority to make changes, and understanding of the proper use of such information should devise appropriate methods to codify their knowledge so that it can be compared with information acquired by others. This requires that the organization's knowledge base be continually changed and updated. In some instances this knowledge may be specific to a machine or process....
>
> Organizational learning is a broad-based strategy for capturing and making available to members of the organization information and knowledge that enable them to benefit from the experience of others–that build on the knowledge of many members of the manufacturing enterprise. In other cases, the knowledge of individuals and groups can be captured by tools, models, formulas, drawings, instruction manuals, and the artifacts of the production process itself. Benchmarking the internal practices of the organization between projects is important.

The performance of an organization can improve through the knowledge that it gains with experience. The lessons learned from mistakes make it less likely that the same errors will be made in the future. Successes encourage people to try the same things again. Learning is the result of using information and the knowledge that is derived from it to the fullest advantage. While learning occurs throughout the entire process of working with and in a system, it is particularly important for accomplishing the long-term improvement that is associated with the "continuous improvement" process. Continuous improvement can be sustained only if the organization learns from past experience and translates that learning into improved performance.

Learning comes from trying new approaches, exploring new ways of doing things, and testing new ideas for improving the various processes that are being used. Experimentation is an important part of learning. It is, of course, impossible to assure that all experiments will succeed. As desirable as it is to experiment, it is important to recognize that there is risk associated with experimentation. If the risk associated with failure is unnecessarily large, it is clear that some other means must be found to obtain the data that are desired. An experiment that risks closing down a factory, or disabling a critical production, or endangering the well-being of the workforce can seldom be justified.

To insist, however, that experimentation take place without risk will reduce the vision of those involved to such small increments that progress will be modest or nonexistent. Recognizing that some experiments will fail must translate into a philosophy that individuals not be judged harshly if they try new ideas that turn out to be unsuccessful. The organization that is afraid to make a mistake or that "shoots the messenger" for telling the bosses that an experiment failed will find that it learns at a painfully slow rate. As Marsing (1992) has aptly observed: "Once the direction is set and the systems are in place to allow a factory to design change, take risk, and position itself to be competitive, it is critical to establish an environment that reinforces the notion that it is good to take risk. Consistency in management's attitude toward change and evolution of the organization to deal with the requirements is paramount."

4.5   INSTITUTIONAL MEMORY AND STRATEGIC OBJECTIVES

To create learning curves, the organization must focus its attention on collection and analysis of the data that are pertinent to its processes. This may be quality data, measures of productivity, data on machine reliability, and so on. Understanding the rate of improvement that has historically been achieved in a system is an essential first step in establishing strategic objectives. The long-term objectives for improvement should not, however, be restricted by these projections. Major deviations from historical trends can occur. New technological developments or new, revolutionary, product designs may make it possible to improve greatly over the historical trends. At the other extreme, it may not be possible to improve the processes indefinitely. Depending on the quantity being considered, the theoretical limits for improvement may be approached, thus leading to a deviation from the historical trends. Any significant deviation, however, should be considered as requiring careful analysis and justification (as in Figure 4.8). Understanding the cause of the deviations can be an important aid in establishing future goals and objectives.

The creation of a learning curve will help the organization develop an institutional memory of the processes and the system that is being measured. While the plotting of a learning curve is not the only way to capture institutional memory about a process–be it the manufacturing process or the product development process–this representation provides a good graphical picture of the progress that has been made in achieving improvement.

The elements of the use of benchmarking information, theoretical limits and the role of experimentation are reflected in the following three foundations. The inclusion of the goals and objectives foundation in both Chapters 3 and 4 is done to emphasize the importance that good data and a good understanding of the system have on the establishment of both longer-term strategic goals and shorter-term operational goals.

4.6   THE *Goals and Objectives* FOUNDATION OF WORLD-CLASS PRACTICE

> FOUNDATION: World-class manufacturers have established as an operating goal that they will be world class. They assess their performance by benchmarking themselves against their competition and against other world-class operational functions, even in other industries. They use this information to establish organizational goals and objectives, which they communicate to all members of the enterprise, and they continuously measure and assess the performance of the system against these objectives and regularly assess the appropriateness of the objectives to attaining world-class status. (Heim and Compton 1992)

4.7   THE *Metrics* FOUNDATION OF WORLD CLASS PRACTICE

Performance evaluation is a process applied throughout the manufacturing organization to measure the effectiveness in achieving its goals. Because of the variety, complexity, and interdependencies found in the collection of unit processes and subsystems that define the manufacturing system, appropriate means are needed to describe and quantify rigorously the performance of each activity.

> FOUNDATION: World-class manufacturers recognize the importance of metrics in helping to define the goals and performance expectations for the organization. They adopt or develop appropriate metrics to interpret and describe quantitatively the criteria used to measure the effectiveness of the manufacturing system and its many interrelated components. (Heim and Compton 1992)

4.8   THE *Experimentation and Learning* FOUNDATION OF WORLD-CLASS PRACTICE

Organizational learning is a broad-based strategy for capturing and making available to members of the organization information and knowledge that enable them to benefit from experimentation and the experience of others. Too often in manufacturing, sources of information become scattered and isolated and individual learning experiences are not automatically converted to organizational memory and made available for all members to draw and build upon. The rate at which an organization improves its performance as a result of learning is perhaps one of the principal determinants of whether it can become best-of-the-best.

> FOUNDATION: World-class manufacturers recognize that stimulating and accommodating continuous change forces them to experiment and assess outcomes. They translate the knowledge acquired in this way into a framework, such as a model, that leads to improved operational decision making while incorporating the learning process into their fundamental operating philosophy. (Heim and Compton, 1992)

REFERENCES

ALLAN, G. G. AND J. S. HAMMOND III. 1975. Harvard Business School. *Note* 175-174.

ARGOTE, L., AND D. EPPLE. 1990. Learning Curves in Manufacturing. *Science* , 247:920-924.

BUTLER, W. E. 1993. Quoted in Blueprint for Best in Class, *Enterprise*, July:13.

CAMP, R. C. 1989. *Benchmarking: The Search for Industry Best Practices that Lead to Superior Performance.*Quality Press, Milwaukee, WI.

CHOOBINEH, F. 1986. FMS: A Totally Programmable Manufacturing Cell, *Conference Papers in Flexible Manufacturing systems: Current Issues and Models*, F. Choobineh and R. Suri (eds.), Industrial Engineering and Management Press, Atlanta, GA.

COMPTON, W. D. 1992. Benchmarking, *Manufacturing Systems: Foundations of World-Class Practice*, J. A. Heim and W.D. Compton (eds.), National Academy Press, Washington, DC:100.

COMPTON, W. D., M. D. DUNLAP, AND J. A. HEIM. 1992. Improving Quality Through the Concept of Learning Curves, *Manufacturing Systems: Foundations of World Class Practice*, J. A. Heim and W.D. Compton (eds.), National Academy Press, Washington, DC:107.

FISHER, P. A. 1992. The Interface Between Manufacturing Executives and Wall Street Visitors-Why Security Analysts Ask Some of the Questions That They Do, *Manufacturing Systems: Foundations of World Class Practice*, J. A. Heim and W.D. Compton (eds.) National Academy Press, Washington, DC:137

FOSTER, R. N. 1982. A Call for Vision in Managing Technology, *Business Week*, May 24:24-33.

GORDON, J. E. 1976. *The New Science of Strong Materials*, Princeton University Press, Princeton, NJ.

GRAHAM, M.B.W. AND S. R. ROSENTHAL. 1986. Institutional Aspects of Process Procurement for Flexible Machining Systems, *Manufacturing Roundtable Research Report Series*, Boston University School of Management, Boston, Sept.

HAMMER, M. AND J. CHAMPY. 1993. *Reengineering the Corporation: A Manifesto for Business Revolution*, Harper-Collins Publishers, New York:194-197.

HEIM, J. A. AND W. D. COMPTON (eds.) 1992. *Manufacturing Systems: Foundations of World-Class Practice*, National Academy Press, Washington, DC:6-7 and 66-68.

HENDERSON, B. AND F. K. LEVY. 1965. Adaptation in the Production Process. *Management Science* 11(6):B136-B154.

KRUPA, D. C. 1992. Time as a Primary System Metric, *Manufacturing Systems: Foundations of World-Class Practice* J.A Heim and W.D. Compton (eds.), National Academy Press, Washington, DC:166.

NEVINS, J. L. D. E. WHITNE, et al. 1989. *Concurrent Design of Products and Processes: A Strategy for the Next Generation in Manufacturing* McGraw-Hill Book Company, New York.

MARSING, D. A. 1992. Taking Risks in Manufacturing, *Manufacturing Systems: Foundations of World-Class Practice*, J. A. Heim and W.D. Compton (eds.), National Academy Press, Washington, DC:189.

SCHNEIDERMAN, A. M. 1986. Optimum Quality Costs and Zero Defects: Are They Contradictory Concepts? *Quality Progress*, Nov.:28.

STALK, G., JR. AND T.M. HOUT. 1990. *Competing Against Time* . The Free Press, New York.

STRATA. R. 1989. "Organization Learning-The Key to Management Innovation," *Sloan Management Review*, Spring:63.

TURNBULL, G. K., E. S. FISHER. E. M. PERETIC, J. R. H. BLACK, A. R. CRUZ, AND M. NEWBORN. 1992. Improving Manufacturing Competitiveness Through Strategic Analysis, *Manufacturing Systems: Foundations of World Class Practice*, J. A. Heim and W.D. Compton (eds.) National Academy Press, Washington, DC:224.

WRIGHT, T. P. 1936. Factors Affecting the Cost of Airplanes, *Journal of Aeronautical Science*, 3:122-128.

QUESTIONS

4.1 Indicate how the learning curve concept can be applied to (*a*) product reliability, (*b*) system response time, and (*c*) machine downtime. What can be learned from a systematic application of the learning concept to these system variables?

4.2 What are the appropriate metrics for examining the performance of (*a*) a library, (*b*) a laundry, and (*c*) an executive search concern? How can these be used to enhance the efficiency and performance of the enterprise?

4.3 What are the appropriate metrics for examining the performance of (*a*) an elementary school with grades K-6, (*b*) a secondary school with grades 7 and 8, (*c*) a high school, (*d*) a public university, and (*e*) a private university?

4.4 Who should be responsible for collecting and analyzing the data that are to be used in constructing the learning curves for the various metrics that are being followed? Why do you identify these particular people?

4.5 How can the risks associated with failure be assessed in determining the value of proceeding with experimentation on a new process for the manufacturing operation? What is the proper level of risk?

4.6 Identify three important pitfalls of benchmarking. What actions can be taken to reduce the severity of these pitfalls?

4.7 Identify the tools that are needed to determine the theoretical limits for (*a*) performance of the product in the hands of the consumer and (*b*) performance of a process that is being used to create the product. Who should be responsible for ensuring that these tools are available and utilized regularly by the activities in the organization?

4.8 How is cost considered when it has been discovered that the theoretical limit of a production process is well below that which is possible?

4.9 What are the theoretical limits of the five most important metrics of performance for (*a*) a library, (*b*) a fast-food restaurant, (*c*) a maintenance service for an electronic instrument company, or (*d*) a local home builder?

4.10 How would you benchmark the performance of each of the enterprises identified in Question 4.9?

4.11 Discuss the distinctions between strategic and operational goals. How do you relate the two and how do you ensure that they are consistent?

4.12 The plant that you manage contains more than 100 machines. Some operate on one shift a day while others operate on three shifts. A number of products are made in the plant, with many of the machines being used for more than one product. Your management has recently become very concerned with the production that is being lost because of machine failure in the plant. You have been asked to prepare a set of charts that will give the management a good view of the impact of machine down time on productivity. What will you choose for your metrics? Make a few sketches of the charts that you will propose and the types of information that can be conveyed by them.

4.13 You have recently returned from a workshop that introduced the merits of continuous improvement. Your factory has operated in the "old mode," by which the employees were told what to do and were expected to comply without questioning the correctness of the directions. Your experience at the workshop has convinced you that your manufacturing enterprise can be much more effective if it can encourage and harness the insights that the employees have. How would you go about introducing this concept to the enterprise? On what timetable would you operate? How would you go about assessing the value of this effort?

4.14 You have just taken over as head of the manufacturing activity of a company that utilizes a well-proven process for making cast metal components. In your last position, you were required to buy components made by similar processes from sources outside the company. In reviewing the current costs of the manufacturing activity, you have strong indications that your processes are not competitive with other manufacturers. How do you go about determining the truth of this situation? If your suspicion proves to be correct, what do you recommend to your management?

4.15 If you had been following the progress of Bell Atlantic for process time, as described in Appendix A4.2, with a learning curve, how would you expect it to look? Remember that the company had a long history of doing things in the old way. What generalizations can you draw from this?

READINGS

ABERNATHY, W. J. AND K. WAYNE. 1974. Limits of the Learning Curve, *Harvard Business Review* Sept.-Oct.:109.

ARGOTE, L., AND D. EPPLE. 1990. Learning Curves in Manufacturing, *Science*, 247:920-924.

COMPTON, W. D. 1992. Benchmarking, *Manufacturing Systems: Foundations of World-Class Practice*, J. A. Heim and W.D. Compton (eds.), National Academy Press, Washington, DC:100.

COMPTON, W. D., M. D. DUNLAP, AND J. A. HEIM. 1992. Improving Quality Through the Concept of Learning Curves, *Manufacturing Systems: Foundations of World-Class Practice*, J. A. Heim and W.D. Compton (eds.), National Academy Press, Washington, DC:107.

FISHER, P. A. 1992. The Interface Between Manufacturing Executives and Wall Street Visitors- Why Security Analysts Ask Some of the Questions That They Do, *Manufacturing Systems: Foundations of World Class Practice*, J. A. Heim and W.D. Compton (eds.) National Academy Press, Washington, DC:137

FOSTER, R. N. 1982. A Call for Vision in Managing Technology, *Business Week*, May 24:24-33.

KRUPKA, D. C. 1992. Time as a Primary System Metric, *Manufacturing Systems: Foundations of World Class Practice*, Heim, J. A. and W. D. Compton (eds.) National Academy Press, Washington, DC:166.

MIZE, J. H. 1992. Constant Change, Constant Challenge, *Manufacturing Systems: Foundations of World Class Practice*, J. A Heim and W. D. Compton (eds.) National Academy Press, Washington, DC:196.

TURNBULL, G. K., E. S. FISHER. E. M. PERETIC, J. R. H. BLACK, A. R. CRUZ, AND M. NEWBORN. 1992. Improving Manufacturing Competitiveness Through Strategic Analysis, *Manufacturing Systems: Foundations of World Class Practice*, J. A. Heim and W.D. Compton (eds.) National Academy Press, Washington, DC:224.

APPENDIX A4.1
EXAMPLES OF THE
LEARNING-CURVE PHENOMENON

A large number of examples of the learning-curve concept have been developed by various authors. The validity of this concept is strengthened as a result of its applicability to such widely varying industries as are represented in Figures A4.1.1-A4.1.4. It is the broad universality of this concept that makes it an attractive tool for estimating the long term costs that a supplier may be able to anticipate and to have sufficient confidence in these estimates that strategic decisions can be made regarding future production levels, marketing strategies, and investments. Figures A4.1.1 and A4.1.2 present the results of early studies of this phenomenon.

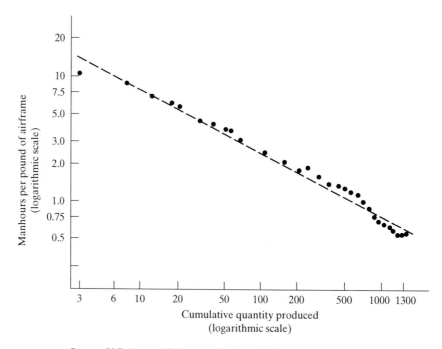

Source: U.S. Army Air Forces, Air Material Command. *Source Book of World War II Basic Data: Air Frame Industry*, vol. I (Washington: 1952).

Figure A4.1.1 Learning curve for B-29 bombers produced at Boeing's Wichita plant. (From Scherer, 1990.)

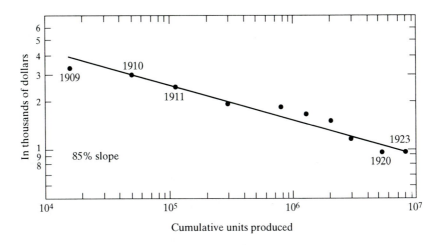

Figure A4.1.2   Price of Model T car, 1901-1923 (average list price in 1958 dollars). (From Aaker, 1984. p215.)

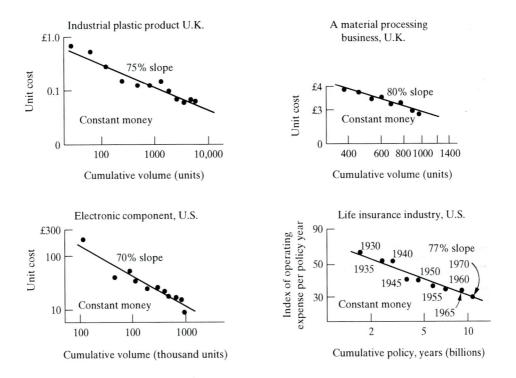

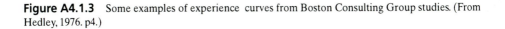

Figure A4.1.3   Some examples of experience curves from Boston Consulting Group studies. (From Hedley, 1976. p4.)

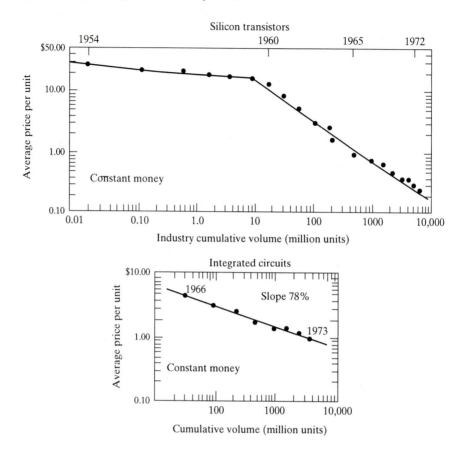

Figure A4.1.4   Price experience curves for semiconductors. (From Hedley, 1976. p6.)

APPENDIX A4.2
TIME AS A METRIC

Commentary by Regis Filtz, Bell Atlantic's newly appointed head of carrier access service (CAS), as reported by Hammer and Champy (1993).

We did what I call a high-level analysis of our work processes–receiving and processing an order for [CAS] service, connecting the service, testing it, and turning it over to the customer. We found, among other things, that from start to finish there were at least thirteen handoffs among different work groups and that some twenty-seven different information systems were involved. Not only was the process slow, it was terribly expensive. Further study showed that while the interval between receiving an order and turning the service over to the customer was fifteen days, our actual work time was only about ten hours. For a thirty-day interval, we did only about fifteen hours of work.

This high-level analysis showed us significant opportunities not only to reduce our process time dramatically, but also to slash our costs by a similar order of magnitude....

We actually set up two different kinds of reengineering teams, one to come up with the ideas and the other to test and refine them in the real world.

The first team we called the core team. To head it, we selected a manager who....was respected by her peer group, and was a good communicator, teacher, and role model. She could and would inspire others.

The core team's leader's first job was to assemble a team of experts from among all the disciplines involved in the fragmented CAS process, being sure that they were competent in their skill areas and that they, too, were respected by their peers and were good communicators. The core team's job was to brainstorm, to redesign, and to blueprint the new process in detail. We gave them a goal. They were to find a way for Bell Atlantic to provide access services to customers in virtually zero cycle time.

We made their goal ambitious for three reasons: First, it's what our customers said they wanted in the long term. Second, meeting it would force a substantive change in the existing process, not just a fix. Third, we figured that zero cycle time was a level of performance that our competitors could never beat.

Frankly, the core team members were apprehensive. At first, they thought their task was impossible, and it took more that a little encouragement to get them to sign on, but they did. They began their work in mid-July of 1991, and within a month, they had designed a new process that physically pulled together under common supervision in one location all of the functions of the old process that had been geographically dispersed, separately managed, and spread among different departments.

As soon as we did a process design, we put the second team, which we called the lab team, to work. Their job was to test the core team's blueprinted design by using it to process real CAS orders. They would try the new process, change it however they liked, and then feed back their results to the core team. Thus, our reengineering process itself was iterative. The lab team became, in effect, a prototype for the case team concept that our core team created.

The lab team was empowered to make whatever changes in work methods and procedures that were necessary to cut the process time, reduce expense, and produce a defect-free output. They were to discard all of the existing functional and departmental measurements and management objectives under which they used to work in their separate departments. Their only concern was figuring out how they could reduce cycle time, cut expense, and improve the quality of the output concurrently.

The lab team took over operational responsibility for servicing customes in a part of central Pennsylvania. Within several months, the team was working with cycle times measured in days instead of weeks. In some cases they had reduced them to hours. The quality of the service improved dramatically, too. Before the lab team began servicing that group of customers, we had four people working full time whose job was to track CAS orders that were not being completed successfully. We have eliminated that group and saved more than $1 million a year on reworks in just one location.

Currently we are extending the case team concept to all of Bell Atlantic's operating subsidiaries. The teams we're installing use the same process and process management systems that were used in the lab team's pilot. We've also identified the culture changes, the new job skills, and the revamped information systems that we'll need.

Bell Atlantic's management systems are being changed as well. We have been and are a hierarchical company that closely supervises individuals and measures their performance by internal criteria. We are moving toward self-managed and cross-functional work teams that are internally motivated to meet customer requirements, and to continuously improve cycle time, cut costs, and improve quality.

The culture and the values of the corporation are also changing. Under the old hierarchy, we could rely upon compliance to the results. With our redesigned processes, we have to rely on commitment. The difference is that in a compliance mode I do what I must do because my boss tells me I must do it. In a commitment mode, I understand what the corporation is trying to achieve and how we're going to achieve it, and I will do whatever it takes to make that happen, including changing the way I do my job if that is what is required.

We previously had individuals performing single functions with absolutely no knowledge of the entire system or of what happened to their work as it progressed downstream. We are moving toward a different arrangement in which tasks are combined, and we are creating systemic knowledge so that each of us understands the process as a whole, what our piece of it is, and how we can interact more effectively with others.

Even as we implement the case team concept, though, the core team is already working on the next iteration of reengineering, in which we will replace the case teams with *one* case worker and some new technology. Essentially, one person will be able to do what a team of people with different specialties does now. Instead of using a team to transcribe the elements of a customer order manually for each of our various systems, we have technology that will allow one person to take a customer's call and use his or her terminal to make electronically all the connections required to set up the service requested. When we reach this iteration, we will essentially be changing the order in which we respond to customer requests. We'll be setting up the service first and then taking the time to figure out how to bill and deep the records we need.

But that won't be the end of it. The final iteration as we see it now is self-provisioning, in which customers will interface directly with our system when they want the service. To the customer, it will seem as if it's working just the way the telephone system works today with routine voice calls. For us, there won't be any manual operations involved, and the cycle time will have become, in accordance with our goal, virtually zero.

REFERENCES

AAKER, D.A. 1984. *Developing Business Strategies*, John Wiley & Sons, New York.

HAMMER, M. AND J. CHAMPY, 1993, *Re-engineering the Corporation: A Manifesto for Business Revolution*, Harper-Collins Publishers, New York. pgs 194-195.

HEDLEY, B. 1976. A Fundamental Approach to Strategy Development, *Long Range Planning*. Dec.

SCHERER, F.M. AND D. ROSS. 1990. *Industrial Market Structure and Economic Performance*, Houghton Mifflin Company Boston.

PART 3

The Pathway to Current Management Practice

In Part III of this book we examine some of the principal events that took place in the development of the philosophy of management as it is practiced currently. In Chapter 5 we provide an historical survey of the many approaches, starting in the late nineteenth century, that have been taken to the management of enterprises. Chapter 6 is devoted to a discussion of the management philosophy that is currently believed to be most effective in achieving world-class performance. Known as participative management, it includes two principal elements: employee involvement and employee empowerment.

As the many aspects of the task of effectively managing an enterprise are being treated, it will be helpful to consider the following synopsis of good management practice. These are described by Yankelovich and Immerwahr (1983) as "do's and don'ts for *stimulating worker commitment and quality performance*." Management seldom faces the task of getting people to agree, in theory, that these are good things to do. Rather, they must confront and find ways to accomplish all of these good things. Helping the manager do this is the intent of the following chapters.

- **DO** tie remuneration directly to performance that enhances the efficiency and effectiveness of the enterprise.

- **DO** give public and tangible recognition to people who keep standards of quality and effort that exceed average satisfactory job performance.

- **DO** accept wholeheartedly the principle that employees should share directly and significantly in overall productivity gains (however defined).

- **DO** encourage jobholders to participate with management in defining recognizable goals and standards against which individual performance can be judged.

- **DO** give special attention to the difficulties that middle managers face in supporting and enforcing programs to restructure the workplace.

- **DO NOT** permit situations to develop where the interests of employees run counter to the well being of the firm–e.g., by introducing new technology in a way that threatens employees' job security or overtime.

- **DO NOT** attempt to improve standards of quality unless you are prepared to accept its full costs–e.g., discarding substandard products, paying more for better components, or transferring or dismissing people who cannot do quality work.

- **DO NOT** permit a significant gap to develop between management rhetoric and the actual reward system. Nothing feeds employee cynicism as much as management blindness or insincerity about the forms of behavior that really "pay off."

- **DO NOT** pretend that programs designed to increase productivity are really intended to enhance job satisfaction and the dignity of work.

- **DO NOT** support special privileges for managers that serve to enhance that status of managers by widening the gap between them and those who do the work–e.g., giving bonuses to managers at the same time that employees are being laid off.

5

Early Management Focus: Historical Highlights

> Rarely, if ever, has a new basic institution, a new leading group, a new central function, emerged as fast as has management since the turn of the century. Rarely in human history has a new institution proven indispensable so quickly. (Drucker, 1985)

While the tasks that management must perform are many and are often dispersed among many different individuals, the principal objective of management is to ensure that the organization is fulfilling its goals and objectives. These will vary, of course, from one organization to another. For a company, the goal may be the development and manufacture of a product at a cost that is attractive to customers and will return a profit to the company and its stockholders. For a not-for-profit activity such as a public library or an educational institution, the goal may be to provide the highest-quality service or education at the lowest possible cost to the user and taxpayer. The importance of clearly stating and communicating the goals of the organization was discussed in Chapter 3. Management must also organize the enterprise, secure the necessary resources, attract the appropriate people to undertake the various tasks, and ensure that the activity operates in a manner that is efficient and effective in accomplishing its goals and objectives. The current chapter begins with a more complete definition of management. The latter sections are devoted to a description of the major contributions that were made to the present fabric of management practice as we know it today.

5.1 ROLE AND DEFINITION OF MANAGEMENT

A bewildering array of activities confronts the manager of any enterprise. He or she must deal with the design and development of the product or service, the manufacture of the product, and the marketing and sales activities, while ensuring that the product or service satisfies the needs of the customer. In addition, relationships with the employees must be addressed and nurtured. There are financial issues, legal issues, the relationship of the enterprise to the community and government, and concerns about future technologies that relate to the enterprise's products and processes. Although no one person below the president or chairman of an organization will have responsibility for all these activities, an awareness of the importance of each and the relationships among them is an important aspect of successful management. As emphasized in Chapter 1, the system of manufacturing embodies all these elements.

The time horizon of concern to the manager cannot be restricted to either the near or the long term. It is not sufficient for the manager to focus exclusively on short-term issues such as inventory control or daily sales while neglecting the long-term needs of the enterprise for new products and processes. Neither can a successful manager concentrate only on the long-term R&D needs of the firm and expect the short-term operational issues to resolve themselves. Whether or not individual managers enjoy dealing with all the issues that confronts them, a successful manager must maintain a balance among them and ensure that neither the short- nor long-term needs of the enterprise are neglected. This requires, of course, balancing the amount of attention given to the details of the operation with the attention that is given to addressing the broad strategic objectives and goals of the enterprise.

Mintzberg has described the manager's many roles with the relationships shown in Figure 5.1 (Mintzberg, 1990:168). Since the title of *manager* conveys a certain formal

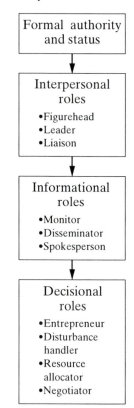

Figure 5.1   The Manager's Roles. (From Mintzberg, 1990, p168.)

status with an accompanying level of authority, the manager is vested with certain powers that he or she is expected to exercise. The three roles that are depicted in the figure flow directly from this formal authority and relate to interpersonal relationships, informational activities, and the making of decisions. As figureheads, managers must perform obligatory functions that are expected of anyone in the position. They must be leaders–the subject of Chapter 8–and maintain contacts with many people throughout the organization in their liaison role. In their informational role, managers must monitor the progress of programs and disseminate information to lower and higher members in the organization. In their decisional role, managers must seek new opportunities, as would any successful entrepreneur, allocate resources for programs, negotiate agreements and program responsibilities, and be the final arbiters in the case of disagreements within their areas of responsibility. In assessing one's strengths and weaknesses in fulfilling these many roles, Mintzberg has developed a list of self-study questions for managers, presented in Appendix A5.1.

It is common to identify someone as a part of management if they have the title *manager, vice-president, president*, or *chairman*. Although this group clearly belongs to management, many people who are not identified by these or similar titles often have important management responsibilities. Line supervisors, team leaders, project leaders, and technical experts all perform functions that are within the purview of management. Using the number of people who report to someone, or the appearance of that person's name on an organization chart, as a principal criterion of whether he or she belongs to management is not appropriate. Since the role of management is to lead and to create an organization and an atmosphere that can fulfill its goals and missions, this is a role that everyone in the organization must play, at least to some degree. As employee involvement and employee empowerment are extended throughout an organization, the demarcation between management and employees becomes more blurred.

This brief description of the tasks confronting the current manager does little to explain why it is that management has come to mean these things. You may even ask why is it that certain elements have come to be emphasized in management whereas others are given low priorities. Why has the practice of mangement resulted in the creation of a discipline of management that is taught in our universities while such topics as manufacturing have not generated a commonly accepted discipline? A first step in answering these questions is to understand some historical events that have influenced our thinking significantly. Elements of the most important historical events persist today and have strongly influenced our approach to many elements of management. To better understand these connections, we examine some of the major events that have shaped thinking about management practice.

5.2 EVOLUTION OF MANAGEMENT PRACTICE

The path that has been followed in arriving at our present understanding of the principles of management has no clearly defined origin and certainly no foreseeable conclusion. To better understand current thinking and to place recent actions in a proper context, it will be useful to review briefly key historical contributions to the philosophy of management. Following a short introduction to the early approaches of management, we concentrate on the period that begins in the late nineteenth century. As in many fields,what will be found is that many of the principal advances have been made by people with a zeal for change. The advocacy that they brought to a particular approach was often controversial and frequently evoked strong reactions. It is, however, through the efforts of "revolutionary" thinkers that large steps are often made in understanding and practice. This is certainly the pattern that has emerged in the practice of management.

As we are unable to devote sufficient time and space to treat the topic in great detail, students are encouraged to expand their insights through perusal of the original papers and the critiques of certain of these papers. The origin of current criticisms of Taylorism, the conflict that developed between Frederick Taylor and Frank Gilbreth, and the impact and later criticisms of the Hawthorne studies–often bordering on personal attacks–amply demonstrate the stresses that have characterized the evolution of the field of management. A deeper knowledge of the conflicts and how they were resolved will help in understanding the source of our present biases, how those biases can be traced to past attitudes and activities, and how new solutions may be sought. The serious student of management practice is encouraged to delve more deeply into the historical context of this topic.

5.2.1   The Artisan/Owner

The early embodiment of the artisan and owner in one person created an enterprise that needed no organizational structure. With all responsibilities for knowing about the business and performing the tasks required, the owners/artisans could follow their instincts and develop the capability to ensure that their products met customer needs. This embodiment of all activities in one person made it unnecessary to consider an organization that would need, even in crude form, management tools. The early enterprise simply did not demand it.

The dominance of the artisan/owner persisted until the middle of the eighteenth century, when the industrial revolution created the capability of transferring work skills from craftsmen to machines. In many ways, the efforts that have been expended to create a discipline of management have had as their objectives–often unstated–the development of tools, structures, and systems for the management of the large enterprises that would enable these organizations to achieve the advantages that artisans/owners have had in terms of understanding, commitment, and capability.

5.2.2   Early Interest in Workshops and Factories

Charles Babbage (1792-1871). Charles Babbage was a scientist interested primarily in mathematics and astronomy. His interest in such practical problems as navigation led him to propose the creation of the *Difference Engine*. It is this invention that has made Babbage of interest to modern generations, for it is remembered as the first proposal for the construction of a mechanical calculator. Babbage proposed this machine as a means of providing high accuracy in calculations that he needed for navigation and astronomy.

Babbage's interests went well beyond that of the difference engine, however. In the period 1823-1827, he became interested in the workshops and factories of England. Toward the end of this period, when a health problem forced him to spend several years on the European continent, he traveled extensively and studied the structure of European industries. On returning to England he summarized his findings in an essay entitled *On the Economy of Machinery and Manufacturers* (Babbage, 1832). Babbage describes his book as follows.

I have not attempted to offer a complete enumeration of all the mechanical principles which regulate the application of machinery to arts and manufactures, but I have endeavored to present to the reader those which struck me as the most important either for understanding the actions of machines or for enabling the memory to classify and arrange the facts connected with their employment.

Urwick and Brech (1956:23) make the following observation about the contribution that Babbage made to this early study of management.

Considering that this [the above] sentence was written more than a hundred years ago, it is astonishing to find scarcely a trace of technical description in the book. Babbage was not concerned with the design or making of machines, but with their use and the organization of human beings for that purpose. His essay is a management study, written from the standpoint of a trained scientist who had had the opportunity of seeing and comparing methods in many factories.

It is this detachment, this use of comparison, his faith in the possibility of applying scientific processes of thought to the organization of industry, which constitute Babbage's unique contribution to the advancement of management. More than half a century before Taylor was to illuminate the same point, with far greater effect because he was a practising engineer, Babbage had stumbled on the underlying truth that there are general principles applicable to the manufacture of products by machinery, and that it is an understanding of these principles rather than the technical knowledge of how to make a particular article which is of first importance.

5.2.3 The Pioneer Proposal: A Recognition of the Need

Henry R. Towne (1844-1924). As co-founder of the Yale & Towne Manufacturing Co., Henry Towne had experience with managing a manufacturing enterprise. He was perhaps the first industrialist to recognize publicly that it was important to train people to manage enterprises more effectively. His early ideas relating to the importance of this were published in a paper presented to the American Society of Mechanical Engineers in 1886. This paper, which is credited with starting the management movement, illustrates Towne's early feelings on this topic (Towne, 1886).

To insure the best results, the organization of productive labor must be directed and controlled by persons having not only good executive ability, and possessing the practical familiarity of a mechanic or engineer with the goods produced and the processes employed, but having also, and equally, a practical knowledge of how to observe, record, analyze and compare essential facts in relation to wages, supplies, expense accounts, and all else that enters into or affects the economy of production and the cost of the product. There are many good mechanical engineers;–there are also many good "business men";–but the two are rarely combined in one person. But this combination of qualities, together with at least some skill as an accountant, either in one person or more, is essential to the successful management of industrial works, and has its highest effectiveness if united in one person, who is thus qualified to supervise, either personally or through assistants, the operations of all departments of a business, and to subordinate each to the harmonious development of the whole.

Engineering has long been conceded a place as one of the modern arts, and has become a well-defined science, with a large and growing literature of its own, and of late years has subdivided itself into numerous and distinct divisions, one of which is that of mechanical engineering. It will probably not be disputed that the matter of shop management is of equal importance with that of engineering, as affecting the successful conduct of most, if not all, of our great industrial establishments, and that the *management of works* has become a matter of such great and far-reaching importance as perhaps to justify its classification also as one of the modern arts. The one is a well-defined science, with a distinct literature, with numerous journals and with many associations for the interchange of experience; the other is unorganized, is almost without literature, has no organ or medium for the interchange of experience, and is without association or organization of any kind. A vast amount of accumulated experience in the art of workshop management already exists, but there is no record of it available to the world in general, and each old enterprise is managed more or less in its own way, receiving little benefit from the parallel experience of other similar enterprises, and imparting as little of its own to them; while each new enterprise, starting *de nouveau* and with much labor, and usually at much cost for experience, gradually develops a more or less perfect system of its own, according to the ability of its managers, receiving little benefit or aid from all that may have been done previously by others in precisely the same field of work.

Surely this condition of things is wrong and should be remedied. But the remedy must not be looked for from those who are "business men" or clerks and accountants only; it should come from those whose training and experience has given them an understanding of both sides (viz.: the mechanical and the clerical) of the important questions involved. It should originate, therefore, from those who are also engineers, and, for the reasons above indicated, particularly from mechanical engineers. Granting this, why should it not originate from, and be promoted by, the American Society of Mechanical Engineers?

The ASME did, indeed, become a forum for the discussion of many of the topics that Towne identified as being essential to the development of an improved understanding of management practice. One can only speculate why it was necessary for a century to elapse between the start of the industrial revolution and the beginning of the management revolution before the need to formalize thinking about management began to be discussed seriously.

5.2.4   Scientific Management

Frederick W. Taylor (1856-1915). Having worked as a laborer, a clerk, and then foreman in a steel mill–the Midvale Steel Company, located on the outskirts of Philadelphia–Frederick Taylor understood many of the practices that limited the productivity of the workers in the factory. Despite his efforts to correct the situation, he found it difficult and often unrewarding. While attending the 1886 meeting of the ASME, he heard the challenge offered by Henry Towne and was encouraged to undertake a more systematic examination of the ideas and principles that were needed to improve the practice of management. It was Towne's speech that launched Frederick Taylor onto the path that was to be the focus of much of the balance of his career. His contributions were many and have led to his being considered the "father of scientific management."

In his 1911 publication entitled *The Principles of Scientific Management*, Taylor summarizes the incentive for much of his efforts (Taylor, 1911).

It is true that whenever intelligent and educated men find that the responsibility for making progress in any of the mechanic arts rests with them, instead of upon the workmen who are actually laboring at the trade, that they almost invariably start on the road which leads to the development of a science where, in the past, has existed mere traditional or rule-of-thumb knowledge. When men, whose education has given them the habit of generalizing and everywhere looking for laws, find themselves confronted with a multitude of problems, such as exist in every trade and which have a general similarity one to another, it is inevitable that they should try to gather these problems into certain logical groups, and then search for some general laws or rules to guide them in their solution. As has been pointed out, however, the underlying principles of the management of "initiative and incentive," that is, the underlying philosophy of this management, necessarily leaves the solution of all of these problems in the hands of each individual workman, while the philosophy of scientific management places their solution in the hands of the management.

Taylor is well known for his focus on identification of the various elements of each task, the determination of the least amount of time needed to accomplish each task, and the construction of a plan to implement the best approach. Much of this work was focused on subdividing the task into its elementary steps and, through the use of the stopwatch, on measuring the time that a skilled worker would take in accomplishing the task. Using these measures as a basis for assessing the most efficient way of accomplishing a task, he was able to demonstrate impressive improvement in productivity. The "scientific" aspect of this approach included developing an understanding of each task, measuring the time that someone skilled in the task would take to accomplish it, and an analysis of how improvements could be made to capture the best possible performance of all workers.

Taylor has been reviled in recent times for his lack of concern for the human being, encouragement in breaking jobs down into specialities so that they could be done more efficiently in a repetitive fashion by someone specially trained, and for the subsequent dehumanization of the workplace. Although much of this criticism has a basis in fact, to focus on this aspect of Taylor's work can lead to overlooking the important and significant contributions by Taylor. Taylor was a champion of the scientific approach. He realized that one could not make good decisions without good data. Good data required careful measurements. Armed with good data, Taylor advocated careful analysis, the objective being to understand what actions could be taken to achieve better performance, usually higher productivity. None of this reasoning is in error. The fact that Taylor repeatedly demonstrated that better understanding leads to better system performance is a tribute to his understanding of the scientific process and the development of steps to implement it.

The Gilbreth's: Frank B. Gilbreth (1868-1924); Lillian M. Gilbreth (1878-1972).

A small group of imaginative people who were contemporaries of Frederick Taylor formed the early cadre of people who believed in and extended the ideas of scientific management. Among these were the Gilbreths, Frank and his wife, Lillian. Frank Gilbreth began his career as a bricklayer. During his early years of apprenticeship, he developed an intense interest in improving the efficiency with which the task of laying bricks could be accomplished. He was very successful in accomplishing this. Not being content, however, to remain a bricklayer, he applied many of the same techniques to establish a successful general construction business. He quickly developed a reputation for completing buildings under budget and in less than the contracted time. The philosophy that dominated Frank Gilbreth's lifelong career is encapsulated in the phrase that came to characterize his work: "one best way of doing work." This interest was not simply limited to increasing the volume of work that a person could do, although this was certainly a result, but also in reducing the fatigue that workers experienced in accomplishing their work. Human capability remained a constant concern to Gilbreth throughout his career.

Gilbreth was strongly influenced by the philosophy of scientific management that Taylor, with whom he first came in contact in 1907, espoused. Although Gilbreth became a strong advocate of measuring the time that each motion required to accomplish a task, he proceeded beyond the simple time-and-motion study that Taylor emphasized to use measurements as a means of analyzing and then improving *how* the worker *performed* each task. Gilbreth pioneered in the use of motion-picture records as a means of analyzing motion. This combination of time measurement and analysis of motions involved in accomplishing tasks were known as *micromotion studies* (Urwick and Brech, 1956:126-147). As Gilbreth's understanding of the best way to approach these issues matured, his approach evolved along a different path from that followed by Taylor and his disciples (Nadworny, 1957), with strong competition ultimately developing between Taylor and Gilbreth.

In their discussion of Frank Gilbreth's career, Urwick and Brech (1956:127) include the following quotation from a letter written by Lillian concerning her husband.

> Mr. Gilbreth felt that the least developed field in industry lay along the lines of discovering and using better methods, and he formulated very early in his life the idea of finding the "one best way to do work." This necessitated finding the best available worker, the best surroundings, equipment and tools, and the best method. The work then developed along two lines. First, an investigation of the worker from every standpoint possible, the physiological, the psychological, and so on, finding out his capabilities, his aptitudes, his aspirations, and his desire for any special type of work. As you can imagine, this led us into all the human sciences with a special stress on advances in education, the new psychology and psychiatry in its various developments.

The concern that the Gilbreth's had for individuals and the capability of individuals to perform the functions expected of them efficiently and effectively led to the association of their names with the study of human factors. Upon Frank Gilbreth's death in 1924, Lillian became the intellectual force behind the studies and the spokesperson for the humanistic approach to managing an operation. As she was trained as a psychologist, this does not seem unusual. What may be viewed as unusual is her melding of concern for the worker with the micromotion studies that had been pioneered by her husband. This focus on the well-being of the worker is amply summarized in a statement that Lillian Gilbreth made in a conference in Sweden in 1949 (Carlsberg, 1962).

> The final point is how can we make the best use of human resources within the individual, within the group and in group interplay. At present the management group confines itself too much to engineers and industrial engineers. Management must not hesitate to look for impasses from outside its own ranks. What is needed here, as in other spheres, is that people with different positions and backgrounds get together and give their points of view: physiologists, anthropologists,–psychologists, sociologists–the

whole set of experts. We increasingly realize that we must make use of human resources. Everybody must contribute, the experts cannot do a constructive job alone. We must make an inventory of our resources and then use them correctly.

Andre A. Potter, former dean of engineering at Purdue University and longtime friend of the Gilbreths, summarized the contribution of these two pioneers in the following way (Potter, 1962):

> Frank Gilbreth built his reputation on 'speed work' made possible by his resourcefulness and the impetus given to workers by his comparative records of achievement designed for each worker. He was an innovator and researcher at heart. He studied and developed the potentialities of mechanical equipment to expand the capacities of the individual. He believed that technology should be the servant of man, and an aid in liberating him from drudgery. He and his wife had a deep appreciation of the supreme worth of each individual and considered the worker of greater importance than the work.
> Dr. Lillian Gilbreth's contribution to the success of the Gilbreth partnership may be traced especially to her keen appreciation of the human factor in industry and to use of psychology as a management tool. She possesses outstanding ability to work with people without causing antagonism, has faith that men and women can be motivated to excel and to find the One Best Way, even when those responsible for a decision have serious differences of opinion. She is a master at the art of conducting free discussion until mutual understanding is achieved and effective agreements reached.

Henri Fayol (1841-1925). While scientific management was developing in the United States, Henri Fayol was stimulating the European community to become more concerned with improving the management of industry. Fayol was a highly respected French industrialist, having served in many capacities, including 30 years as general manager of the French coal and iron combine Société de Commentry-Fourchambault. The most complete discussion of his ideas are to be found in his publication *Administration Industrielle et Générale* (Fayol, 1916). In discussing Fayol's attitudes regarding administration, Brodie (1967) notes that Fayol commended those in attendance at the International Congress of Mines and Metallurgy on the technical progress that had been achieved in the mining and metallurgy industries but admonished them to think more broadly about the administration of activities.

> He wanted to make the point, however, that the Congress had been devoted almost exclusively to technical matters, yet it seemed to him that a similar approach would be no less beneficial if applied to administrative problems. As he saw things, everyone in an undertaking played a part in administering it. He wanted to see an interchange of experience and of ideas, analogous to that which took place on technical matters. He also urged the importance of observation and the placing of personal experience on record, so that a body of precepts on administration could be built up.

Fayol identified a group of principles of good management and a set of rules of good administrative practice. The principles of good management that Fayol identified are:

1. Division of labor
2   Authority and its corollary, responsibility
3. Discipline
4. Unity of direction
5. Unity of command
6. Subordination of private interest to the general good
7. Staff remuneration
8. Centralization
9. Orderliness
10. Equity
11. Stability of tenure
12. Initiative
13. Unity

The five rules of good administrative practice that were identified by Fayol are included in each of these practices.

- **Foresight.** A plan of action is a schema for the future, based on the resources of an undertaking, the particular characteristics of its activities and the possibilities that the future promises.
- **Organization.** Organization deals with the provision of those physical and human resources which an undertaking needs for its effective operation.
- **Command.** Every man in a position of leadership is responsible for getting the best out of those who work under him.
- **Coordination.** The numerous individual transactions which take place in an undertaking must be coordinated in such a way as to achieve smooth and successful running.... It means getting things in the right proportion and in balance, subordinating the adventitious to the essential.
- **Control.** Control is a process of verification, to check that things happen in accordance with policy, plans and orders and that mistakes or failures will be exposed and will be followed by appropriate action.

One can only be struck by the reasonableness of these lists and the insight that Fayol expressed as a result of his experience in managing a large organization.

5.2.5   Human Relations in Management: The Hawthorne Studies

Elton Mayo (1880-1949) and Fritz J. Roethlisberger (1898-1974). The Hawthorne studies were initiated in 1927 at the Hawthorne Works of Western Electric Company in Chicago. They were led by Elton Mayo, then head of the Department of Industrial Relations Research of the Graduate School of Business Administration, Harvard University, and Fritz J. Roethlisberger, Wallace Brett Konham Professor of Human Relations of the Harvard Business School. These studies had as their genesis an earlier group of experiments done at the Hawthorne Works between 1924 and 1927 of the effect of plant illumination on employee efficiency. These studies were carried out by the National Research Council of the National Academy of Sciences with participation by the Western Electric Co.

The objective of the early studies on illumination was to establish a quantitative understanding of the effect of room illumination on productivity in the test room of the works. Considerable care was taken to ensure proper establishment of a control group against which the performance of the test group could be evaluated. The results of these extensive studies were surprising. Contrary to all expectations, no clear correlation could be found between performance and the level of illumination of the workplace. Not only did performance improve when the level of illumination was increased, it also improved when illumination was decreased. The performance of the control group also increased at more or less the same pace as for the test group. Two additional experiments under more controlled conditions were carried out, with similar results.

The conclusion that was drawn from these three experiments was that illumination was not the most important (perhaps was even unimportant, except in the extreme) factor influencing productivity in the test room. This conclusion contributed to a growing realization that many aspects of human behavior must be explored if employee reactions to the workplace are to be understood. This realization led to the development of a subsequent set of experiments that were directed by Mayo and Roethlisberger at the Hawthorne Works. A detailed discussion of the later experiments and their results can be found in a book by Roethlisberger and Dickson (1950).

The first of the Mayo/Roethlisberger experiments was undertaken in an effort to better understand the factors that affect productivity. Because of the apparent complexity of the issues, it was decided to isolate a small group of workers in a separate room removed from the regular operations. In this way it was hoped that more careful and systematic studies could be done. A group of six female workers in the relay assembly test room were chosen as the test group. The investigators originally identified six questions that they hoped to answer (Roethlisberger and Dickson, 1950:28).

1. Do employees actually get tired out?
2. Are rest pauses desirable?
3. Is a shorter working day desirable?
4. What are the attitudes of employees toward their work and toward the company?
5. What is the effect of changing the type of working equipment?
6. Why does production fall off in the afternoon?

Experimental data were collected on this test group over the ensuing 26 months (13 experimental periods). The results showed a general increase in productivity under a variety of changing work conditions.

In looking back it is clear that two essentially different sorts of changes occurred in the first seven periods of the experiment. There were those changes introduced by the investigators in the form of experimental conditions; these were well noted and recorded. There was another type of change, however, of which the investigators were not so consciously aware. This was manifested in two ways: first, in a gradual change in social interrelations among the operators themselves, which displayed itself in the form of new group loyalties and solidarities; secondly, in a change in the relation between the operators and their supervisors. The test room authorities had taken steps to obtain the girls' cooperation and loyalty and to relieve them of anxieties and apprehensions. From this attempt to set the proper conditions for the experiment, there arose indirectly a change in human relations which came to be of great significance in the next stage of the experiment (Roethlisberger and Dickson, 1950:58-59).

The following summarizes the conclusions that were reached at the end of the 13 experimental periods:

On the basis of the studies reported in this chapter, the investigators concluded that these operators were at all times working well within their physical capacity. There is no evidence in support of the hypothesis that the increased output rate of all these operators during the first thirteen experimental periods was due to relief from fatigue. This hypothesis, which at first seemed most plausible, had to be abandoned....Although the observer's records gave little evidence of boredom or apathy, it was believed that more exacting inquiries into the operators' attitudes should be made in order to warrant treating the available evidence as conclusive....A careful statement of the kinds of changes which had taken place in the attitudes of the operators, and of the relation of such changes to alterations in the work situation, seemed more urgent and more important than a decision as to whether or not monotony was being experienced in this or that degree. (Roethlisberger and Dickson, 1950:127).

The findings in this experiment were as surprising as in the earlier illumination experiments–*improved output did not appear to arise from factors that were commonly held to be important in motivating a worker or detracting from a worker's performance.* As a result, additional testing was undertaken with a second relay assembly group and with a mica splitting test room group. The conclusions that were reached relative to these tests were as follows:

At least two conclusions seemed to be warranted from the test room experiments so far: (1) there was absolutely no evidence in favor of the hypothesis that the continuous increase in output in the Relay Assembly Test Room during the first two years could be attributed to the wage incentive factor alone; (2) the efficacy of a wage incentive was so dependent on its relation to other factors that it was impossible to consider it as a thing in itself having an independent effect on the individual. *Only in connection with the interpersonal relations at work and personal situations outside of work, to mention two important variables, could its effect on output be determined* [emphasis added]. (Roethlisberger and Dickson, 1950:160)

One of the important personal situations outside work, referred to in the above conclusions, was the prevailing concern among many workers that general economic circumstances were such that the company was going to be forced to make substantial layoffs.

The next phase of the studies was an extensive set of interviews with approximately 21,000 employees–nearly 70% of the workforce at the plant–over a period from 1928 to 1930 that were intended to determine employee attitudes to their employment.

> The investigators felt that they had finally developed an adequate conceptual scheme for understanding factors determining satisfaction or dissatisfaction at work....In the early stages of the interviewing program, statements reflecting employee satisfaction and dissatisfaction were referred to the employees' physical environment....The more the investigators worked with this interview material, the clearer it became that they were dealing with essentially two kinds of statements: factual and nonfactual....Before nonfactual statements could be used they had to be...related to some context other than the explicitly stated one. The first context suggested to the interviewers was the personal situation of the employee.... The limitation of this approach, the investigators found, was that it led them farther and farther away from the immediate industrial situation....It became clear that many employee comments which had formerly been interpreted in terms of the interviewee's personal situation could be better understood if they were interpreted in the light of the employee's existing social relations within the plant: the social organization of the group with which he worked and his position in that group. (Roethlisberger and Dickson, 1950:373-374).

As a result of the interview study, in 1931 and 1932, a series of observations was undertaken of the *social organization* of workers using the bank wiring observation room. Relative to the bank wiring observation room studies, it was concluded that (Roethlisberger and Dickson, 1950:523-524):

> The social organization of the bank wiremen performed a twofold function: (1) to protect the group from internal indiscretions, and (2) to protect it from outside interference....
>
> It may be seen, therefore, that nearly all the activities of this group may be looked upon as methods of controlling the behavior of its members. The men had elaborated, spontaneously and quite unconsciously, an intricate social organization around their collective beliefs and sentiments. The question as to what gave rise to those sentiments and beliefs, whether they arose from actual or potential threats to their security, as the operators claimed, is an important one....

The bank wiring observation group not only exhibited the internal social organization described above, but had developed a complex relationship with the wider company organization.

> It has been shown that the internal function of this organization was to control and regulate the behavior of its members. Externally, however, it functioned as a protective mechanism. It served to protect the group from outside interference by manifesting a strong resistance to change, or threat of change, in conditions of work and personal relations....Had it been explicitly stated, their behavior could be said to have been guided by the following rule: "Let us behave in such a way as to give management the least opportunity of interfering with us." (Roethlisberger and Dickson, 1950:525)

In a period of six years, the Hawthorne experiments had developed a wealth of data, challenged accepted concepts, identified new areas of study, and contributed greatly to understanding many of the factors that influence behavior in the workplace. The literature abounds, however, with articles that are critical of the methods used in the Hawthorne experiments. Mayo and his colleagues have been criticized for the implications that they drew of the results and the use that they made of the conclusions in broader contexts. Henry A. Landsberger of the Cornell Social Science Research Center has summarized and further analyzed the reasons and rationale for much of this criticism (Landsberger, 1958). The insights offered in this monograph are particularly useful to students in understanding the sociological factors that influence the doing and reporting of research. In the present context, only a few quotations of the view that Landsberger has derived of the Hawthorne experiments themselves and the recounting of the results by Roethlisberger and Dickson are offered.

> The significance of the authors' conclusions is not that they demonstrated to students of industrial relations that they were being insufficiently sociological and psychological, but instead, one must see the authors as demonstrating to industrial psychologists like themselves (as well as to noninstitutional economists) that industrial relations factors could not be ignored even in the heartland of industrial psychology: in the investigation of reactions to lighting and rest pauses. (Landsberger, 1958:87)

> Viewed in the context in which we have sought to place it, the book's contribution is, we believe, profound and quite unrivaled. Nowhere will the researcher who delves into the more microscopic social phenomena find a better illustration and explanation of how even his data are deeply affected by certain larger variables, such as the *nature of economic institutions, the state of technology, urban integration, ethnic feelings, size of organization, and so forth and so on.* [emphasis added] (Landsberger, 1958:89)

> The Hawthorne Studies gain their importance from the fact that they were comprehensive and encompassed a wide range of empirical data. They contained a rich description of an industrial situation at a time when the new fields of industrial sociology and industrial social psychology were at least as short on data about specific situations as they were on theory or theories. (Landsberger, 1958:92)

> The study of supervision merged into a study of industrial bureaucracy as a whole, once it was recognized that: (a) analytically speaking, problems of supervision could arise at any level in the hierarchy and were equivalent to problems of authority with which bureaucratic theory has long concerned itself; and (b) the foreman's style of supervising others depended not only on his personality, but also on the supervision he received from his superiors. This resulted, in the first instance, in directing attention toward the top of the industrial hierarchy. (Landsberger, 1958:105)

5.2.6   Theory X and Y

Douglas M. McGregor (1906-1964).   In 1957, McGregor presented an address to the Fifth Anniversary Convocation of the MIT School of Industrial Management entitled "The Human Side of Enterprise." In this address he offered suggestions for new approaches to managing enterprises. One can properly view these suggestions as integrating many of the ideas developed in the Hawthorne studies into a more sociological framework. He described the conventional views of management, as contained within Theory X, to have the following attributes (McGregor, 1957):

1. Management is responsible for organizing the elements of productive enterprise–money, materials, equipment, people–in the interest of economic ends.

2. With respect to people, this is a process of directing their efforts, motivating them, controlling their actions, modifying their behavior to fit the needs of the organization.

3. Without this active intervention by management, people would be passive–even resistant–to organizational needs. They must therefore be persuaded, rewarded, punished, controlled-their activities must be directed.

McGregor's analysis of the inadequacy of management is reflected in Theory X. Adopting the approach of a social scientist, he chose worker motivation as the goal that should be optimized by management. Drawing heavily on the works of Abraham Maslow, he argued that human motivation is determined by responses to a group of needs, which he identified as physiological needs, safety needs, social needs, ego needs, and self-fulfillment needs.

> Man is a wanting animal–as soon as one of his needs is satisfied, another appears in its place. This process is unending. It continues from birth to death.
>
> Man's needs are organized in a series of levels–a hierarchy of importance. At the lowest level, but pre-eminent in importance when they are thwarted, are his *physiological needs*. Man lives for bread alone, when there is no bread. Unless the circumstances are unusual, his needs for love, for status, for recognition are inoperative when his stomach has been empty for a while. But when he eats regularly and adequately, hunger ceases to be an important motivation. The same is true of the other physiological needs of man-for rest, exercise, shelter, protection from the elements.

> *A satisfied need is not a motivator of behavior*! (McGregor, 1957:22-28)

McGregor notes that the next level of needs that a person addresses after his or her physiological needs are satisfied are their safety needs: protection against danger, threat, and deprivation. Unless people are in a dependent relationship where they fear deprivation, this need is often expressed in the form of "the fairest possible break....Arbitrary management actions, behavior which arouses uncertainty with respect to continued employment or which reflects favoritism or discrimination, unpredictable administration of policy–these can be powerful motivators of the safety needs in the employment relationship at every level, from worker to vice president (McGregor, 1957:22-28).

At the next level, humans beings are motivated principally by their social needs: needs for belonging, association, acceptance by their associates, and giving and receiving friendship and love. Group associations and permanent teams that develop a cohesive relationship often display these characteristics. Assuming that these needs present a threat to the organization, management frequently takes actions that reduce or eliminate the capability of people to achieve these needs in the work environment. Workers who are motivated by social needs will probably react negatively to these restrictions, often becoming resistant, antagonistic, uncooperative, and noncommunicative. If the workplace does not provide satisfaction of the person's social needs, they will satisfy those needs outside the workplace through associations in churches, lodges, investment groups, gourmet groups, and so on. In such circumstances the workplace is no longer satisfying the social needs of the employee, and the motivation to excel in the workplace is greatly diminished.

At a still higher level, humans become concerned with their egoistic needs: the needs for self-respect, independence, achievement, competence, and knowledge combined with the needs for status, recognition, and appreciation. The typical organization has traditionally offered little satisfaction to the worker in terms of these needs.

The uppermost motivational level for the person is self-fulfillment. It is at this level that the person seeks to realize his or her greatest potential for continued self-development and for being creative in the broadest sense.

Since "a satisfied need is not a motivator of behavior," when a person's needs are satisfied at one level, motivation will be achieved only by looking to the next higher level. Thus safety is a proper motivator only after a person's psychological needs have been fulfilled. Social needs will become a motivator only after physiological *and* safety needs have been met. Egoistic needs will become a motivator only after physiological, safety, *and* social needs have been met. If a person suddenly finds that a lower-level need is not being fulfilled, efforts to motivate with higher-level schemes will be unsuccessful. Figure 5.2 illustrates the pyramid effect of these various levels of motivation.

Since most organizations have attempted to motivate their employees using salary, benefits, holidays, and so on–essentially concentrating only on the physiological needs of the worker–when most workers are searching for fulfillment of their safety, social, egoistic, or self-fulfillment needs, it is not surprising that management and workers have often reached an impasse in their relationship. McGregor viewed this as a lack of understanding and willingness by management to recognize the inherent needs of individuals and the means by which these needs could be satisfied to the benefit of the organization and the individual. In an effort to find ways to correct this situation, McGregor suggested that Theory Y be adopted by management. The elements of this theory are (McGregor, 1957:22-28):

1. Management is responsible for organizing the elements of productive enterprise–money, materials, equipment, people–in the interest of economic ends.

2. People are *not* by nature passive or resistant to organizational needs. They have become so as a result of experience in organizations.

3. The motivation, the potential for development, the capacity for assuming responsibility, the readiness to direct behavior toward organizational goals are all present in people. Management does not put them there. It is a responsibility

Figure 5.2   McGregor's hierarchy of worker motivation.

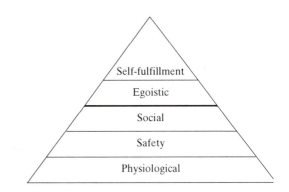

of management to make it possible for people to recognize and develop these human characteristics for themselves.

4.   The essential task of management is to arrange organizational conditions and methods of operation so that people can achieve their own goals *best* by directing *their own* efforts toward organization objectives.

While this statement of Theory Y does not prescribe how management is to accomplish these objectives, the objectives are clear and persuasive. Perhaps the comments by Edgar Schein in the introduction to *The Professional Manager* (McGregor, 1967) best characterize what McGregor was trying to convey.

> In *The Human Side of Enterprise,* McGregor gave to the academic and industrial community a set of values about people in organizations which were widely held but had not been clearly stated or examined....the essence of this attitude is to trust people, to believe in their capacity to integrate their own personal values with the goals of the organization. Doug believed that individual needs can and should be integrated with organizational goals. In the extreme, Theory Y has meant democratic processes in management, giving people a greater voice in the making of decisions and trusting them to contribute rationally and loyally without surrounding them with elaborate control structures.

Much of our future discussions concerning the management and treatment of people will build upon and reflect the objectives that McGregor included in his Theory Y.

5.2.7   The Scanlon Plan

Joseph Scanlon (1899-1956). Joseph Scanlon was a labor leader and a contemporary of McGregor. Although his concerns were primarily for the laborer, he sought to identify a means by which labor and management could cooperate to the benefit of both. In describing the elements of the Scanlon Plan, McGregor states the following (Lesieur, 1958):

> The plan implements Scanlon's underlying belief by establishing three broad conditions within which it becomes possible and natural for all members of the firm to collaborate in contributing to its economic effectiveness. These conditions are:
>
> 1.   A formally established "area of collaboration" and machinery (production and screening committees) for coordinating such collaborative efforts throughout the whole organization. This is accomplished without undermining collective bargaining or weakening the local union.
>
> 2.   A meaningful, realistic, common objective (the "ratio") in terms of which such collaborative efforts can be objectively measured.
>
> 3.   A psychologically adequate system of rewards (non-economic as well as economic) for a wide range of contributions to the effectiveness of the enterprise. (Traditional incentive wages, profit sharing, and suggestion-system awards are quite inadequate in terms of modern psychological theory.)
>
> As a consequence of establishing these three conditions, the employees and the managements of Scanlon Plan companies literally discover a new way of life. The process is not easy; some of the learning is rough indeed.

McGregor offers the following insight into the Scanlon Plan (McGregor, 1966):

> If you as management are considering the development of a new product or process in industry, you expect without question to devote a lot of time, money, and energy to turning your initial idea into a working process or a finished product that can go on the market. The initial idea is very remote from the final product that appears on the market....when we turn to look at the problems of managing people or to matters having to do with the organization of human effort, we find management attempting (quite unconsciously, I think) to shortcut this process completely and to assume that you can go from the original idea directly to the sale or use of the product without any intervening development process. For me, the growth of the Scanlon Plan since the middle thirties represents a kind of development research that is still far from completion.

5.2.8 Theory Z

William Ouchi (1943-). Ouchi has been a strong advocate of the practices that are followed in Japan and that have proven particularly valuable in providing the Japanese with much of their competitive advantage in manufacturing. In their studies of management practice, particularly in Japanese plants located in the United States, Johnson and Ouchi identified five important aspects of the managerial style that had been transferred from Japan to the United States. These were the following (Ouchi, 1981; Johnson and Ouchi, 1974):

1. *Bottom-up process.* The Japanese manager believes that initiative, change, and problem solving is accomplished best by those who are closest to the problems (i.e., those who are at the lowest level of the organization). The Japanese manager has the task of improving on the initiatives of others and creating an atmosphere that will motivate all members of the organization to seek better solutions.

2. *The senior manager as a facilitator.* The senior manager facilitates inquiries and encourages his or her subordinates to search for solutions to the problems that they have identified.

3. *The middle manager as an initiator and coordinator.* Middle managers are expected to understand the nature of the problems, to formulate tentative solutions to them, and to coordinate with managers in other functional areas to generate a composite plan for presentation to senior management.

4. *Decision by consensus.* The consensus-building process attempts to reach a common decision based on discussions and the exchange of views among all functional areas affected by an issue. "Contrary to what many Westerners think, the Japanese system does not demand that all participants 'sign off.' Those who affix their seals to the document containing the decision are indicating their consent, which is not the same as their approval. Each participant is indicating that his point of view has been fairly heard.... In this manner, the Japanese sidestep the nearly impossible task of attaining unanimity."

5. *Concern for the employee.* Concern for the whole employee–not for just his work performance–is a characteristic of the Japanese company. This concern, the Japanese are convinced, is essential to attaining productivity and high standards.

Ouchi concludes that corporations should follow the following four principles in their operations (Ouchi, 1981; Johnson and Ouchi, 1974). These constitute the Theory Z of management practice.

1. Give employees lifetime employment.
2. Promotions should be based on length of service as distinct from evaluation of immediate job performance.
3. Individuals should not be specialized in their positions but should move throughout the company.
4. Decisions should be made through a collective decision process.

While many of the practices that the Japanese have determined to be beneficial can be endorsed and encouraged as important practices for management, it is worth noting that the actions contained in items 1 and 2 of Theory Z are not the only ways of achieving the security that McGregor identifies as being important to achieving employee motivation. Furthermore, all Japanese companies do not subscribe to principle 1, as is evidenced by the statistic that the number of workers in Japan who have the benefit of lifetime employment is a small fraction of the total workers. Thus, in examining the applicability of Theory Z to current situations, it is important to ask whether it has applicability only to selected segments of industry or perhaps only to certain companies. While offering some benefits, strict application of principles 1 and 2 can create a number of unfortunate circumstances that can serve to the detriment of the organization. These are discussed further in subsequent chapters.

5.3 A CONJECTURE CONCERNING FUTURE OPPORTUNITY

It is worth noting that the Japanese management style that has been so discussed and lauded by many Westerners and is the center of Ouchi's Theory Z is focused on satisfying the first three levels of human needs identified by McGregor: the psychological, safety, and social needs. The traditional Japanese system of management does not go beyond this level. It does not offer significant support for satisfying the ego or self-fulfillment needs of the individual. It does not seek to motivate through either the egoistic or self-fulfillment needs. While much of Western philosophy–particularly as practiced in the United States – encourages individualistic behavior and thus recognizes the need to support the ego and self-fulfillment needs of people, the management approaches that we have frequently adopted do not provide a mechanism for accomplishing this in much of our industry. It is worth speculating whether recognition of the individual, as encouraged in Western cultures, will offer these cultures an inherent advantage over Eastern cultures *if* the management of Western enterprises can only bring themselves to recognize the importance of fulfilling the safety and social needs of their employees. With this as a base, it seems possible that Western cultures can develop a system that has the inherent advantage of providing individual motivation at the highest level. If so, it is not farfetched to imagine that a system could develop that is stronger, more productive, more responsive, and more efficient than any that has been seen to date.

SUMMARY

It is striking that this chapter began with an account of the role of the artisan/owner, who had essentially all the knowledge about an enterprise and its customers within one person, moved through a period in which human influence was deemphasized, and has concluded with an account of the rediscovery of the importance of involving all employees in the discovery and elimination of problems. The route that has been followed in this evolution can be summarized as starting with the humanized management practice of the artisan, turned to a dehumanized system for the large complex enterprises, and more recently has attempted to return to the humanized approach. While the manufacturing system that exists today is much more complex than was experienced by the artisan of early years, the structure currently being introduced into many world-class companies seeks to provide the large enterprise with many of the benefits enjoyed by the artisan.

A question was raised early in the chapter as to how the discipline of management developed given that topics such as manufacturing had not generated a well-defined structure. Beginning with Henry Towne, there developed an awareness of the importance of management. Furthermore, the belief arose that management techniques are transferable to essentially all situations and that a well-defined collection of knowledge is needed for a practitioner of the art of management. With these recognitions, management schools evolved that attempt to communicate the needed collection of facts and theories to those who would practice the art of management. In contrast, manufacturing has only recently assumed a position of sufficient prominence to be regarded as of strategic importance to corporations. In early years, schools known as A&M's (arts and mechanical schools), many of which were created by the land-grant college acts, included manufacturing in the list of courses taught. This process gradually disappeared and it is only in recent years that manufacturing has become an accepted topic of study in engineering and, to a lesser degree, in some schools of management. Furthermore, the techniques used in manufacturing are far from universal. The techniques and processes used by the semiconductor industry have little in common with those of the automotive industry, for example. This lack of universality of techniques continues to be a major barrier to the development of a true discipline of manufacturing.

REFERENCES

BABBAGE, C. 183.2. *On the Economy of Machinery and Manufacturers*, Cary and Lea, Philadelphia.

BRODIE, M. B. 1967. *Fayol on Adminstration* . Lyon, Grant and Green, London:7.

CARLBERG, B. C. 1962. *The Journal of Industrial Engineering*, XIII:192.

DRUCKER, P.F. 1985 *Management: Tasks, Responsibilities, Practice*, Harper&Row, New York.

FAYOL, H. 1916. *Administration Industrielle et Générale,* Bulletin de la Société de l'Industrie Minerale, (3).

JOHNSON, R. T. AND W. OUCHI . 1974. Made in America Under Japanese Management, *Harvard Business Review,* Sept-Oct. 61.

LANDSBERGER, H.A. 1958.. *Hawthorne Revisited: Management and the Worker, Its Critics, and Developments in Human Relations in Industry*, Cornell University, Ithaca, New York.

LESIEUR, F. G. (ed.). 1958. *The Scanlon Plan: A Frontier in Labor-Management Cooperation.* New York, Technology Press and John Wiley & Sons, Chapter 2. Reprinted as Chapter 8 in *Leadership and Motivation*, W. G. Bennis and E. H. Schein (Ed.) MIT. Press, Cambridge, MA. 1966, p. 121.

MCGREGOR, D. 1960.. The Human Side of Enterprise, *Management Review,* 56..

MCGREGOR, D. 1966.. *Leadership and Motivation*, the MIT Press, Cambridge, MA.

MCGREGOR, D. 1957.. *The Professional Manager,* W. G. Bennis and C. McGregor (eds.) McGraw-Hill Book Company, New York:xi.

MINTZBERG, H. 1990. The Manager's Job: Folklore and Fact, *Harvard Business Review,* Mar-Apr:163.

NADWORNY, M. J. 1957. Frederick Taylor and Frank Gilbreth: Competition in Scientific Management, *Business History Review* 31:22-34.

OUCHI. W. 1981. *Theory Z: How American Business Can Meet the Japanese Challenge* , Addison-Wesley Publishing Company, Reading, MA.

POTTER, A. A. 1962. *The Journal of Industrial Engineering*, XIII:190.

ROETHLISBERGER, F. J. AND W. J. DICKSON. 1950. *Management and The Worker*, Harvard University Press, Cambridge, MA.

TAYLOR, F. W. 1911. *The Principles of Scientific Management*, Harper&Brothers, New York: 103(reissued 1934).

TOWNE, H. R. 1886. The Engineer as an Economist, *Transactions of the American Society of Mechanical Engineers*, 7: 428.

URWICK, L. AND E. F. L. BRECH. 1956.*The Making of Scientific Management*, Sir Isaac Pitman & Sons, London.

QUESTIONS

5.1 Many agreements between union and management involve explicit statements about the duties that each job classification can perform. In some factories there may be tens of job classifications. If he were alive today, would you expect Fred Taylor to embrace this structure? Why?

5.2 The Human Factor Society encompasses more than a dozen sections, includung man-machine interactions, training, safety, and ergonomics, each of which emphasizes a particular activity that relates to how people interact with each other and with their environment. Although the existence of these can be traced to the early work of Frank and Lillian Gilbreth, they go well beyond the work of these early researchers. In thinking about what the field of human factors might encompass in the year 2025, what topics would you expect it to cover? Would these maintain a connection that could be traced back to the Gilbreths?

5.3 In McGregor's Theory Y, it is stated that people are not by nature passive or resistant to organizational needs. It is suggested that they have become passive and resistant as a result of their experiences within the organization. Identify some of the experiences that led to the creation of these attitudes in workers.

5.4 Can the concept of lifetime employment be used as a motivating factor and not simply taken for granted by a workforce?

5.5   In Ouchi's Theory Z, it is stated that the length of service should be considered to be distinct when evaluating an employee's job performance. How should a manager balance the effects of length of service and job performance when considering who should be promoted into a recently created new position?

5.6   How can U.S. companies expect to gain employee confidence and commitment to growth while they continue to engage in the practice of laying off employees when the economic structure of a company changes?

5.7   Do time-and-motion studies conflict with worker involvement and empowerment?

5.8   How would you apply McGregor's Theory Y in a enterprise that uses a large number of temporary employees? Are there limitations to the application of this theory to the type of organization?

5.9   McGregor identified a hierarchical structure of needs that dominate human motivation. Assume you can be assured that your physiological needs are going to be met throughout your working life.

   (a)   What are some of the choices in jobs or professions that you would consider if you were to be motivated principally by safety needs (using safety in the sense that McGregor uses it)?

   (b)   Now assuming that your physiological and safety needs are being met, what are some of the choices in jobs or professions that you would consider if you were to be motivated principally by social needs?

   (c)   With your physiological, safety, and social needs met, what are some of the choices in jobs or professions that you would consider if you were motivated by ego needs?

   (d)   With your physiological, safety, social, and ego needs met, what are some of the choices in jobs or professions that you would consider if you were motivated by self-fulfillment needs?

   (e)   For parts (a), (b), (c), and (d), give your reasons for identifying the particular jobs or careers that you have listed with the particular level of needs.

5.10   Your personnel officer has just completed a lengthy discussion of the advantages and disadvantages of McGregor's Theory Y and Ouchi's Theory Z. You conclude that both theories have some validity and some limitation and that you need to undertake some experimentation to determine what can best be done with both theories. Since your manufacturing enterprise is effectively divided into small groups, you conclude that you can try some things in one group prior to taking any actions throughout the plant. What kind of experiment would you carry out to determine how you might use egoism and self-fulfillment as effective motivators of performance? What kind of experiments would you do to evaluate the importance of lifetime employment? How would you evaluate these experiments?

READINGS

GIBSON J. 1992. Taylorism and Professional Education, *Manufacturing Systems: Foundations of World-Class Practice,* J. A. Heim and W.D. Compton (eds.), National Academy Press, Washington, D.C.:149

MCGREGOR, D. 1957. The Human Side of Enterprise, *Management Review,* 46 (Nov.): 22-28; 88-92.

APPENDIX A5.1 SELF-STUDY QUESTIONS FOR MANAGERS*

1. Where do I get my information, and how? Can I make greater use of my contacts? Can other people do some of my scanning? In what areas is my knowledge weakest, and how can I get others to provide me with the information I need? Do I have sufficiently powerful mental models of those things I must understand within the organization and in its environment?

2. What information do I disseminate? How important is that information to my subordinates? Do I keep too much information to myself because disseminating it is time consuming or inconvenient? How can I get more information to others so they can make better decisions?

3. Do I tend to act before information is in? Or do I wait so long for all the information that opportunities pass me by?

4. What pace of change am I asking my organization to tolerate? Is this change balanced so that our operations are neither excessively static nor overly disrupted? Have we sufficiently analyzed the impact of this change on the future of our organization?

5. Am I sufficiently well-informed to pass judgement on subordinates' proposals? Can I leave final authorization for more of the proposals with subordinates? Do we have problems of coordination because subordinates already make too many decisions independently?

6. What is my vision for this organization? Are these plans primarily in my own mind in loose form? Should I make them explicit to guide the decisions of others better? Or do I need flexibility to change them at will?

7. How do my subordinates react to my managerial style? Am I sufficiently sensitive to the powerful influence of my actions? Do I fully understand their reactions to my actions? Do I find an appropriate balance between encouragement and pressure? Do I stifle their initiative?

8. What kind of external relationships do I maintain, and how? Do I spend too much of my time maintaining them? Are there certain people whom I should get to know better?

9. Is there any system to my time scheduling, or am I just reacting to the pressures of the moment? Do I find the appropriate mix of activities or concentrate on one particular function or problem just because I find it interesting? Am I more efficient with particular kinds of work, at special times of the day or week? Does my schedule reflect this? Can someone else schedule my time (besides my secretary)?

10. Do I overwork? What effect does my work load have on my efficiency? Should I force myself to take breaks or to reduce the pace of my activity?

11. Am I too superficial in what I do? Can I really shift moods as quickly and frequently as my work requires? Should I decrease the amount of fragmentation and interruption in my work?

Source: MINTZBERG, H. 1990 The Manager's Job: Folklore and Fact, *Harvard Business Review*, Mar.-Apr.:174.

12. Do I spend too much time on current, tangible activities? Am I a slave to the action and excitement of my work, so that I am no longer able to concentrate on issues? Do key problems receive the attention they deserve? Should I spend more time reading and probing deeply into certain issues? Could I be more reflective? Should I be?

13. Do I use the different media appropriately? Do I know how to make the most of written communication? Do I rely excessively on face-to-face communication, thereby putting all but a few of my subordinates at an informational disadvantage? Do I schedule enough of my meetings on a regular basis? Do I spend enough time observing activities first-hand, or am I detached from the heart of my organization's activities?

14. How do I blend my personal rights and duties? Do my obligations consume all my time? How can I free myself from obligations to ensure that I am taking this organization where I want it to go? How can I turn my obligations to my advantage?

6
Current Management Focus: Participative Management

> The basic philosophy, spirit, and drive of an organization have far more to do with its relative achievements than do technological or economic resources, organizational structure, innovation and timing. All these things weigh heavily on success, but they are transcended by how the people in the organization believe in its basic premises and how faithfully they carry them out. (Thomas J. Watson).

The rejuvenation process that began for U.S. industry in the late 1970s and early 1980s was stimulated by intense competition from overseas manufacturers. The U.S.-based producer was forced to consider steps that would improve the productivity of the enterprise while improving the quality of the products being produced. Since the principal competition was being generated by companies in Japan, it was natural that the U.S. companies would explore the techniques the Japanese were using. Benchmarking of Japanese firms became a serious goal of many companies.

The first impression-usually before in-depth visits were made to observe the performance of the Japanese competitor-was that the Japanese were using much more modern technology in their manufacturing facilities than were available in comparable U.S. plants. It was suggested that there were more robots, more automation, and more modern equipment than was available to the U.S. manufacturer. If the U.S. manufacturer could only spend enough on automation, it was reasoned, it would be possible to match the Japanese performance.

The first series of in-depth visits usually demonstrated that the technology in use in Japanese plants was not significantly different from that in use in U.S. plants. A few more robots might be seen and somewhat more automation might be used, but the differences in the use of new technology was far less than expected. These early visits usually generated large amounts of data on the performance of Japanese firms. On subsequent analysis, reaction by U.S.-based firms was often that something was wrong with the data. It seemed impossible that statistics on such items as the number of items being built per employee, or the time taken to bring a new product to market, or the number of engineering change orders that were needed during the development and introduction of a new product could be accurate. They simply seemed too good to be true. The reaction was disbelief that the Japanese could be that good. Either the data were being falsified intentionally or there were other factors that were not being included in the analysis.

In these early visits it was common for there to be considerable discussion of the *quality circles* prevalent in Japanese industry. It was generally concluded by visitors that these circles enabled the Japanese to achieve the high quality that was evident in their products. It seemed straightforward, therefore, that U.S. industry would move to adopt quality circles in all their plants as a way of improving quality. When early attempts to introduce quality circles failed in the United States, it was often stated that this was a result of the "cultural" differences that existed between the Japanese and U.S. workforces.

Other reactions were derived from analysis of financial statistics. One was that the principal benefit for the Japanese was from their low cost of labor. Another was that if the yen/dollar relationship could only be made more attractive, the entire competitive problem would disappear and U.S. industry would be able to compete. It was also concluded that the low cost of capital in Japan compared with that in the United States made it easier for the Japanese to justify investments.

To a certain degree, all of these early reactions had validity. It is certainly true that the cost of capital was less in Japan than in the United States, thus making it more difficult for U.S. manufacturers to modernize their plants and improve their facilities. It is also true that labor was, in the 1970s and early 1980s, less expensive in Japan than in the United States. Although it was difficult to generate comparable numbers for the cost of products made in Japan with those made in the United States, it is quite clear that the Japanese had a cost advantage for a wide variety of products (e.g., estimates of approximately $1000 to $3000 per car delivered to the United States was generally acknowledged). While only a portion of this resulted from lower labor costs, wages were a contributor to the differential. Although the quality of Japanese products were improved as a result of quality circles, it is now clear that this was only one of many factors that had given the Japanese an advantage in terms of quality. Finally, as noted above, although some Japanese factories were more automated than those in the United States, it became clear that the level of technology being employed in Japan was not greatly different than in the United States.

What became clear, ultimately, is that while all these factors were, and to some extent always will be important, they explain only a small part of the advantages that Japanese industry has enjoyed. The important question became: What was being missed as exploration continued of the various means by which the Japanese were able to develop, manufacture, and deliver products of very high quality to their customers at costs that were significantly below that of U.S. producers?

6.1   THE CONCEPT OF EMPLOYEE INVOLVEMENT

The factors that had been missed earlier in attempts to understand the Japanese industrial success were found to be more complicated than those suggested in early studies. Although such factors as disparities between the rates of interest between the two countries and the dollar-to-yen ratio were important, the differences that were most important related to the management approach employed by the Japanese in their enterprises. It is important for those who aspire to manage world-class institutions to understand the management philosophy the Japanese used and to search for ways of adapting it and including in their management practices the best of its attributes. With this in mind, it is important to appreciate the importance of the practice of employee involvement and employee empowerment.

Employee involvement refers to the *practice of providing a structure* that involves the employee in the search for the best way of accomplishing the tasks that are being done. As Welliver (1992) writes:

> As the flow of information increases, people with the practical knowledge will regularly help form organizational goals and objectives in a give-and-take process, which the Japanese call "catch-ball". The manager would say, "Here's what I'd like you to do". And the employee would respond by saying how he thought he could do it. They would toss ideas back and forth until they finally agree. In America, we are more apt to play hardball. Top management says, "do it or else," and if accomplishments fall short, the employee is hit with demerits.

In the practice of employee involvement, the philosophy of continuous improvement is foremost. Each employee is encouraged to offer suggestions as to how to continuously improve the operations for which they are responsible and to seek ways to improve the overall system while also helping others improve their operations. While the longtime practice of using a suggestion box as a means of encouraging employee suggestions is one step in accomplishing this, an effective employee involvement program is much more. A box misses the personal interactions that are needed to accomplish this. Achieving continuous improvement through involvement of the employee requires many changes in the operating systems. Consensus must be reached by the employees regarding actions that affect all of them. Information is needed by all of the employees if they are to have the basis for making good judgments regarding operating changes. Resources must be available for implementing the ideas that are forthcoming from the employees.

It is for these reasons that *providing a structure* was emphasized above. It suggests that the structure must be such that employee involvement is actively encouraged. The organization must be committed to the involvement of all employees. Not only must it encourage it, if must reward those who engage in it. Everyone must understand that it is expected practice. In order for these practices to flourish, the management must be a strong and visible advocate.

6.2   THE CONCEPT OF EMPLOYEE EMPOWERMENT

If employee involvement is to be effective, the employee must see that their efforts provide benefits to the organization. They must know that their suggestions will be given proper consideration; that good suggestions will be provided necessary resources for implementation; and, that they will be rewarded for seeking to improve. The best way to accomplish these things is to give the employee the *responsibility and the authority* to proceed with making improvements without requiring a review of each action by someone at a higher level. This results in the practice of employee empowerment. For *employee empowerment* to be successful, the organization must have confidence in its employees; it must respect their judgment; and, it must expect a high level of responsibility from the employees.

It was the creation of a system in which the practices of employee involvement and employee empowerment could flourish that enabled the "quality circles" of the Japanese companies to be so effective. The failure to recognize that a commitment to a totally different philosophy of management was necessary if "quality circles" were to be successful was the hallmark of the early attempts by U.S. companies to copy these practices.

The combination of employee involvement with employee empowerment is the management philosophy that is encapsulated in the term *participative management*. As described above, it is clear that the employee is expected to be more than an idle observer of the processes with which he is dealing. They are expected to be active and to contribute to the success of the enterprise.

A. T. Kearney (1989), the international management consulting firm, has published a brief profile of nine companies that have achieved competitive success through a renewed emphasis on manufacturing. It is informative to see what some of these companies said about the importance of employee involvement and employee empowerment.

Harley-Davidson

Harley-Davidson discovered, after floundering with quality circles for three years, that the key to employee involvement was giving up traditionally held control and realizing that 'the guy on the shop floor knows more about solving the problem than you do.' The goal–not yet achieved in all cases, but steadily being realized–is to enable a person on the line who has a problem 'to pick up the phone, call the supplier, and get it fixed.' ...

To achieve such a level of employee involvement, the usual hierarchy of responsibilities was replaced with a system in which each employee has "ownership" over a particular operation.

Oxford Speaker Company

To stay competitive in manufacturing, Oxford also needs reliable low-cost labor. After a wide search, Oslac and Carnevald found it in an unusual place. On an Indian reservation in Mississippi they established a joint venture with the Mississippi Band of Choctaw Indians. Starting from ground zero, the Mississippi plant reached high-quality production in a little more than two years and now supplies loudspeakers directly to Chrysler and Ford.

Every employee, from the Choctaw Indians in Mississippi to the workers at the main plant in Chicago, is intimately involved in Oxford's operations-as a producer, a quality controller, and, ultimately, as a recipient of the rewards that involvement creates.

All employees are encouraged to make recommendations to their supervisors concerning quality and cost savings. Weekly meetings are held to determine the cause of rejects, break them down, and figure out why they happened. And, in the ultimate empowerment of production personnel, any employee on the line has the authority to shut it down if he sees it producing speakers of substandard quality.

Du Pont Corporation

To make its new orientation work, Du Pont has had to make equally sweeping changes in its relationship with its employees. This was an area, according to Blanchard, where much needed to be done. In the past, he pointed out, 'each department was pretty much left to its own devices' and from the standpoint of the Chemicals and Pigments Department, which he headed before his current position, "the situation was so desperate...that we were determined we were going to increase the participation and involvement of our people".... Blanchard said he was also absolutely convinced that our people have so much to contribute, and we had the kind of environment, with so many controls, that we had stifled their enthusiasm.'

> ...According to H. Gordon Smyth, senior vice president of employee relations, "Du Pont believes that the time is right to institute a more open approach to employee communications–to ask employees for their opinions about the company's operations instead of taking the more traditional approach of conducting a study and implementing the prescribed course of action."

The Eaton Corporation

Eaton Corp. has had the following experience at its Watertown, Wisconsin, plant (1993).

> Their reasons ...go beyond better customer contact. They have created a broad program to foster interaction among people both inside and outside the organization. Managers at Watertown ...are striving to fulfill a goal that eludes most companies: to get the most out of what people know, and can do, by getting them to act in concert, as one.
>
> That approach has paid off handsomely. Productivity at Watertown is growing 10 percent a year. Both net income and return on investment more than doubled in the last five years, as revenues reached $38 million, mainly from sales of counters and printed circuit boards.
>
> Watertown uses a variety of means to create an integrative work force-extensive teaming, disciplined planning, and widespread sharing ofknowledge-all built on a foundation of empowement of its 335 employees.

6.3 INTRODUCING PARTICIPATIVE MANAGEMENT INTO AN EXISTING SYSTEM

> Recognition that the employee, irrespective of level, can identify the source of problems and offer important insights into solving them is not revolutionary. These ideas are not new to those who have studied and examined the elements of successful management practice. In 1958 Douglas McGregor wrote (Lesieur, 1958):

> > First consider with me the knowledge and the insights that we have acquired within the last fifteen or twenty years about delegation and decentralization within industrial organizations. We have learned that, if we push decision making down in an organization as far as we possibly can, we tend to get better decisions, people tend to grow and to develop more rapidly, and they are motivated more effectively. Most companies today of any size at all are persuaded that the principles of decentralization and delegation-applied with wisdom-are fundamental to the successful operation of their organizations. We recognize

that no small group of management or no single manager can have all the answers; even if he does have them, he will lose a great deal if he attempts to make all the decisions. He will never have an organization that grows and becomes healthy in its own right.

Unfortunately, McGregor's declaration that most companies-in 1958-had accepted the principal that employees should be empowered at all levels was very premature. With few exceptions, industrial organization in the 1960-70's did not follow this pattern. The content of many jobs was narrowly circumscribed. Job classifications in the plants prevented cooperation between tradesmen and between workers and management. Sections or divisions were rewarded on the basis of their local performance, not their contribution to the performance of the total system. The industrial enterprise was often a collection of units that talked little to each other, competed vigorously for resources, and felt little commitment to the well-being of the other units. In recent times such systems have been described as organizations of "silos."

Adopting policies and processes that encourage the practice of including the employee in the processes of management was the first step to be taken by most companies. Although beset with some difficulties, this was generally easier than the next step, the giving of authority, to the employees, indeed the responsibility, to take the actions that were necessary to implement their ideas.

To change a philosophy of management that starts from the precept that a supervisor or manager knows what is best and that the workers are to comply and not question-because the manager has the authority to enforce all instructions-to one in which people are expected to offer suggestions and to help find solutions can be traumatic to everyone. Imagine the problem for the supervisor, having been subjected as a worker to manage-by-authority, in accepting that he must now turn to workers, who were viewed previously as not being capable of thinking, and asking them to tell him/her better ways to do their job better. Or consider the dilemma for the workers who were previously chastised by their union for attempting to improve on techniques for doing the job, in responding to the request that they tell the management how productivity can be improved with the result that fewer workers may be needed. Consider the manager's reaction, who previously demanded that he/she "review and sign-off" on all matters effecting operations, upon being told that this practice must be eliminated as the workers must be given the authority to make the decisions without his/her prior approval.

It is not surprising that all sorts of reasons and excuses were given why these changes would not work. The suggestion that employees should be involved in many of the decisions led to an early claim that management was being asked to abrogate its responsibilities and to turn over management of the plant to the employees. However, nothing could be further from the truth. While the roles of management and worker must both change with the introduction of these new processes, this does not mean an abrogation of final authority on the part of management. A broad class of decisions must still be made by management. The new processes, however, recognize that detailed decisions about how a task can best be done is best made by those who are doing the task. Many decisions fall outside this category.

Another criticism that was often heard concerned the need to reach consensus agreements. What was lost in these debates was the true meaning of consensus. Among the definitions of *consensus* is the phrase "feeling of agreement". Synonyms include *peace, harmony*, and *accord*. Consensus building does not mean that everyone must agree before actions can be taken. It does mean that a process is in place that is designed to offer the opportunity to all who are involved to offer suggestions, ideas, and opinions. In many matters, management must continue to make decisions even if some of the participants in the discussion continue to disagree with them. In addition, the manager or supervisor is expected to explain the basis for their decisions carefully so that everyone will understand and be willing to accept the decisions, even if they do not agree with them. As (Ouchi, 1981) reminds us in section 5.2, consensus does not necessarily imply unanimity. Communications becomes a very important part of the manager's tasks in the new system.

Everyone's role must change in making the transition from the authoritarian to the participative management philosophy. The manager must change from that of a giver of tasks, the reviewer, and the approver to that of coordinator, communicator, setter of objectives, and generator of resources. The role of the employee must become less that of a receiver of orders, passive follower, and robot worker to more that of active participant, creator of ideas, implementer of opportunities, and responsible decision maker. To assume that the changes would be easy for either of these groups as they move from the earlier pattern to the latter, would be to grossly underestimate human nature.

To change an existing organization can be an exceedingly difficult task. With the great inertia that large organizations develop, employees can find all sorts of reasons to try to protect the current status. Rules and guidelines that have been created over the years cannot easily be dropped or drastically altered. An organization learns how to do things and it is very difficult for it to learn how to do them differently. It is not surprising that a large organization is seldom able to accomplish radical changes without the presence of strong external forces. Examples of the difficulties and efforts that are needed to accomplish change are given in section 6.4 for Ford Motor Company and in Appendix A6.2 for Lexmark International, Inc. *peace, harmony*, and *accord*. Consensus building does not mean that everyone must agree before actions can be taken. It does mean that a process is in place that is designed to offer the opportunity to all who are involved to offer suggestions, ideas, and opinions. In many matters, management must continue to make decisions even if some of the participants in the discussion continue to disagree with them. In addition, the manager or supervisor is expected to explain the basis for their decisions carefully so that everyone will understand and be willing to accept the decisions even if they do not agree with them. As Ouchi (1981) reminds us in Section 5.2.8, consensus does not necessarily imply unanimity. Communications becomes a very important part of the manager's tasks in the new system.

Everyone's role must change in making the transition from the authoritarian to the participative management philosophy. The manager must change from that of a giver of tasks, the reviewer, and the approver to that of coordinator, communicator, setter of objectives, and generator of resources. The role of the employee must become less that of a receiver of orders, passive follower, and robot worker to more that of active participant, creator of ideas, implementer of opportunities, and responsible decision maker. To assume that the changes would be easy for either of these groups as they move from the earlier pattern to the latter would be to grossly underestimate human nature.

To change an existing organization can be an exceedingly difficult task. With the great inertia that large organizations develop, employees can find all sorts of reasons to try to protect the current status. Rules and guidelines that have been created over the years cannot easily be dropped or drastically altered. An organization learns how to do things and it is very difficult for it to learn how to do them differently. It is not surprising that a large organization is seldom able to accomplish radical changes without the presence of strong external forces. Examples of the difficulties and efforts that are needed to accomplish change are given in Section 6.4 for Ford Motor Company and in Appendix A6.2 for Lexmark International, Inc.

6.4   THE PROCESS OF CHANGE AT FORD

The Ford Motor Company was one of the first large organizations in the United States to make the change successfully from being authoritarian to being participative. It is instructive to examine in detail some of the lessons that were learned in the process of accomplishing this. In the late 1970s and early 1980s, Ford was suffering from a combination of forces that threatened its very existence. Two oil embargos by OPEC, a high inflation rate in the United States, an unusually small annual U.S. market for vehicles, a burgeoning number of Japanese car imports, products whose quality was inferior to that of the Japanese, and a product line that needed drastic renewal all combined to produce operating losses of billions of dollars. Although it was not widely recognized, the weight of these circumstances could have caused the company to fail. It is within this context that the discussion by Nancy Badore in Appendix A6.1 should be read. This is a how-to description of why and how a large company undertook to change its philosophy of operation. It is an account of the enormity of the task that was undertaken and the lessons that were learned in accomplishing it.

Badore (1992:90) listed the early lessons that help generate the confidence that was needed to undertake the further stages.

1.   A business focus is essential in the early stages of a process this size–in our case the focus was on quality.

2.   The goal must be viewed as worthy of achievement by all involved–it was this worthiness, not loyalty to management or union, that earned the time and commitment from the initial volunteers.

3.   Every level of the organization has a crucial role–from the chairperson to the average worker on the factory floor. Leaving levels out can subvert the process.

4.   "Early experimenters" (the first volunteers) in these change efforts are crucial to success. It will be their results, not the concept itself, that will attract the interest and support of a critical mass of the organization.

5.   The organization must have the capability to learn from its early successes and failures and to transfer this knowledge to other elements of the organization.

6.   Such efforts are sustained when they are tied to competitive improvement–ours has lasted through changes in union leadership, plant managers, vice presidents, and business cycles. "Cultural change for the sake of change" would not have lasted this long.

The combination of people who understood the severity of the problems, the willingness to experiment and try new things, and early successes that could be used to build confidence enabled the company to accomplish things that few others large organizations had been able to do at the time. Badore's account of some of the unexpected and severe difficulties that were encountered can be sobering. Making the decision to undertake these changes should not be taken lightly. It is clear that the changes can be successful only if everyone in the organization is willing to support the changes, because they believe in them, not because someone tells them to believe in them. Success will not come easily, for there are simply too many barriers to overcome. An organization must have the commitment to change if it is to accomplish change.

While actions were under way to change the management philosophy of the company, several key decisions were being made with regard to the product. The design group at Ford was asked to redesign the Thunderbird according to *their* intuitions, not those of management. The result was the first U.S. design of an "aerodynamically pleasing" vehicle. Substantially increased sales for the new design demonstrated that the concept was valid. Employee empowerment *had been successful.*

The development of an all new midsize car was the next major task that the company was forced to undertake. Because of the Thunderbird's success, it was to be designed using the aerodynamically pleasing form with a quality level that was to be as near that of Japanese products as was possible. Known later as the Taurus, it was decided that it should be created by a design team that included all key activities, including design engineering, product engineering, manufacturing engineering, marketing, and finance–the first major product design team for Ford. Because of the market environment and the high cost of an all-new car, it was recognized from the outset that this product had to be "right" in terms of quality, features, and appearance or the company would probably fail. In other words, this was a "bet your company" product.

Even at this stage many employees did not believe that the company had changed from a philosophy of production first and quality second. It was only after the company delayed introduction of the Taurus by several months to ensure that the quality was as it should be, thereby foregoing millions of dollars in sales, that they began to become believers. When the same action was taken with the delayed introduction of the Aerostar, for the same reasons, people in the company began to accept the concept that quality had moved into a top-priority position. The next priorities were the development of processes such that neither quality nor production would need to be sacrificed–that both would be correct the first time. More is said regarding this in later chapters.

Throughout this time, the company was proceeding with the introduction of employee participation. Off-site sessions were held, encouragement was given by the chairman, president, and vice-presidents, consultants were employed, organizations were changed, the corporate goals and objectives statement (shown in its 1984 form in Figure 3.1) was developed and distributed, and a cascading of this statement was made by each group. It was a time of turmoil, exhilarating change, painful experiences, and new hope. As viewed by many outside the company, it was an incredible process. As Nancy Badore is so fond of saying: "My friends would just turn away and giggle when I told them I was working to help make Ford a *people-friendly* company." While the process of change is never complete and must constantly be encouraged, the strong financial position that Ford enjoyed in the early 1990s must be attributed to its willingness to change and success in accomplishing it.

SUMMARY

The encouragement of employee involvement and employee empowerment in large U.S. companies is a phenomenon of the 1980s and 1990s. Finding approaches to the effective use of the knowledge and experience of the employee has been critical to many companies in their efforts to improve efficiency and enhance the quality of their products and services. It is being shown in many circumstances that achieving a world-class organization can occur best if systems such as these are implemented. As Badore (1992) described, enormous effort must be expended to introduce this concept into a large company. It requires the commitment of many people at all levels. Further, once introduced, it must be nurtured and encouraged.

Clearly, the challenge to the manager of the 1990s and beyond is to find the means to meld the best of all approaches into a new form of management. The scientific method of measurement and analysis must not be abandoned. The recognition that human motivation is driven by a variety of needs must be allowed to enter the thinking and planning of management. As we attempt to learn and apply the principles of sound management that have been developed over the past century, sight must never be lost of the fact that specific tools and general theories must each be adapted to particular circumstances. Participative management and employee empowerment have been demonstrated to be essential elements in allowing a large enterprise to retain the advantages that the artisan/owner had in terms of understanding, commitment, and capability. It is hardly a surprise, then, that this latest development in our evolutionary thinking would focus on the value of the individual to the enterprise. These practices are entirely consistent with the propositions offered by McGregor and several of those offered by Ouchi.

The successful manager is one who can use the tools for a particular situation, who will learn from past successes and failures, and who will be willing to tackle new situations in new ways. While management philosophy is strongly influenced by the size and philosophical orientation of an enterprise, the successful manager must be adaptable to circumstances and act accordingly, taking the best of all approaches and melding them into an effective philosophy.

REFERENCES

BADORE, N. L. 1992. Involvement and Empowerment: The Modern Paradigm for Management Success, *Manufacturing Systems: Foundations of World-Class Practice occur*, J. A. Heim and W.D. Compton (eds.), National Academy Press, Washington, D.C.:85-92.

EATON CORP. 1993. Quoted in "Making Factories Hum", *Enterprise,* July:15.

KEARNEY, A. T. 1989. *U.S. Manufacturing Competitiveness–Profiles in Competitive Success.*

LESIEUR, F.G. (ed). 1958. *The Scanlon Plan: A Frontier is Labor-Management Cooperation.* Technology Press and John Wiley & Sons, New York, Chapter 2; reprinted as Chapter 8 in Leadership and Motivation, W.G. Bennis and E.H. Schein (eds.) MIT Press, Cambridge, MA, 1966, p.121.

OUCHI, W 1981 *Theory Z: How American Business Can Meet the Japanese Challenge*, Addison-Wesley Publishing Company, Reading, MA.

WATSON, T.J. 1963. Quoted from R.P. Hummel, 1987. *The Bureaucratic Experience*, St. Martin's Press, New York: p. 104-105

WELLIVER, A. D. 1992. Going to the Gemba, *Manufacturing Systems: Foundations of World-Class Practice*, J. A. Heim and W. D. Compton (eds.), National Academy Press, Washington, DC:238-244.

QUESTIONS

6.1 "Management by walking around" refers to the practice of managers making themselves visible to and available to employees by being in the workplace on a regular basis. Can this practice be consistent with the practice of delegating decision making to lower-level employees? How would you go about enhancing the value of management by walking around?

6.2 How can you as a manager "go to the gemba" and then conclude that actions should not be taken as suggested by the line workers without losing credibility with them?

6.3 Over the past three years, your company has been actively engaged in the process of introducing participative management. Your assessment is that this has generally been successful. Operating results justify this optimism. What future actions will you undertake to ensure that the progress that you have made is not lost? What actions will you take to further enhance the adoption of participative management?

6.4 You are the manager of a plant that makes machine tools, most of which are sold to companies that manufacture farm equipment. What are some of the decisions that are made in the plant that will not be reviewed with all employees and will therefore not be included in the process of participative management? If some employees object to this exclusion of some decisions from participative management, how can you justify them to the employees?

6.5 You are the manager of the manufacturing activity for your company. You have begun the process of changing the management philosophy of your plant from being authoritarian to that of participative management. Over the past four weeks you have held numerous meetings, attended off-site group discussions, and provided encouragement and support for the principles of employee involvement and employee empowerment. This morning you were confronted by one of your excellent long-term employees who stated that she did not want to accept the responsibility associated with employee empowerment. This employee stated that she was comfortable with being told what tasks to do, she did not want to feel responsible here as well as at home for decisions, and she wanted to continue doing things as she had always done them. What would be your response? How would this response be viewed by other employees? What actions would you take?

6.6   Your manufacturing plant regularly operates using two shifts. Maintenance is done routinely during the off-shift when there is no production. Plant management has been very active in promoting employee involvement and employee empowerment. A problem has arisen in the actions proposed by the quality team on shift 1 and that proposed by the quality team on shift 2. The actions proposed would appear to generate results for one shift that will negatively affect the other shift. How do you resolve this? How do you reduce the likelihood that similar events will occur in the future? The quality group for the maintenance shift sees very different opportunities for improvement than are seen by the production shifts. How do you handle these different approaches when the groups are meeting at very different times and do not overlap?

READING

WELLIVER, A. D. 1992. Going to the Gemba, *Manufacturing Systems: Foundations of World-Class Practice*, J. A. Heim and W.D. Compton (eds.), National Academy Press, Washington, DC:238-244.

APPENDIX A6.1 INVOLVEMENT AND EMPOWERMENT: THE MODERN PARADIGM FOR MANAGEMENT SUCCESS*

The history of industrial development is replete with challenges that have changed the direction of individual companies as well as entire industries. Firms and institutions that were operating successfully were forced to find imaginative and innovative new ways of operating or face the prospect of extinction. New ways or new technologies were needed to accomplish what had previously been thought impossible. The past decade–or perhaps decade and a half–will undoubtedly be noted as one of these critical periods in industrial history. Industry groups and firms that had dominated the scene since World War II were suddenly confronted with a new and very different challenge–worldwide competition. Reasonably inexpensive transportation and highly reliable communication systems, combined with new approaches to managing and controlling the manufacturing enterprise, enabled manufacturers in all segments of the globe to compete in markets that had previously been reserved to those who manufactured where the product was marketed. Furthermore, the plateau on which the competitive battle was joined focused not just on costs but on quality, responsiveness, and flexibility–all in the name of satisfying a reawakened interest in providing customers with what they needed or demanded.

While no unique set of elements properly describes all the companies or industries that were challenged by this wave of world competitiveness, many firms found that they suffered from some of the following characteristics. The customer was not recognized as having the determining influence on product attributes or performance; tensions existed among various units and between various levels of the enterprise; no mechanism existed for setting priorities among the many desirable corporate objectives; suppliers were treated as a necessary evil to be tolerated but not trusted; and management and labor were confrontational in attitudes and objectives.

Firms that manifested these characteristics, even though they were common industry practice, found themselves at a distinct disadvantage to competitors who emphasized the customer, provided a product of high quality, and maintained an internal working environment that was stimulating and cooperative. It is not an overstatement to say that many U.S. firms were confronted with the highly unpleasant alternative of changing their ways of doing business or facing extinction. It was in this environment–the new competitive environment of the mid-1970s to early 1980s–that U.S. firms turned their attention to the development of employee involvement.

What is Employee Involvement?

The term *employee involvement* means inclusion of the employee in the operation of the system. But it is the creation of the process–the tool and the means–by which this is accomplished that is the topic of this discussion. The two principal objectives of employee involvement are as follows:

1. To create, share, and make "real" for all employees a vision of the goals for the overall enterprise as well as for each organizational unit.

2. To seek and share the knowledge possessed by individual employees in achieving that vision.

*Source: Badore, N. L. 1992. Involvement and Empowerment: The Modern Paradigm for Management Success, *Manufacturing Systems: Foundations of World-Class Practice*, J. A. Heim and W.D. Compton (eds.), National Academy Press, Washington, D.C.:85-92.

This cannot be viewed as a grand philosophic statement of principle. Rather it must be an operational statement of a process that calls forth a new level of participation by the employees. This statement must serve as a guide by which each employee can support the shared goals of the enterprise while at the same time serving as a standard against which the appropriateness of alternative actions can be assessed.

Relationship of Employee Involvement and Empowerment

Employee involvement recognizes that individual employees have the best opportunity to understand and appreciate the problems that are unique to their positions; that the employees also have the greatest insight and experience in suggesting ways of solving those problems. It does not directly follow, however, that a mechanism is available by which that knowledge and experience can be put to use in solving those problems. This is the function of "employee empowerment." If proper advantage is to be taken of the knowledge that the employee possesses, it is necessary for an organization to empower the employee to implement the improvements that they know to be necessary. By so doing, the enterprise is making the employee an integral part of the process of staying competitive.

Why was "Employee Involvement and Empowerment" Deemed So Critical?

Although size of the enterprise is not a prerequisite for being successful, certain aspects of "smallness" clearly make some tasks easier to achieve. Among these are the ability to achieve good communications among all employees, a common understanding of corporate and organizational objectives and goals, and a greater chance that the managers know personally the employees who must be involved in problem solving. The role of employee involvement and empowerment is, in a sense, the means by which a large organization attempts to achieve many of the benefits that are generic to the small organization. Although certain organizational structures and systems are required in larger organizations, the effort to accomplish meaningful employee involvement and empowerment is directed at preventing the organizational structure and systems from providing barriers to finding the best solutions to problems. Furthermore, the involvement process provides a means of humanizing the organization and maintaining participation by individuals at all levels–a process that is intended to lift the organization to new heights of competitive performance through the best use of the skills and interests of the individual. It is the means by which continuous improvement can be made an operating goal for all levels of an organization.

One might question, of course, the nature of the stimulus for developing employee involvement and empowerment. It would be pleasing to assert that recognizing the importance of employee involvement and empowerment was a natural part of organizational maturation and that the organizations saw it as a natural way to improve their performance. In fact, this was not the case and it was embraced, sometimes reluctantly, often belatedly, when the competitive crisis became so stark that management and workers alike were willing to take what was viewed as drastic action in order to survive. Industry appears to be no more immune to crisis management than any other segment of our society.

The process by which firms and enterprises achieved these changes has not been easy. It has been a difficult transition from the hierarchical structure, dominated by managers who "know what is best" for the organization and who expect others to do whatever task is asked of them, to a more open and sensitive environment in which people are willing to listen and respond to what they hear.

Accomplishing The Change

In discussing some of the issues and barriers that exist in accomplishing the transition to a more "people-oriented" company, I will draw on my experiences at Ford Motor Company. For simplicity, these experiences are presented as occurring in three distinct phases, but it should be realized that no such simple demarcation actually existed. In fact, some element of each phase was under way at all times.

I will try to offer some insight into the problems we experienced and the lessons we learned. The process of change is never easy for an organization. For a large, complex organization, it can be particularly traumatic. A retrospective assessment of the process and its results–after many of the frustrations have diminished in importance with the passage of time–can only be viewed as immensely valuable to the organization and to the people who have participated in it.

Phase I—Employee involvement: change at the plant level (1979-1982). One should recall that the late 1970s was a period of economic downturn, in part a result of OPEC actions. It was also a period of severe labor stress with frequent confrontation between labor and management. This was the environment that provided the stimulus for experimentation and change. The change sought was originally cultural-improvement of the relationship between union and management within the plants. The vehicle to achieve the change was to be the establishment of quality circles, a somewhat unoriginal concept at the time, given that the quality circle movement was being imported widely from Japanese manufacturers.

What differentiated the Ford experience from other quality circle efforts at the same time–efforts that consistently ran out of steam–was the fortuitous combination of several elements. These included timing–the near-depression that hit the manufacturing sector at the same time the project was launched; the care taken to prepare each plant, one at a time, to install quality circles; and the vitality that was inadvertently built into the process when age-old adversaries–union leaders and line managers–were "married" to provide joint oversight to the change efforts. Perhaps the most critical element of all was the fact that quality, selected as the overarching goal for these joint efforts largely because it was noncontroversial, would turn out to have important consequences for the competitiveness of the company and the union alike.

In 1979, Ford's U.S. manufacturing organization consisted of some 80 plants and parts depots varying in size from several hundred to several thousand employees. Organizing to manage change on this scale meant applying concepts devised by researchers and thinkers who dealt with much smaller organizations. We could find no precedent for a change effort as extensive as the one we contemplated.

In this first phase of change at Ford, a formal organizational development model was created centrally to be implemented in a decentralized fashion–that is, on a plant-by-plant basis. This model drew heavily on the theoretical work of Walton (1969) at Harvard and of Beckhard and Harris (1987) and Schein (1969) at Massachusetts Institute of Technology. It drew also on our analyses of why so many of the attempts to install quality circles in the United States were failing in the early 1980s. The model presented, almost simplistically, a series of five steps to be taken by the leadership of an organization before embarking on a significant change, that is, before launching quality circles or "Employee Involvement Groups." We defined "leadership" as the senior-most positions in the plant on the company and union sides. That is, the leadership of the employee involvement efforts was to be provided only by a pairing of the plant manager and the union leader.

Creating a model, however theoretically valid and practical in its intent, was one thing; getting people to use it was another. "Marketing" the concept of employee involvement–and the business and political rationale underlying it–was a crucial part of the change process. The job of speaking on behalf of the change effort was undertaken by two people with high visibility and credibility in the plants: Peter Pestillo, Ford's vice president for labor relations, and Don Ephlin, vice president of the United Automobile Workers' Ford Department. For several months, on request, they jointly visited plants to hold informal meetings with line managers and union officials. They accomplished two things at these events: they modeled an unprecedented camaraderie and they educated their audiences about the link between business outcomes and the ways business is conducted.

At the end of 1979, several plants had volunteered to sponsor the first pilot employee involvement projects. These living laboratories were to teach the rest of the company about the real issues underlying a major change effort. We found, for example, that it was not difficult to enlist volunteers from management to improve quality or from the union's political hierarchy; nor was it difficult to enlist volunteers from the ranks of the people working in the plant. What *was* difficult was handling the sheer workload associated with implementing improvement ideas once the problem-solving groups began to meet and work.

Only belatedly did we discover that quality circles reverse the work flow of the traditional hierarchical organization: ideas for implementations were suddenly flowing upward from the broad base of our plant pyramids. There is much talk about management sincerity as the crucial factor in efforts to change quality. We were to learn that it is even more a question of being organized to handle the incremental work flow associated with putting new ideas to work: reviewing group recommendations and assigning priorities–including capital allocations when necessary–and responsibilities for implementation. Communicating back to the Employee Involvement Groups the final determination on suggestions, the status and timing of projects stemming from the accepted suggestions, and why certain suggestion would not fly, also took a tremendous amount of coordination and (for us) new sensitivities and skills.

Nonetheless, these early projects prevailed; they achieved remarkable results in improved quality and various organizational (and union) measures of payoff. By the end of the first three years, quality had been improved up to 40 percent by external industry measures. The projects, moreover, have persisted and have expanded and spawned many other significant initiatives.

It might be instructive to enumerate some of the lessons we learned from this early experience:

1. A business focus is essential in the early stages of a process this size–in our case the focus was on quality.

2. The goal must be viewed as worthy of achievement by all involved it was this worthiness, not loyalty to managment or union, that earned the time and commitment from the initial volunteers.

3. Every level of the organization has a crucial role–from the chairperson to the average worker on the factory floor. Leaving levels out can subvert the process.

4. "Early experimenters" (the first volunteers) in these change efforts are crucial to success. It will be their results, not the concept itself, that will attract the interest and support of a critical mass of the organization.

5. The organization must have the capability to learn from its early successes and failures and to transfer this knowledge to other elements of the organization.

6. Such efforts are sustained when they are tied to competitive improvement–ours has lasted through changes in union leadership, plant managers, vice presidents, and business cycles. "Cultural change for the sake of change" would not have lasted this long.

Phase II—participative management: the change in middle management (1982-1985). The early experience at the plant level demonstrated a number of things. First, it was clear that an immense reservoir of good ideas resided with the employees. Second, it was necessary to modify significantly the existing relationships among management and labor if we were to be successful in tapping this resource. Third, the process of change that was being pursued could not be successful if it was limited to the local plants and concerned only with the unionized workers and their management.

The second phase in the process concentrated on bringing the middle management group of the operation into the process. With some successes to point to at the plant level, we found it possible to move somewhat faster in the second phase. The techniques were not all that different, however. A series of business conferences and workshops was organized for teams that made up the total business group. These meetings involved middle managers who shared common issues, for example, management teams from plants that produced similar products and the entire executive body from division headquarters who oversaw that constellation of plants. This provided an opportunity–even the necessity–for a role reversal, with the plant managers now modeling and sharing with the headquarters "bosses" what they had learned at the plant level. The headquarters personnel, in turn, found themselves responsible for articulating the business vision and competitive outlook for the enterprise as a whole, thus binding the plant management teams to the goals of the larger enterprise.

The meetings were held off-site in an environment of open discussion, free exchange, and constant debate–always with a coordinator who had an earlier successful experience in the process. Typically, five days of continuous immersion in the business problems was required. The results of these meetings might be briefly summarized by one phrase–improved understanding. The participants came away with an improved understanding of the business issues, their customers, the competition, the problems and constraints faced by others in the organization, and the potential benefits of the employee involvement process. Further, the participants discovered that the opportunity to discuss and debate the issues in this setting opened a wealth of new communication channels.

More lessons were learned from these experiences. Two were perhaps most important:

1.   Involvement of the entire "system" in the process greatly reduces the time needed to accomplish a change in culture.

2.   Debate is necessary to accomplish a change–it both clarifies the issues and is an essential element in obtaining genuine commitment from everyone as to future actions. The building of this genuine commitment and understanding stood in marked contrast to the shoulder-shrugging compliance that had, in the past, been mistaken for agreeing with "the boss."

Phase III—senior management change (1985-present).   As the process proceeded and included more and more people throughout the organization, it became clear that a continuing mechanism was needed to provide a meaningful involvement of the senior management in this process. This led to the establishment of the Ford Executive Development Center. The center has the responsibility to provide an environment for bringing together the company's executives from around the world to develop a consensus on company strategies and to focus on the customers and their needs and wants. The format of the group sessions is similar to that used in the earlier stages of the program–a site away from the office, groups of 50 or less, a mixture of people from various levels of the organization worldwide, participation through the five days by at least three of the corporate vice-presidents, and a concluding session in which company issues are debated and discussed with the chairman or the president of the company. During the week, speakers from outside the company are also invited to challenge the group to think differently about particular problems.

As we found in the earlier phases of this effort, the lessons we learned are not revolutionary. It is critically important, however, that we continue to operate in such a way that those lessons are not forgotten. Perhaps the three clearest lessons from Phase III are as follows:

1.   The executive development center provides a straightforward means by which senior managers can acquire a worldwide outlook about the business.

2.   A greater openness has developed among Ford employees to new, even disturbing concepts necessary to our success in the future.

3.   Debate and questioning of the status quo with the senior management is an important part of understanding and developing support for the worthiness and rightness of corporate strategic objectives.

Conclusions

An effective work force that is encouraged to search for ways of achieving continuous improvement is a key foundation of the modern manufacturing enterprise. Employee involvement in problem identification and employee empowerment that encourages the employee to take actions in creating a more efficient and effective system are critical to becoming world class. While it is not easy for large organizations to be "people oriented," their ultimate success in meeting the demands of the marketplace requires a commitment and a willingness to search for new ways of solving problems. "Employee involvement" is one of the keys to achieving this. The process is never ending. Employee involvement must be constantly stressed, continuously practiced, and regularly evaluated if it is to become a foundation of the company's operating philosophy.

REFERENCES

BECKHARD, R. AND R. T. HARRIS. 1987. *Organizational Transitions: Managing Complex Change*, Addison-Wesley Publishing Company, Reading, MA.

SCHEIN, E. H. 1969. *Process Consultation: Its Role in Organizational Development*, Addison-Wesley Publishing Company, Reading, MA.

WALTON, R. E. 1969. *Interpersonal Peacekeeping: Confrontations and Third Party Consultation*, Addison-Wesley Publishing Company, Reading, MA.

APPENDIX A6.2   CASE STUDY: LEXMARK INTERNATIONAL, INC.

In the late 1980s, IBM was struggling with the strategic issue of how to position itself in the laser printer, typewriter, and office supply business. While there was no question that this business group, having annual sales of over $2 billion, supplied components that were critically important, it became increasingly clear that these could not be considered as part of the core business. IBM was, after all, in the business of selling computers and computer systems. Since the core businesses received the attention and the largest share of the resources, the components group was not able to do many of the things that were necessary to ensure that the business was healthy and truly competitive.

IBM did not want to divest the laser printer, typewriter, and office supplies business and simply step away from it. It was important that a strong future supplier of these important components be assured. Further, they were reluctant to sell the businesses to a competitor. Finally, their commitment to lifetime employment was genuine and any actions that they might take regarding this business had to be consistent with this practice.

In March 1991, IBM concluded its effort to move its laser printer, typewriter, and office supplies business from beneath the IBM umbrella. An unusual arrangement was developed in which these businesses were bought by Lexmark International, Inc., the owners of which were a venture capital firm–Clayton, Dublier & Rice–employees of Lexmark and IBM. The agreement allowed Lexmark to sell all current and pipeline products under the IBM logo for five years, thus maintaining a tie that provided a continuing supply of important components to IBM. Agreements also provided Lexmark with access to certain future technical developments at IBM and encouraged the joint licensing of patents.

Before divestiture, IBM reorganized the activities and reduced the workforce from just over 5000 to 4100. Generous benefit packages that encouraged voluntary separations, retirement, or transfers to other IBM activities resulted in no layoffs, consistent with IBM policy of lifetime employment even in the presence of such a monumental change as a divestiture. The final arrangements satisfied the main concerns of IBM while giving Lexmark the flexibility that it needed to proceed to develop a strong independent entrepreneurial business.

The upper management of Lexmark, drawn from the ranks of IBM, had more than 90 years of experience with IBM. Named as CEO was Marvin Mann , who had been with IBM for 32 years in positions such as general manager of service industries and in charge of manufacturing for the information products division. He fully understood the IBM environment. Lexmark began with an employee base that was talented but constrained through its experience with the operating philosophy of Big Blue. It inherited an organizational structure that was characteristic of IBM. The task was how to make a workable company of this entity. Clearly, this was no simple matter, as the conversion from one of the most conservatively organized organizations into an entrepreneurial organization had rarely been accomplished.

Marvin Mann understood the importance of employee involvement, team participation, responsibility, and accessibility, even though these practices were not common in the IBM environment. If he was to implement these at Lexmark, many things must change. Two important steps were taken. First, the complex hierarchial organizantion that was characteristic of IBM was abolished. Having experienced the frustrations of an organization with eight layers between top management and the assembly workers, he understood the need to reduce the structure. The number of management layers in Lexmark was reduced to four. Second, a change in the process of making decisions was made. IBM used a contention policy that was intended to guarantee that decisions were optimal because all altenatives were examined. Department managers from neighboring areas were expected to challenge each other's proposals. Members of the central staff were expected to offer altenatives and to insist on their being analyzed. The intricacy of the process created confusion about who was responsible for the actions that were to be taken. The result was delays, controls, and interference, with little assurance that final decisions were closer to optimality than was the original proposal. Mann abolished the contention policy. Staffs were disbanded. Decision responsibility was placed with the originator of the proposals. The ultimate desire was to have an atmosphere in which people were innovative, trusted their own judgments, and were allowed to learn from mistakes.

Finding ways to convince people that a new mode of operation must be achieved proved to be more difficult. Being willing to accept personal responsibility for decisions came slowly. An employee who requested approval to spend $26,000 on a new piece of equipment was taken aback when told that his judgment was trusted and that he should go ahead and sign the requisition. After further study that employee concluded that a $6000 expenditure would be sufficient. In another instance, a team of employees spent days redesigning an assembly line to provide enhanced efficiency. After presenting it to management and being told that management would sign for it only after *each* team member *signed* the proposal, the team spent another month deciding that they had enough confidence in the plan to put their names on it.

Gradually, the operating philosophy of the employees changed. Teams became very aggressive in setting their operating objectives. This could never have occurred in the old organization. An operating objective of 90 percent for the fraction of orders that were filled properly would have seemed appropriate a few years ago. The team concluded that this was inappropriate when viewed in the current environment. With 50 million orders filled annually, 10 percent of orders improperly filled translated into too many unhappy customers. The team concluded that operating objective could not be less than 100 percent. The ensuing effort to accomplish this was large, but successful-all because the employees saw the need and insisted on fulfilling it.

Sources:

1. IBM One Day, Lexmark the Next. *Management Review*, Jan. 1994:38.
2. Culture Shock: Story of an IBM Unit That Split Off Shows Difficulties of Change. *Wall Street Journal*, July 23 1992:A1.
3. The Tying on the Wall. *The Economist*, Oct. 3, 1992:74.
4. Laser Printer Assembly is a Snap at Lexmark International, *Industrial Engineering*, Aug. 1993:37.
5. Printer at Work, *Byte*, May 1994:149.
6. IBM's Sell-Off of Lexmark: Linchpin of an Alliance. *Mergers & Acquisitions*, Nov. 1991:38.

Issues for Discussion:

1. The relationship that Lexmark and IBM developed is not common in most divestitures. What are some of the circumstances that would make an arrangement of dubious value or could even jeopardize the success of the new entitity? Why?

2. If IBM had chosen not to proceed with divestiture, what actions could have been undertaken? What might the consequences have been?

3. Could the changes that were undertaken by Marvin Mann in the Lexmark organization have occurred while the organization was a part of IBM? Why?

4. Might it have been better to have accomplished an even stronger separation from IBM through the selection of a management team that had never been associated with IBM?

5. Lexmark found that achieving an atmosphere of empowerment requires ownership, patience, and support. How would you describe what they did in terms of these three actions?

APPENDIX A6.3 CASE STUDY: UNITED ELECTRIC CONTROLS CO.

In late 1986, Bruce Hamilton was appointed vice-president of manufacturing for United Electric Controls, Inc., a company that makes industrial temperature and pressure sensors. It was not the most enviable timing for a promotion. The 60-year-old family operated company was losing money on sales of $28 million. The 450 or so employees were suffering with the loss of bonuses, overtime and raises. The manufacturing system was old and out of date. The company was organized and managed in ways that were common for the time. Decisions were made by managers, employees were largely viewed as commodities, and problems were allowed to fester. Bruce Hamilton was not experienced in manufacturing when he was appointed to this job. But it was clear that many things had to change.

By 1991 lead times for filling orders had been cut from 16 weeks to a few days, 60% of orders were built and shipped within three days, average lot sizes had been reduced from 200 to 10, and work-in-process had been reduced by 80%. Hamilton's goal by year end 1991 was to build and ship customer orders the day following receipt. Sales increased to $36 million in 1991 and the company was profitable.

Did a unique set of circumstances allow United Electric to achieve such dramatic change, or could the actions be duplicated by other companies seeking rejuvenation? Somewhat greater detail concerning the events that took place at United Electric may help answer these questions.

The first important step was the recognition by Bruce Hamilton that changes were needed in the way products were manufactured. But this recognition did not demonstrate an easy path for accomplishing the changes. In fact, this recognition just emphasized the enormity of the task.

As any good manager would do, Hamilton started by searching for information about how others had accomplished this. He began by reading many of the books that discussed the issues that affect manufacturing productivity. At about this time someone gave him a copy of the Goldratt and Cox book *The Goal*. In it he found many of the issues identified that were troubling him about operations at United Electric. Thinking that others in the organization might find the book interesting, he began sharing it with them. With the support of David Reis, company president, Hamilton purchased and distributed about 150 copies of the book to all employees in the production area. Reis later expanded the number of recipients to include all employees.

A number of people recognized that the actions that Goldratt and Cox suggested for improving competitiveness had a good change of helping solve many of the problems that existed at United Electric. Informal discussions were followed by more formal meetings, which were followed by seminars and then with off-site meetings. A consensus regarding the need for change and the possibility of being able to accomplish it began to emerge. A second important step had been taken in the process of change.

In assessing the importance of the various subsequent actions that were taken in the process of achieving the turnaround, Bruce Hamilton identifies the key as the *employees*. It was they who made the many suggestions about improving processes, simplifying operations, building only what was needed to fill orders, and so on. It was they who had the ideas of how to improve the way that they did things, including some seemingly mundane, but important improvements. These included such ideas as the use of pegboards in innovative ways, the creation of a simple visual device for testing systems for leaks, the reduction in complexity of parts procured from vendors, and the reduction in the forms used for procurement

APPENDIX A6.4 A WORD OF CAUTION ABOUT EMPLOYEE PARTICIPATION PROGRAMS[*]

A legal challenge has been created to the form and substance of some employee participation programs. These challenges have been made before the National Labor Relations Board (NLRB) by members of the International Brotherhood of Teamsters–in the case involving Electromation–and by the Chemical Workers Association, Inc.–in the case involving E.I. du Pont de Nemours & Company. In both cases the NLRB ruled that the form in which the employee participation programs were structured made them a labor organization. Under NLRB legislation, management cannot interfere with a labor organization (i.e., management must not interfere with the workers' right to join a labor union and to engage in concerted union activities).

A two-part test is applied to determine if management is exercising improper influence on a labor organization. The first is to determine whether the entity is a labor organization and under the jurisdiction of the NLRB. The second is to determine if the entity is dominated improperly by management. Under NRLB §2(5), an entity is considered to be a labor organization if all of the following conditions are satisfied:

- Employees participate.
- The organization exists, at least in part, for the purpose of dealing with employers.
- Dealings between the entity and the employers involve conditions of employment, including such matters as grievances, wages, incentives, and so on.

*Source: Fink, R. L., R. K. Robinson, and A. Canty. 1994. Dupont v. Chemical Workers Association: Further Limits on Employee Participation Programs, *Industrial Management*, Mar. Apr.3.

Clearly, most employee participation programs satisfy the first. If the deliberations of the entity lead to *recommendations* that deal with conditions of employment, the third condition is also considered to be satisfied.

If the entity is found to be a labor organization, the second test must be applied to determine whether it is dominated by management. Under NLRB §8(a)(2), a labor organization is considered to be dominated by the employer if any one of the following three conditions are met:

- The employer interferes with the formation of the entity.
- The employer interferes with the administration of the entity.
- The employer provides the entity with financial or other support.

The latter condition can be claimed to have been met if the employer allows the employees to engage in these activities during working hours or supplies facilities and supplies. The former condition can be claimed to have been met if the employer is involved in establishing the entity.

As onerous as these rulings appear, they should not be viewed as preventing all employee participation programs. First, not all employees are covered by the NLRB (e.g., domestic workers, agricultural workers, independent contractors, employees of federal, state, and local governments, and employers subject to the Railway Labor Act). Second, in a company with a union, agreement by the union, as a part of collective bargaining with management, to the establishment of these entities removes the concern about management domination. For nonunionized companies, the problem can be more difficult.

The issue remains cloudy at the time of this writing. Although it is possible that the laws under which the NLRB operates will be modified or that other alternatives will be made to provide companies with freedom to establish employee participation programs, no clear direction has been taken.

PART 4

Achieving an Effective Organization

In Part IV we discuss a variety of actions that can be helpful in achieving an effective organization. With the recognition that employees are the most important asset of an organization, this part begins with a discussion of employees. The topics that are discussed are the issues of motivation, teams, the setting of objectives for each employee, the appraisal of their performance, the various means of providing reward for good performance, and the responsibilities that the employee has to the organization. Chapter 8 focuses on the all-important issues of leadership for the organization, emphasizing the role of leadership for various levels of the organization. In chapter 9 we discuss the formal organizational structure. Particular attention is given to the importance of the relationships that exist between the elements of the enterprise and the suppliers and vendors that support them.

7

The Employees

There is growing recognition that for enterprises in advanced industrial societies to compete effectively in a world economy while maintaining and improving the standard of living, firms must develop and fully utilize their human resources (Kochan and McKersie, 1992).

While much has been written of the *lights out factory*, a factory that operates without human involvement, it is increasingly clear that the modern manufacturing enterprise will continue to require highly talented, dedicated people who are committed to continuous improvement in the performance of the system. There is every reason to believe that this will continue for the foreseeable future. Assessing the needs of the enterprise for talented people, recruiting those people, nourishing and fostering their development, and encouraging them to make their best contribution is one of the most critical elements of good management. As Badore (1992) notes: "An effective workforce that is encouraged to search for ways of achieving continuous improvement is a key foundation of the modern manufacturing enterprise." It is not an exaggeration to conclude that the employees are an enterprise's most important asset.

If employees are of such importance to an enterprise, it is helpful to try to understand how *they* view *their* place in the organization. A particularly useful insight can be gained from a study that was carried out in the early 1980s by the Public Agenda Foundation under the direction of Daniel Yankelovich and John Immerwahr (1983). Although this survey is more than 10 years old, there is considerable evidence that it continues to reflect the view of many employees in many companies. The following is extracted from their report.

A number of findings suggest that the American workplace is currently structured in ways that undermine the strong work ethic values that people bring to their jobs. Four specific problem areas emerged in our research.

Disincentives. One of the most startling findings of the national survey of working Americans is the degree to which the American workplace has undercut the link between a jobholder's pay and his or her performance. This situation represents a sharp departure from the traditional American value of individualism. A central theme of our cultural heritage supports the idea that individuals will fail or succeed through their own effort and hard work.... Our survey provides evidence that many people in the workforce are deeply concerned about the lack of connection between their effort and their pay.... [It] also shows that a close relationship exists between the failure of the workplace to reward effort and the tendency of people to hold back on the effort that they put into their jobs.

Lack of good fit	Describes Present Job %	Want More of on Job %
Pay Tied to Performance	22	61
Good Pay	39	77
Not too much Strain and Stress	20	56
Chance for Advancement	29	65

Managers who don't motivate. A second problem is the failure of managers to motivate people to perform effectively.... Most jobholders hold a generally positive attitude toward their managers. They like them personally and respect their dedication....[They] also have very positive feelings about the treatment that they receive from their managers.

Incidence of bad experiences with managers

	Agree (%)
Management doesn't know how to motivate workers	75
Everyone gets the same raise regardless of how hard they work	73
People don't respect authority in the workplace anymore	70
Today people want more of a challenge on the job	67
People don't see the end result of their work	68
Labor unions don't encourage their members to work hard	64

A mismatch between people's values and the existing reward system. A gap between what people value and the rewards and incentives that they receive also weakens work behavior and undercuts the work ethic. For much of our recent history, managers have assumed (reasonably enough) that most jobholders sought material success and an improving standard of living.... Today's work force brings a remarkably pluralistic set of values to the workplace.... Jobholders who work primarily for self development have a very different set of needs and expectations from people who work for material success and for an increasing standard of living.

Mismatch between jobs and values

	Good Match %	Bad Match %
Work to survive (38%)	22	16
Work to improve living standard (43%)	15	28
Work for self development (17%)	12	5

[The percentages beside the statements of motivation indicate the percent of respondents that indicated this as their principal expectation.]

A persistent confusion between job satisfaction and productivity. Intuitively, it seems plausible that a satisfied jobholder will also be an effective worker....This superficially plausible approach is largely incorrect.... Some work characteristics enhance both motivation and job satisfaction. Our research suggests that this is true of interesting work....A convenient location or good fringe benefits or low amounts of rush and stress make [the] job more agreeable ... but these factors do not enhance [the] motivation to work more effectively.... Most jobholders make a surprisingly sharp and clear distinction between factors that primarily enhance motivation and those that primarily are satisfiers. The features that enhance motivation center on such work ethic values as responsibility, a chance for advancement, and a challenging job.

Factors that enhance productiveness

	Would work harder for %	Makes job more agreeable %	Both %
Good chance for advancement	48	22	19
Good pay	45	27	22
Pay tied to performance	43	31	16
Recognition for good work	41	34	17
Job enables me to develop abilities	40	27	21
Challenging job	38	30	15
Job allows me to think for myself	37	33	17
A great deal of responsibility	36	28	14
Interesting work	36	35	18
Job requires creativity	35	31	20

Factors that enhance job satisfaction

	Would work harder for %	Makes job more agreeable %	Both %
A job without too much rush and stress	15	61	13
Convenient location	12	56	12
Workplace free from dirt, noise and pollution	12	56	12
Working with people I like	17	54	13
Get along well with supervisor	19	52	12
Being informed about what goes on	21	49	16
Flexible work pace	20	49	12
Flexible working hours	18	49	15
Good fringe benefits	27	45	18
Fair treatment in workload	24	45	18

Job satisfaction and productiveness

	Want more of on Present Job %	Motivator %	Satisfier %
Good pay	77	45	28
Recognition of good work	70	41	35
Good fringe benefits	68	27	50
Chance for advancement	65	48	22
Job security	65	29	43
Interesting work	62	35	35
Pay tied to performance	61	43	31
Job allows me to learn new things	61	34	36

In sum, we conclude that the American economy is failing to utilize one of its most powerful resources-a widespread commitment to the work ethic. Although many people want to work hard and do good work for its own sake, the workplace is structured in ways that discourage rather than support this norm. As a result, people work below their potential and do less than they want to. The demands of jobholders for managers who know how to motivate, for pay tied to performance, and for other changes in the workplace that increase motivation reflect people's desire to give more to their jobs than they are currently giving.

7.1   MOTIVATING THE EMPLOYEE

Strictly speaking, the answer to the question managers so often ask of behavioral scientists–"How do you motivate people?"–is, "You don't." Man is by nature motivated.... His behavior is influenced by relationships between his characteristics as an organic system ... and the environment.... Creating these relationships is a matter of *releasing* his energy in certain ways rather than others. We do not motivate him, because he *is* motivated.(McGregor, 1966:208)

The message seems clear. The task is to develop the circumstances under which each person will find it valuable to focus his or her efforts in directions that are of mutual benefit to the individual and to the organization. As McGregor has stated: "We have learned that, if we push decision making down in an organization as far as we possibly can...people tend to grow and to develop more rapidly, and they are motivated more effectively." The study by Yankelovich and Immerwahr suggests that employees are eager to be challenged and motivated. Management's task is to create an environment that encourages decision making at all appropriate levels. This is the essence of employee involvement and employee empowerment.

McGregor's pyramid of motivation must be recognized as a key to accomplishing this enhanced motivation. Once the physiological needs are satisfied, safety needs become the motivator. Stability in employment and fair treatment on the job are essential in satisfying this need. Although Ouchi's advocacy of lifetime employment may not be embraced, it is certainly true that U.S. companies could adopt employment practices that would offer the employee more job security without restricting management prerogatives unduly. It should not surprise management that employees will be concerned with the safety of their employment when management reacts to each economic fluctuation by adjusting levels of employment, requesting salary concessions from employees, and undertaking to reduce employee benefits, often while giving themselves large pay bonuses. Fairness is the issue. If the enterprise is really in dire financial straits, most people would agree that something will need to be done. The pain that the actions inflict must be shared equally among all employees if they are to believe that they are being treated fairly.

The perception of fairness is not focused only on job security. Equal treatment and equal opportunity, irrespective of gender, ethnicity, or age have been addressed in legislation, largely because fairness was not being achieved in its absence. The existence of special perks for management-reserved parking spaces, special dining rooms, carpeted offices-help create an atmosphere that leads to group stratification. If people are to experience security, they must have an environment that treats them with fairness. Fairness dictates that everyone understand that benefits are derived primarily from the quality of the work they do, not the fact that they have a particular title behind their name or a particular ethnic heritage, or are of a particular age or gender. As U.S. industry is successful in providing a more fair and equitable environment, enhanced mutual trust among all employees will develop and increased loyalty to the company will develop.

The next level of motivation is social. Recall the results of the Hawthorne experiment (Section 5.2.5) that a group of people working closely together for a common goal develops a social structure. This is the essence of the success of *teams*. Teams have been found to be effective in implementing what McGregor advocated–pushing decision making down to lower levels. While teams offer a means of social motivation, they also have other benefits for an organization. It is often easier to identify a problem, explore various solutions to it, and offer a possible solution while working within a small team than it is while working alone. Consensus decisions made by a team tend to protect the individual from criticism and scorn. While difficulties can arise in team decision making, as discussed in Chapter 9, the team that overcomes these difficulties can offer the organization alternative solutions to problems that may be very informative and imaginative. Not only does the organization benefit, but the team members can savor the realization that they made an important contribution to the organization.

Pushing the decision making to lower levels implies that employees at these levels have the information that they need to make decisions. Noting that the Yankelovich and Immerwahr survey found that a large fraction of workers complained that they did not have sufficient information and could not see the end result of their work, a particular effort must be made to offer employees what they need to make sensible decisions. Communication of the proper information is a key aspect of organization leaders. This is discussed further in Chapter 8. Assuming that employees have the necessary information on which to base their decisions, they must be empowered to do so. Motivation will be destroyed if the teams are asked to identify problems and offer solutions but are restricted in moving forward in implementing their ideas. Active, effective teams can lead to improved performance and to increased motivation. They can be of great benefit to both the organization and the employee.

> The values in "participation" as a tactic of management do not lie merely, or even primarily, in the fact that "people like to be considered" about decisions affecting them. It is that participation, when it is sincere and genuine, is an open recognition of the interactional character of influence. (McGregor, 1966:215)

> When people understand the vision, or larger task, of an enterprise and are given the right information, the resources, and the responsibility, they will "do the right thing"... Empowered people-with good leadership, empowered groups-will have not only the ability but also the desire to participate in the decision process. This level of involvement will enable and encourage the individual to make decisions rather than adopt a passive or reactive attitude, waiting to be told what to do.(Hanson, 1992)

Not every activity in an organization is amenable to a team approach. The Yankelovich and Immerwahr survey suggests some of the important factors that motivate people individually to improve their performance. People responded that their productivity would be enhanced if their jobs enabled them to develop their abilities, was challenging, allowed them to think for themselves, was interesting, and required creativity. It may seem unrealistic to attempt to create these features in many of the jobs in the manufacturing area. Management is challenged, therefore, to find new and usual ways to make routine positions more interesting. Some have found that the rotation of people among jobs helps encourage an employee who is assigned temporarily to a job that is less interesting. People will make the effort to think about their job and try to make it more interesting if management demonstrates that it is prepared to listen and help them accomplish this. Sometimes jobs can be restructured or grouped in ways that offer more interesting aspects to all who are involved. When the well-being and motivation of employees becomes an important element, Gilbreth's admonition of the "one best way of doing work" takes on a meaning that transcends the mechanical aspects of the tasks.

7.2   OBJECTIVES, APPRAISALS, AND REWARDS

7.2.1   Management by Objectives

Employees who are unclear of the expectations of their management are almost certainly going to find reasons to debate the content of their performance appraisals. A successful appraisal system depends on creating an appropriate mechanism for identifying, discussing, and agreeing on what is expected of an employee. Various means of accomplishing this have been developed. *Management by objectives* (MBO) is such a system. The intent of a MBO system is to establish a set of objectives for each person in an organization against which his or her performance will be measured in the coming months. These may be initiated by management, reviewed with the employee, and modified as appropriate. Alternatively, the action might take place in reverse order, with the initial drafting being done by the employee. The intent is to develop a clear statement of the objectives that the individual *and* management agree is appropriate. To ensure that both agree, some organizations have adopted a policy of having both the employee and the appropriate level of management sign the statement.

An important element of MBO is the identification of an action plan that is capable of delivering the objectives. If to accomplish the objectives, management must provide certain resources, it must be clear that the objectives are contingent on obtaining those resources. If circumstances within the organization or the market change, the objectives may need to be revised. Some organizations require that a new MBO be created if circumstances demand it. Others retain the original but include an appropriate modifying statement. However it is accomplished, the evaluation of how well the individual accomplished his or her objectives must reflect the circumstances within which the employee was working.

There are concerns with a MBO system that need to be addressed. Since it must be responsive to the changing circumstances and to the environment, the objectives may require frequent alteration. Maintaining an up-to-date system can be time consuming. Frequently, people do not make the necessary changes with the result that the statements are largely meaningless when the time comes to use them. This creates a situation that can be difficult for both the employee and the management.

Another problem can occur if the objectives are quite specific. While detailed items may allow more quantitative determination of performance, they may ignore some of the most important aspects of the position. How, for example, should an objective be stated when the intent is to encourage the continuous improvement of a process? The establishment of a quantitative goal that can be measured may have the result of not encouraging an employee to exceed that goal. How can an ill-defined objective, such as exploring the applicability of a new technology, be stated such that it can be used in a realistic assessment of performance? How should the objectives differ for individuals and for a team? If written for both, does the existence of objectives for individuals detract from efforts to encourage team participation? How does the organization assure that there is consistency among activities at all levels? There are no universal answers to these and other equally important concerns. They must be addressed, however, by each organization in such a manner that the MBO system is tailored for the organization's needs while providing an effective statement of the expectations of each employee's accomplishments.

7.2.2 Appraisals

Most enterprises require that all employees receive an appraisal of performance on a regular basis. Not only do these provide a means of assessing strengths and weaknesses of performance, they are useful in helping both the employee and the enterprise determine how to improve performance. They are also used to determine the rewards that each employee receives in terms of pay and bonus. In addition, the appraisals create a historical record of employees' accomplishments in a variety of job assignments.

Although many organizations expect appraisal of performance to be done annually, others do this semiannually. New employees often are provided an evaluation after a probationary period of a few months. The frequency that is chosen depends on the needs of the organization.

Although the format of the appraisal will vary among organizations, the information that is collected is rather generic. The intent of all is to evaluate performance and relate this to the competency level of the individual. Some organizations require that a fixed set of items be evaluated. In an effort not to restrict the reviewer, some organizations prefer that a paragraph be written that discusses the individual's performance. In other cases these two methods will be combined and may also include a list of critical incidents that have occurred since the last appraisal, both good and bad. For a team effort, a collective judgment may be required of the performance of the group. When a statement of objectives has been created for each employee, this becomes the basis for the evaluation of performance.

Many organizations require that the appraisals be reviewed and concurred in by one or more levels of management above the person who prepared it. This process is expected to do two things: first, it helps eliminate the possibility of personality conflict or favoritism between a supervisor and employee; second, it helps assure uniformity of approach across segments of the organization. After the appropriate levels of the organization have agreed on the content of the appraisal, the employee will be presented with the appraisal. It is usually the case that the employee is asked to signify agreement with the appraisal by signing it. In the event that the employee disagrees with the appraisal, an appeal process must be available and everyone must know how appeals are to be handled prior to the initiation of discussions. The appeal may involve discussions with a higher level of management or with a representative of the human relations department, it may involve a written objection, or if the immediate supervisor agrees, it may generate a modification of the original appraisal.

Because of the appraisal process, there must be a way of resolving disputes.

In addition to the specifics included in the individual performance appraisal, there is often a requirement that the appraiser offer some quantitative measure of the individual's performance relative to other members of the group. There are, again, many ways of doing this. Some use the levels of outstanding, excellent, good, fair, and poor. Some will use a numerical ranking of 1 through 5 or 1 through 10. Whatever the measure, it is common that over time, the fraction of people who are classified as outstanding or who fall in the ranking of 1 (where 1 is highest) will be high. Thus the ranking system becomes skewed. There is a natural tendency for this to happen. First, people dislike giving members of their group, with whom they must continue to work, a low rating. Second, supervisors and managers will tend to rationalize a high ranking by arguing that *their* group is above the average of the people in the organization, so everyone in their group deserves to be near the top. Third, if an employee truly performs at the top level one year but at a somewhat lower level the next year, it becomes difficult to go up and down in ratings. The result is that people become locked into their level of rating.

There is no easy solution to this set of problems. Some organizations have attempted to reduce this tendency of "ranking creep" by demanding that each organization generate a normal distribution in the rankings. Although this will reduce the creep, it suggests to employees that an arbitrary scale is being superimposed on them. This can discourage further improvement. It is like having to wait for someone to die or retire before being promoted. Some organizations attempt to correct this problem by requiring that higher-level personnel who review the appraisals serve to normalize the ratings. Although this may help assure a more uniform distribution, this places the responsibility for the assignment of performance ratings at a level that has less personal knowledge of individual performances.

An alternative approach to solving this problem is to create a forced ranking of all people in the unit. The immediate concern of those being asked to do the forced ranking is how a highly productive engineer can be ranked against a highly productive technician or an indispensable secretary. Although this is indeed difficult, answering the following two questions can be very helpful. As the supervisor (manager) of this unit, who would I most hate to lose as I attempt to fulfill the mission that has been set for this unit? The corollary question then is: As the supervisor (manager) of this unit, who could I most easily do without as I attempt to fulfill the mission that has been set for this unit? Answers to these two questions will have identified the names of the individuals who appear at the top and bottom of the forced ranking list. The same questions are then asked for the second level, and so on. Although it is clear that there is some arbitrariness in the ranking of nearest neighbors on the list (e.g., should names 9 and 10 be interchanged), it is not likely that there will be any uncertainty between who is listed in the top and bottom third of the list. It should be noted that the rankings will be sensitive to the objectives of the unit as well as the individuals in the group. If a unit's objectives are changed drastically and the talents of certain individuals in the group ceased to be well matched to the tasks, this creates a serious problem. If a person ranked at the top of the list last year falls to a low level this year because of a change in mission for the group, it is incumbent on management to find an alternative assignment for the person. The organization must provide the person with the opportunity to adjust or to find a new position.

Irrespective of the system of appraisal and ranking that is adopted, the appraisal offers an opportunity for the employee and the person responsible for evaluating performance to discuss issues that influence performance. It offers an opportunity to identify strengths and weaknesses, to discuss corrective actions for the weaknesses, and to seek to enhance the person's strengths. It offers an opportunity to discuss career objectives and the progress that is being made in achieving those objectives. Unfortunately, these opportunities are often misused or dismissed. People find it unpleasant to discuss another person's limitations, particularly if the person disagrees–which is often the case. The result is that the discussion of performance may be perfunctory or even uncritical. Often, the appraisal that is given is more positive than the performance warrants since the person giving the appraisal may wish to avoid confrontation. This, of course, destroys the integrity of the process.

The skillful manager/supervisor will find a means of communicating to the person the sense that improvement is needed and that certain actions will probably assist in improving performance. Offering ways to improve performance rather than simply criticizing helps reduce the tendency of most people to defend their current actions. Helping individuals overcome their natural tendency to justify their current actions is a critical first step in achieving improved performance.

The survey results presented in the introduction to this chapter clearly indicate that people want to be recognized for good performance. While the appraisal system is an opportunity to do this in a formal sense, recognition should not be delayed until the annual or semi-annual meeting between the employee and the person giving the appraisal. Their is ample evidence that delaying recognition of a job well done or delaying a discussion of poor performance reduces the impact of the discussion.

While frequent discussions and continuous feedback can be a very effective means of encouraging good performance, they can, if done improperly, destroy the atmosphere of openness, trust, commitment, and sharing that is important for a productive workplace. Giving an employee major credit for a successful development that should be shared with others will destroy the sense of fairness and lead to charges of favoritism. Giving a critical performance appraisal in front of co-workers can demean and discourage an otherwise productive employee. Discussing the faults of one worker with another can lead to mistrust. An organization that treats people with respect and dignity will find that those people will be willing to exert extra effort for the organization when it is needed.

Appendix A7.1 contains a description and copies of the performance management process material that was used by Ford Motor Company in its appraisal system. It is important to note the emphasis that is given to team work, employee involvement, and participative management. They found that it is important to separate these activities from the technical and business accomplishments that are a part of the position.

Caterpillar Co. has taken the evaluation process one step further. In addition to the review of the employee by management, illustrated in Appendix A7.1, they found it useful to make the process of evaluation symmetrical by also having employees comment on and rate the effectiveness of their supervisors. The form that they have developed to accomplish this is also contained in Appendix A7.1. This is designed to give the supervisor information that will help improve their relationships and overall capabilities. This procedure is not unlike that followed in many universities when students are asked to rate their professors on effectiveness.

7.2.3   **Rewards**

Not only are appraisals used to evaluate and encourage improved employee performance, they are used to help determine the rewards that each employee receives in salary increases, promotions, or job rotations. The results of the Yankelovich and Immerwahr survey suggest, however, that many people are unhappy with their organization's approach to determining rewards. For example, many people want a stronger coupling between pay and performance. They acknowledge that this would enhance their job satisfaction and productivity.

With the forced ranking list, it is possible to have a nearly continuous distribution of rewards that goes from zero, or near zero, to an award for the highest rank that is roughly twice that received by the person in the middle of the list. The concern that is likely to be voiced by the manager (supervisor) is that the person at the bottom of the list may be performing to his or her limit, that it is a task that must be done, that not everyone can be outstanding, and that it is unfair that this person's pay not keep pace with the average. The response to this might be that perhaps this person should not have been placed at the bottom of the list if they are doing a task that must be done. This is just one of the issues that must be taken into account when setting the upper and lower limits for rewards. Rather than having a continuous distribution of merit raises, it might be preferable to divide the employee list into three or five groups and give everyone in each group the same increase, with the amount for each group determined by the relative ranking of the group. This will make it somewhat less critical whether a person is ranked 9 or 10 in a group of 30.

If people are urging a stronger connection between pay and performance, why is it so difficult to accomplish? Perhaps the most direct answer is that it forces the manager/supervisor/leader to rank each employee's performance against that of other employees. If there is a fixed amount of merit funds, as is usually the case, it becomes necessary to decide the amount that each employee will be given. Because of the unpleasantness associated with ranking one employee against another, there is a natural tendency to want to give everyone the same merit increase. Various means of ranking employees were discussed above. Whatever the mechanism, this process should serve as a basis for distributing the rewards.

The handling of rewards for team effort is particularly difficult. A team that is working well together and whose members are trying to carry equal responsibilities must be considered as a unit. Although it is not uncommon for the team leader to have a somewhat, but only slightly, higher salary or wage scale than the members, care must be taken not to suggest through the merit-raise program that the leader is being rewarded more than is justified. A reasonable approach is to treat the team as a unit. The various teams in an activity can and should be compared, with the team having the best performance receiving the most reward. If the distinction between the best- and worst-performing team is small, the distinction given in terms of merit rewards should be modest.

In addition to salary rewards for performance, there are many other ways in which it is possible to recognize unusual contributions. Some companies hold an annual awards event–a banquet, picnic, or retreat–in which unusual performance will be recognized by the highest levels of the company management. This may be for an outstanding technical achievement, an unusual contract performance, and so on. Recognition by the chairperson or president can be an exhilarating experience for someone who never has an opportunity to meet these individuals directly in the normal course of business. Some companies give awards for patents that are issued; others, for important publications. All of these are efforts to recognize performance and to acknowledge publicly that the contribution is highly appreciated.

7.2.4   Potential for Promotion

Opportunities for advancement arise during organizational growth and through the replacement of people who depart through normal attrition. Choosing the proper person for each position is critical. The performance appraisal system frequently has limited value in assessing the capabilities of people to assume significantly enlarged responsibility. As an evaluation of performance regarding a particular set of objectives, a performance appraisal is not often structured to examine the long-term capability of the person for different roles. This has led many companies to introduce a management development analysis that regularly requires an assessment of the level of responsibility that employees may be able to assume in future years. Predicting the future for individual progress is, at best, difficult. It has the benefit, however, of helping identify the key people who may become future leaders and whose progress needs to be monitored closely. It may also identify limitations that certain people may have that are amenable to correction through training or counseling.

In far too many instances, promotions have been made as a result of good performance on earlier assignments. In such instances, a person who is ideally qualified to accomplish one type of task may find himself or herself promoted into a position for which he or she is not prepared or motivated to excel. This can be a disaster for both the organization and the individual. It is critical, therefore, that choices be made for promotion in terms of the talents and abilities for the new position, not as a reward for good performance at former and current positions. The management development process is intended to emphasize the potential for the future rather than dwelling on performance in the past.

In some instances, individuals have declared that they do not wish to move into management positions. Those who prefer and are best suited to remain in specialist positions may be as or more valuable to the company than are those who move through the ranks of management. For such individuals, alternative ways need to be found to reward and to recognize their contributions. This has led some organizations to create what is known as *dual ladders* for their technical people. For those who want to make a long-term commitment to the application of technology, the dual ladder offers a means to advance and be rewarded for achievements. The steps in the dual ladder are made parallel to the steps of middle-management positions. Titles such as senior scientist and fellow are often given to these levels. In those organizations that offer this choice, it is necessary for the individual to determine which path provides the most motivation and to determine whether his or her talents favor one path more than another. Either choice can offer opportunities for leading–one through people and the other through technology.

7.2.5   The Experience of Capital Holding Corporation

A good summary of the approach to appraisals, promotions, and rewards is given by Pamela Godwin of Capital Holding Corporation's Direct Response Group (DRG) (Hammer and Champy, 1993). This description is characteristic of the process currently being followed by many companies.

The ... audit that we did has made a big contribution to some major changes underway in human resources management–for instance, in the redesign of our promotion and reward system.

Our new philosophy is that we will reward for performance and promote on ability.
In the old system, a promotion was a reward for having done a good job. That doesn't
make sense. People should be rewarded for doing a good job. We'll do that with compensation. Promotion, on the other hand, should depend on people's ability to do the job
to which they're being promoted, and it has little to do with their performance in the jobs
they have now. So, we are redesigning the performance development program.

We've separated the results review and the performance development review. We
give people their results review and adjust their compensation. Then, several months
later, we conduct the performance development review. We separate the two reviews so
that we can acknowledge and reward the results they turned in and still be clear about
their need for additional growth and development. Promotion is not a reward, and by not
treating it as such we hope to eliminate the Peter Principle in our company.

A measure of an employee's willingness to learn and develop as well as a measure
of his or her ability to function in a team environment will be parts of our compensation mix. We're replacing the current job evaluation system that focuses on the performance of a single task with one that evaluates employee performance in developing a
wider array of the more highly evolved skills that our business model demands. I guess
this is pretty radical. We're going to pay people in part for the job they've done-for performance, in other words–and in part for getting ready to do the job we'll ask them to
do in the future–for learning.

Also, in 1992 we launched goal-sharing program that is different. Merit increases
have always been awarded based on individual performance. This year we're cutting
the individual merit increases by half. We tied the other half to team and corporate performance on all levels, not just for managers. In 1993, all normal merit raises will be
based on team and corporate performance, and we will have a separate recognition
program for individual stars. In general, we want the individual to prosper when the
team and the corporation do.

Because we're reworking our training and development program and because we've
placed a new priority on the wide and candid sharing of information, within two years
we hope to be well along the way to establishing what I call a "thinking culture" in the
company. By thinking culture, I mean a work environment in which employees, especially those closest to the customer, have the autonomy to make decisions and the training and information they need to make the right ones.

7.3 WHAT TALENT DOES THE ENTERPRISE NEED?

The usual practice in filling vacancies is to identify the talents needed to meet the
short-term objectives of the job and to look for someone who can fill those needs.
The long-term needs of the organization are frequently given lower priority. Although clearly the short-term needs must be satisfied, a dominant focus on these
may create a situation in which the organization is not prepared for the new directions that the market may demand. A broader view of the needs of the enterprise
for talented people can be determined through benchmarking (Compton, 1992). A
significant deficiency may require a vigorous recruiting effort or an effort to re-
train current employees.

7.3.1 Benchmarking Employee Talents

Assessing the talents of the current workforce requires good records and a consistent effort at maintaining their currency. To assess the relative capability of your personnel, it will be necessary to collect the following information for comparison with that acquired for world-class competitors:

- Identification of those who have professional degrees in fields that are relevant to the business (e.g., in science, engineering, management, finance, and technology).
- Identification of those who have an advanced degree in fields that are relevant to the business.
- Identification of those who are regularly enrolled in continuing education courses that are relevant to their positions or to the future needs of the company.
- Identification of those who have a special expertise that might be useful in selected circumstances (e.g., proficiency in foreign language or experience with unusual computer languages).
- Identification of those who have been recognized by outside organizations (e.g., awards by professional societies, officers of professional societies, appointment to membership on national committees, invitations to present lectures at important regional or national meetings).
- Identification of those who have been awarded a patent.
- Identification of those who have published learned papers in refereed archival journals.
- Identification of those who are specially skilled in the trades.

The list above has emphasized those fields relevant to the business. Many people enjoy taking courses, expanding their horizons, and learning about areas and fields well removed from the interests of their company. Although there is no reason to discourage this, the emphasis on evaluating the capability of the workforce must focus on the areas that are relevant to the companies needs.

The collection of this information is important in obtaining a profile of the talent possessed by the organization's personnel. It does not, of course, identify whether these capabilities are adequate for the tasks that the organization is facing. This matching of the current profiles of the workforce with the tasks that must be accomplished is a continuing problem for the management. It must be addressed each time a program is initiated or a new direction taken. Sometimes the trends that are evident in the marketplace demand that people with vastly different backgrounds be brought into the organization. Examples abound where this has occurred: the application of integrated circuits in products that were basically mechanical, the need to develop new computer-aided design tools, the substitution of materials that must be processed in ways that are quite different from those used for currently used materials, and the need to create and expand computer communication networks.

While the changes in the marketplace are often gradual, allowing the enterprise to adjust gradually, some changes are so pronounced that drastic action is required. Introduction into the marketplace of a product including a technology that is new to the company may create a crisis for a field service activity that must maintain and repair the product. In some cases this may dictate that a different strategy be developed for servicing in the field. An example of this is the repair procedure for the microcomputers now used in automobiles. Repairs are not accomplished in the field. Any problem with the microcomputer results in replacement of the unit in the field, with repairs being made at regional centers.

The concern regarding the acquisition of needed new skills must not overshadow the risks associated with the loss of highly important skills through retirement, death, or resignation. To an increasing extent, companies are being forced to deal with the retirement of many of their skilled tradespeople. With the disappearance of many apprentice programs throughout the country, companies are finding it necessary to make extra efforts to ensure that they do not lose the skills needed to operate and maintain their facilities.

With a knowledge of the profile of the skills of their workforce, a company is now in a position to ask how their capability compares with that of their competition. This step in benchmarking presents some serious difficulties. A survey of the competitor's organization chart may indicate that they have identified broad activities that are being emphasized (e.g., the existence of a computer science department or a social science research department suggests that certain disciplines will probably be included). While information on the number of patents, number of publications, amount of involvement that a competitor's workforce has in national organizations and so on, can be obtained and is of some value, it would be of more interest to know the number of employed engineers or scientists that are trained in a particular academic discipline. Contacts with academic placement offices can often provide an early indication that a competitor is hiring in a new field, emphasizing one discipline more than another, or simply not hiring. Some information can be obtained through individual professional contacts. As your employees come in contact with those of the competitor, they learn things about programs, levels of funding, and people. This can be valuable information in assessing the capability of a competitor if there is someone who is striving to assemble and analyze it. The usual response by the U.S. firm to this need for fact finding is that there is no time and resources available to accomplish this. This may be a short-sighted response to an important problem.

The difficulties in obtaining good data can be partially overcome through cooperative arrangements in which data are shared among companies. In some instances this can be done through trade associations. In others it is accomplished by agreement among a few companies. Care must be taken during this sharing of information to ensure that none of the provisions of the anti-trust laws are violated.

Benchmarking has shown that Japanese employed in manufacturing have, on average, a higher level of academic achievement than that of comparable U.S. workers. A substantially larger fraction of the Japanese have been found to have engineering degrees than will be true in the United States. This is a source of concern, since the level of technical sophistication in the manufacturing sector is continuing to climb. The comparison of U.S. and Japanese manufacturing companies also finds a smaller fraction of the Japanese workforce that is occupied with financial, legal, and audit functions. Such information is of enormous importance to an enterprise in assessing its performance relative to its competitors.

7.3.2   Training

Just as the organization must strive continuously to improve the efficiency of each of its operations and processes, so must it encourage each employee to strive to improve his or her capabilities. This frequently occurs through participation in in-house training courses, attendance at special seminars, and enrollment in continuing education courses provided by local colleges or universities. The rapid expansion of technology is creating a situation in which people can quickly lose their technical competency, by some estimates in a time as short as seven years. The only alternative is to continue to explore new areas, to enroll in courses that upgrade skills, and to continue to maintain an intellectual curiosity about the world and the directions of evolution of technology.

Motivating employees to engage in continuing education demands that the management demonstrate the value that it places on this. This can be done by giving recognition to those who complete a significant new level of education, by offering to pay all or a major part of tuition costs associated with enrollment in classes, and by offering important classes at the manufacturing location with some part of the time required to complete the course being absorbed by the company. If continuing education is in the best interest the employee *and* the company, both should share in the time committed to it and in the cost of completing it.

Industry takes the issue of training very seriously. The total investment that U.S. industry makes in education and training is staggering. Eurich and Boyer (1985) estimated that the amount spent annually by industry on in-house and company-sponsored education was about $60 billion, an amount more or less equal to that annually spent by *all* of higher education in the United States. There is no reason to believe that the relative importance of industry-sponsored education has decreased significantly since 1985.

Establishing a good working relationship between local colleges and universities, contracting with them to develop special courses that introduce the new technology at a level that is appropriate to employees, and providing the necessary resources for equipment and space to accommodate the classes is an important way of building a strong connection between the local technical and academic communities. Not only does this provide a source of expertise, but the interaction can enhance the understanding of industry's needs and encourages the academic community to alter its curriculum in ways that will be responsive to the needs of industry.

For the continuing education experience to be most useful, serious consideration needs to be given to the format that is used in presenting the material. The traditional classroom may not be the most appropriate. Flexibility in the content and the rate of presentation is also important, inasmuch as students may have a wider range of backgrounds and be less uniformly prepared than is found in the normal academic environment. Particular care must be taken to relate the material to the environment in which the employee is working. A topic that appears unrelated to the job, to the talents that are needed, or to the company's products or processes will be more easily dismissed than will a topic that is related directly to the student's immediate environment.

7.4   RECRUITMENT

The successful recruitment of highly talented people into an organization is difficult but essential if the organization is to prosper and remain competitive in the world marketplace. The people being recruited must feel the excitement of the organization, must understand the opportunities that are available, must recognize that the organization needs them, and must feel that they can prosper if they work hard. The people best able to convey these feelings are those with whom they will work. Although it is important to focus on the immediate position that is to be filled, along with the tasks that must be done, the organization must look beyond this immediate need. People should be recruited who are flexible and who have the potential to do a variety of tasks and to become future leaders in the organization.

Prahalad and Hamel (1990) describe the human talents that a corporation must have in terms of the core competencies that the enterprise requires to maintain its competitiveness. They define core competencies in the following way:

> Core competencies are the collective learning in the organization, especially how to coordinate diverse production skills and integrate multiple streams of technologies.... Core competence is communication, involvement, and a deep commitment to working across organizational boundaries. It involves many levels of people and all functions.

From this definition they proceed to describe the impact that this has on the types of people who should be employed by the company.

> The skills that together constitute core competence must coalesce around individuals whose efforts are not so narrowly focused that they cannot recognize the opportunities for blending their functional expertise with those of others in new and interesting ways.

In attempting to assess the candidates potential for long-term development, it is important that they meet and talk with people somewhat removed from the immediate job opening and with people having various levels of responsibility in the organization. A consensus needs to be reached in terms of the technical capabilities, willingness to work in a team, the level of commitment to excellence, and the long-term potential of being successful in management. No organization is capable of achieving a perfect record in recruiting. The more information that is acquired, the more contacts that are made, the more likely are the members of the organization to make the correct recommendation concerning the candidate.

Giving recruitment this prominence raises the question of how high-quality people are identified as potential candidates. Recognizing that certain rules and regulations must be followed in recruiting, during interviews, and in making selections, it is important to attempt to have as highly qualified a review group as possible. If the position requires someone with a professional degree, frequent visits to the campuses of leading schools can lead to early identification of the best students. The interest of these students can be nurtured and they can be made acquainted with the opportunities that exist in the organization. The best people to handle these contacts are usually recent graduates of that institution. They know the institution and what is important to students searching for a position.

A high standard of excellence must be established. Managers who are recruiting for a new member of the group need to be sufficiently secure in their position that they will seek to recruit people better than themselves. Rather than viewing such people as a threat, the leader must recognize that performance will be enhanced by having the very best people in the group. The successful leader understands that he or she will look good in the reflected glory of the outstanding members of their group. Recruitment of new team members involves special effort, inasmuch as the other members of the team must be enthusiastic with the choice of new members. Before bringing in new team members, consensus must be assured among all existing members. For the reasons mentioned above, the team must seek constantly to enhance its capabilities while searching for people who will maintain the camaraderie that successful teams generally possess.

7.5 EMPLOYEE RESPONSIBILITY

Just as the successful manager must develop a strong responsibility for the well-being of employees, employees must recognize that they have an equally important responsibility to the enterprise and to the people with whom they work. If the enterprise is to be successful and prosper, it is necessary that this shared responsibility be recognized and that everyone accept their role in making it succeed. Past feelings of mutual distrust between management and employees must be eliminated. Sabotaged products by workers retaliating for management actions deemed to be unfair must disappear, just as the lack of sensitivity of managers to the importance of the employee must cease. While advocating that different actions are needed on the part of management, it is important to emphasize that employees have important responsibilities that must be fulfilled if participative management is to be successful.

7.5.1 Participation

Earlier we noted steps that management must take to create an environment that encourages continuous improvement, employee involvement, and employee empowerment. While these actions can offer encouragement, support, and resources to employees, these desirable attributes of management practice can be achieved only with the active involvement of the employees. The actions of management are necessary but not sufficient to guarantee success. Employees must also participate if the process is to be successful. Everyone must actively seek to improve the environment. They must observe conditions, suggest alternatives, discuss opportunities, and assume responsibility for improving. Participation cannot be passive. Each employee must become active and must join with others in assuring that participation is positive and ongoing.

It is not unusual to find that some employees relish this environment and that others resist it. Those who have never experienced a work environment that requires personal initiative can find it threatening. People cannot be expected to assume an unfamiliar role without assistance. Training may be needed and can usually be accomplished with group sessions that are led by experienced coordinators. Encouragement can be provided by co-workers who have experienced such an environment. It is seldom sufficient simply to declare that suggestions are invited. It is better to help people understand the types of suggestions that are expected (e.g., productivity-enhancing actions), and the topics that are not open to suggestion (e.g., the salary structure of the company). The creation of quality circles is an attempt to bring people together into a team and to provide the opportunity for collective decision making to occur. Since the members of these teams are generally volunteers, it is essential that people take advantage of the opportunity to join and to participate.

It is not unusual, of course, to find some people who adamantly refuse to join in any of these activities. Although people should not, indeed cannot, be forced to become involved in participative management, their reluctance cannot be allowed to inhibit the efforts of others. It may be necessary to find assignments for these people that are more isolated and less involved with others. In this way, the noninvolvement of such people in quality circles or similar activities can be less detrimental.

7.5.2   Communications

While the management has been admonished to communicate more effectively with employees, employees have the responsibility to communicate effectively with each other and with management. For people to transmit accurately to each other information that has been received from management, the information that they receive must be timely and reliable. The same is true when entering into discussions for improvement with fellow workers. When discussing the importance of safety as a motivator of people, we concentrated on the issues of safety from loss of job and from unfair treatment. In an analogous manner, safety can be associated with information. If employees can be certain that information that they are receiving is accurate and timely, they will be more motivated to respond. Uncertainty in the correctness and value of information creates a serious motivational barrier for people. Rumors will have drastically reduced influence in an environment that is characterized as being candid and open and possessing timely and accurate information.

Many communications contain elements of fact and expressions of opinion. There is no reason to fear opinion or to attempt to eliminate it, as opinions often are the stimulus for change and improvement. It is important, however, that it always be clear which part of the communication is fact and which is opinion. People can have varying opinions about the interpretation of the facts, but every effort should be made to minimize debate about facts. This applies to both employees and management. Great care should be used to obtain, validate, and communicate facts. Everyone should understand that opinions may need to be changed and modified as a result of additional facts and further discussion.

7.5.3   Being a Good Team Member

Being a good team member requires that each member of the team commit to the success of the team. Gratification in accomplishing a task must be shared with other team members. McGregor argues that a properly motivated team can overcome many of the deficiencies that have come to symbolize group efforts and group decision making.

> It is demonstrably possible to create relationships between individuals that make up a face-to-face group such that the group exhibits properties almost diametrically opposed to those observed in the typical committee or staff group or task force in everyday organizational life.

Such groups make decisions that are effectively implemented, they are creative and innovative, they operate efficiently, they are not crippled by disagreements nor hampered by dominant personalities. Pressures for conformity are minimal, the knowledge and skills of each member are effectively utilized. The outputs of the group need not be mediocre or least-common-denominator compromises, but can often yield decisions and solutions at a level of performance superior to the sum of the outputs of the individual members operating separately. Finally, the members see the group as a setting within which there are attractive opportunities to achieve many of their individual goals and to gain intrinsic rewards. (McGregor, 1966:229)

Northwest Airlines (1991) describes the desirable characteristics of a team player as follows in its training session on the art of giving service.

A Team Player Is

- *Conscientious.* Someone who is always thorough in their work. They pull their own weight because they don't want to make it harder on their teammates. They have a personal commitment to quality customer service.
- *Informed.* Someone who keeps up on current developments regarding the "rules of the game." They stay on top of things to be the most effective player they can be. People who are in the know deliver better service.
- *A communicator.* Someone who keeps information flowing. They contact the appropriate team members when circumstances arise that might be important to the team effort. Team players listen as well as talk. Communication equates to quality customer service.
- *A networker.* Someone who realizes the value of having a professional relationship with their fellow team members. They work to nurture a spirit of camaraderie on the team. People who work well together are better able to serve the customer.
- *A cheerleader.* Someone who encourages their teammates and tells them they are doing a great job! They aren't fair weather players, only for the team when they are winning. Their team spirit and pride is obvious to the customer.
- *Human.* Someone who realizes that no one is perfect. We all make mistakes. They are comfortable with mistakes and treat them as learning experiences. A team player never belittles fellow teammates for making a mistake. Our customers are human too. They appreciate candor.

Using the foregoing characteristics of an effective team player, Northwest Airlines (1991) constructed a self-evaluation quiz entitled "Are You a Team Player?" The intent of this quiz is to encourage their employees to carefully consider their individual strengths and weaknesses and to take such actions as might be appropriate to improve their performance in a team. "Are You a Team Player?" appears in Appendix A7.3.

7.5.4 Ethical Conduct

Employees at all levels, from the lowest to the highest levels, are expected to engage in ethical conduct. While "ethical conduct" often is interpreted to mean refraining from stealing company property, it has much broader implications. Proper utilization of resources, careful treatment of sensitive information and data, accurate reporting of results and observations, and the exercise of restraint in passing along rumors and innuendos are all examples of the practice of ethical conduct.

While it may seem straightforward to determine what actions are ethical in any situation, it is often more complex. As Andrews (1989) remarks in his discussion of ethics:

> Making ethical decisions is easy when the facts are clear and the choices black and white. But it is a different story when the situation is clouded by ambiguity, incomplete information, multiple points of view, and conflicting responsibilities. In such situations-which managers experience all the time-ethical decisions depend on the decision-making process itself and on the experience, intelligence, and integrity of the decision maker.
>
> Responsible moral judgment cannot be transferred to decision makers ready-made. Developing it in business turns out to be partly an administrative process involving: recognition of a decision's ethical implications; discussion to expose different points of view; and testing the tentative decision's adequacy in balancing self-interest and consideration of others, its import for future policy, and its consonance with the company's traditional values. But after all this, if a clear consensus has not emerged, then the executive in charge must decide, drawing on his or her intuition and conviction. This being so, the caliber of the decision maker is decisive-especially when an immediate decision must arise from instinct rather than from discussion.

Although no specific path can assure that ethical decisions are always made, the following guideline can be useful in determining what actions are appropriate in a particular situation: "If the actions that I am contemplating are made known to people outside the company, can they ever embarrass my employer, my fellow employees, or myself?" If the answer is yes, it is usually important to seek other alternatives.

7.6 THE *EMPLOYEE* FOUNDATION OF WORLD-CLASS PRACTICE

> Accomplishing the objective of creating a world-class manufacturing organization must begin with recognition that the most important asset of an enterprise is its employees. When properly challenged, informed, integrated, and empowered, the employees can be a powerful force in achieving the goals and objectives of the organization....
>
> Creation of the environment in which employees can participate in the activities of the organization demands a change in the thinking of many people. It is not just the supervisors, managers, vice presidents, the president, and the chairman that must be willing to participate, but also the employees on the plant floor.... The rationale for employee involvement is predicated on the assumption that individual employees have the best opportunity to understand and appreciate the problems that are unique to their positions. They know their jobs and they know what limits their performance.
>
> FOUNDATION: Employee involvement and empowerment are recognized by world-class manufacturers as critical to achieving continuous improvement in all elements of the manufacturing system. Management's opportunity to ensure the continuity of organizational development and renewal comes primarily through the involvement of the employee. (Heim and Compton, 1992)

REFERENCES

ANDREWS, K. R. 1989. Ethics in Practice, *Harvard Business Review*, Sept.-Oct.:100.

BADORE, N.L. 1992. Involvement and Empowerment: The Modern Paradigm for Management Success, in *Manufacturing Systems: Foundations of World-Class Practice.* J. A. Heim and W. D. Compton (eds.) National Academy Press, Washington, DC: 85-92

COMPTON, W. D. 1992. Benchmarking, in *Manufacturing Systems: Foundations of World Class Practice*, J. A. Heim and W.D. Compton (eds.) National Academy Press, Washington, DC: 100.

EURICH, N. P. AND E. L. BOYER. 1985. *Corporate Classrooms: The Training Business*, Carnegie Foundation for the Advancement of Learning, Princeton, NJ.

HAMMER, M. AND J. CHAMPY. 1993. *Re-engineering the Corporation: A Manifesto for Business Revolution*, HarperCollins Publishers, New York, 1993:188-189.

HANSON, W. C. 1992. The Integrated Enterprise, in *Manufacturing Systems: Foundations of World Class Practice*, J. A. Heim and W.D. Compton (eds.) The National Academy Press, Washington, DC: 161.

HEIM, J. A. AND W. D. COMPTON (eds.) 1992. *Manufacturing Systems: Foundations of World Class Practice*, National Academy Press, Washington, DC:35-37.

KOCHAN, T. A. AND R. B. McKERSIE. 1992. Human Resources, Organizational Governance, and Public Policy: Lessons from a Decade of Experimentation, in *Transforming Organizations*, T. A. Kochan and M. Useem (eds.), Oxford University Press, New York:169.

McGREGOR, D. 1966. *Leadership and Motivation*, MIT Press, Cambridge.

NORTHWEST AIRLINES. 1991. *The Art of Giving Service: Phase Two* (Private Communication).

PRAHALAD, C. K. AND G. HAMEL. 1990. "The Core Competence of the Corporation," *Harvard Business Review,* May-June:82.

YANKELOVICH, D. AND J. IMMERWAHR. 1983. *Putting the Work Ethic to Work: A Public Agenda Report on Restoring America's Competitive Vitality*, Public Agenda Foundation, New York:26-30.

QUESTIONS

7.1 How can a company encourage the individual creativity of employees when operating in a management by objectives (MBO) environment? Does MBO inhibit employee creativity?

7.2 In preparing the forms for performance evaluation, input from the employee is desirable. What information would you request from the employee in preparing for a performance evaluation?

7.3 In an environment in which individuals are encouraged to perform as team members and in which team effort is rewarded, how can promotional choices be made among individuals? How is the team spirit maintained when individuals are the ultimate recipients of promotion?

7.4 Some companies require that employees who receive special training at the company's expense enter into a contractual agreement to stay with the company a number of years or repay the company if they leave. Is this unduly restrictive, or is it justifiable? Why?

7.5 Should the approaches that are taken to recruit talented individuals be applied to blue-collar employees as well as to white-collar employees? What limitations, if any, are there to application at all ranks?

7.6 In an environment is which employee involvement is emphasized, how do you handle an employee who prefers to work in a structured environment and receive explicit instructions in terms of assignments rather than actively participating in the decision-making process?

7.7 What are the advantages and disadvantages of the dual-ladder approach? Are there limitations to the usefulness of this system?

7.8 It is 7:00 A.M. You have spent the past 30 minutes going over yesterday afternoon's mail. Expecting another 30 to 45 minutes of quiet work time, you are surprised to have two of your junior technical staff come in and ask to discuss a problem with you. Although you know the two women only slightly, their serious composure leads you to agree to talk with them. They bring you a complaint about their manager and one of your immediate sub-ordinates-discrimination and favoritism in the treatment of people with whom they work and possible improper use of company facilities for personal gains. What do you say to them? What are your subsequent actions? How would you resolve this ultimately?

7.9 Two years ago you joined the company as supervisor of a line that reworks and repairs units that fail the quality tests at the end of the production line. You have a dedicated group of workers who are anxious to participate in improving the firm's performance in the marketplace. Unfortunately, they feel helpless in accomplishing this in as much as they can only take products that are given to them and attempt to put them in working order. You have found it difficult to be responsive to their frustration since the manufacturing floor is organized in areas that report to three different managers. As one of a group of 10 first-line supervisors, you have just been invited to have lunch with the vice-president of your division. The list of people being invited includes two supervisors whose lines manufacture the components that are responsible for the largest fraction of defects in the final product. The invitation indicated that the principal topic for discussion would be ways of improving the performance of the division. What will you tell the VP about the situations on the plant floor? What suggestions will you make concerning ways to improve the overall performance of the division?

7.10 Each month your company chairman convenes a meeting of senior managers and asks one of the executives to summarize the plans for his or her area of responsibility for the next 12 to 36 months. You, as senior vice-president for manufacturing, will be addressing this group at 10:00 A.M. this morning and will present the plans for the manufacturing group. You are prepared to summarize the plan that was reviewed and approved by the finance committee of the company a month ago. This plan calls for an expansion of existing facilities by 25% to meet an expected increase in market demand for existing products. Total cost of the expansion will be about $10 million. Your request for an expansion of 50% was scaled back because of limited funds for capital expansion. You are surprised to find in the morning newspaper an interview with the company president that announces that the company is going to accelerate a long-range development that it has been pursuing and that it will be striving to introduce an all-new class of products in 2-1/2 years. You know that these new products will use a material handling system that is not compatible with existing manufacturing facilities. A previous estimate indicated that a $30 to $50 million investment would be required to produce these products. Since the president is out of town and cannot be reached, you are unable to talk with him before the meeting. What will you report to the management group later this morning concerning your plans? What will you say regarding the impact of announcement by the president on the manufacturing group's plans?

7.11 You have been assigned the task of recruiting new graduates at Purdue University for your organization. You have discovered over the past three years that an appreciation of the importance of team participation and a willingness to work closely with others in a team environment is critical to being successful in your company. Prior to arriving on campus for the interviews, you have assembled a list of eight key questions that you will use to determine which candidates have the greatest likelihood of success in a team environment. What eight questions will you bring to campus?

APPENDIX A7.1 PERFORMANCE EVALUATION

Performance evaluation is most effective when it helps management and employees enhance the performance of all participants and to encourage improved development. While the evaluation serves to provide a record of performance, which can be very important in instances in which disagreements arise between management and the employee, it should be viewed as a positive means of communicating to members of the organization. To accomplish this, the manager must assure that evaluations are accurate, credible, timely, and reflect the performance relative to the tasks that were expected to be fulfilled. Furthermore, the employee must clearly understand the basis on which they will be judged.

In the Ford example the organizational objectives of participative management, employee involvement, team efforts, goal setting, and so on, are identified as important. If performance relative to these objectives is to be evaluated explicitly, it will be essential that all members of the group/team/enterprise understand this at the beginning. Accomplishing this may require constant reminders. Some organizations accomplish this through the display of banners announcing these objectives. Others use organization meetings, off-site sessions, discussions during the setting of MBOs, and so on. Whatever the mechanism, employees must not be uncertain about the basis on which they will be evaluated.

Performance evaluations are easy and pleasant when the manager has good things to say to employees. They can be difficult when the performance has been less than desired. In either case, the employee should not be surprised by the evaluation. Interactions since the last evaluation should provide many opportunities to praise good performance, encourage improvement, and identify troublesome areas in which expectations are not being met. It is far better to be able to tell an employee: As we talked last month, your performance in this area is not as good as we expect, but it is gratifying to see that you have taken steps to improve. If the progress continues over the next six months at the same rate, I think we will be able to consider this to be a very satisfactory area of performance." Although the employee will still be disappointed with an unsatisfactory rating in this area, the fact that you were able to offer encouragement is very important.

Caterpillar uses a similar evaluation procedure and then carries it one step further to provide a means by which employees evaluate their managers. Although this procedure is used by a limited number of companies, it has benefits. It offers clear evidence of the needs that the employee is experiencing in his or her relation with the boss. Objections to this procedure include concerns about the objectivity of people relative to their boss, concern that a boss who is meeting resistance while trying to change the way things are done will receive very negative reviews, and concern that a boss who receives negative scores will become vindictive. Although all of these are valid, an enterprise that recognizes these possible pitfalls and prevents them from damaging the procedure can find that this process leads to improved relationships among employees and their bosses.

Consistency of evaluations among different groups is always a problem in a large organization. Although all may use the same form, various people will evaluate various questions differently. Consider question 38 of the Caterpillar questionnaire. While the desirability/capability of the employee taking risk is likely to differ greatly between an employee on the producing line and one in the research laboratory, is the manager less *effective* while limiting the capability of a line worker to take risks? Or consider question 42. If priorities change for the group as a result of customer demands, new opportunities, and so on, is it improper for the boss to change the assignments of the people in the group? If so, will his or her evaluation be lower because of this?

PERFORMANCE MANAGEMENT PROCESS: FORD MOTOR COMPANY

The forms that follow were used by the Ford Motor Company in its Performance Management Process. Instructions concerning the use of these forms include the following statements:

> The Performance Management Process links goal setting, coaching for performance, and performance evaluation as integrated elements in a continuous process. Its main purpose is to develop people and improve performance through a clear understanding of goals and regular coaching. In addition, its purpose is to provide candid and accurate formal evaluations to support rewards for performance practices. Managers and subordinates have a mutual responsibility for making the process work. Both have a role in setting goals, coaching, and evaluation.
>
> Under the Performance Management Process, formal evaluation is based on progress toward key goals, human resources management and teamwork, and accomplishment of key position responsibilities.
>
> All three areas are important; there are no predetermined weights for each area. Any weighing and the factors to be considered in each area are subjects to be agreed upon in the goal setting discussions. It is essential, however, that evaluation take into account:

- *short*-term results
- *long*-term effects
- *how* results were achieved

> The Performance Management Process differentiates between key goals and key position responsibilities. Key goals are differentiated for on-going position responsibilities in that key goals will likely be a greater departure from past practices, be more difficult to achieve, command more time than in the past, or command more time than other responsibilities.
>
> Key goals may be team goals, i.e., goals which represent the integrated activities of more than one person. The process of setting and clarifying goals is basically the same whether they are team or individual goals. The major difference is that where team goals are involved, the discussion of content and process expectations should be team discussions.
>
> Human resources management and teamwork is being set apart to highlight its importance. All employees should be evaluated in part on the basis of their performance in human resources management and teamwork regardless of whether or not they supervise others. Included in this heading is:

- Participative management practice
- Employee involvement
- Self-development and the development of others
- Support of team efforts
- Respect for others
- Goal-setting, coaching and evaluation

In addition to key goals and human resources management and teamwork, most jobs have a number of other duties which may demand an individual's time. These are the ongoing position responsibilities of the job. Position responsibilities are those business plan objectives for which most managers are responsible, e.g., quality, cost, scheduling, and timing. They may also include important, but routine, project responsibilities.

PERFORMANCE MANAGEMENT PROCESS

NAME _____   POSITION _____

KEY GOALS/OBJECTIVES	CONTRIBUTION TOWARD GOALS/OBJECTIVES

HUMAN RESOURCES MANAGEMENT AND TEAMWORK (Comment on the individual's interpersonal skills, participative management skills, and contribution to team efforts.)

ACCOMPLISHMENT OF KEY JOB RESPONSIBILITIES (Comment on the individuals performance of continuing job responsibilities.)

PERFORMANCE RATING
    Current
    Year
    Rating 199X [FM]

Appraisal prepared by _____ Date _____

Appraisal reviewed & concurred by _____ Date _____

Employee Acknowledgement _____ Date _____

PERFORMANCE MANAGEMENT PROCESS (Continued)   Page 2

NAME _____

PART II - COACHING/COUNSELING FOR PERFORMANCE IMPROVEMENT

A Coaching/Counseling discussion was held on _____
                                                     (Date)
SUMMARY OF PROPOSED DEVELOPMENTAL ACTIONS

Employee's signature_____

Supervisor's signature _____

EMPLOYEE'S COMMENTS   (On performance appraisal, coaching/counseling discussion and/or career interests.)

PERFORMANCE MANAGEMENT RATING SCALE

(EX) Exceeds Requirements/Expectations - Individual clearly exceeds the level of performance required by the job. Even in senior management positions, this category should not be the average rating. This rating should be reserved for individuals who made an extraordinary contribution to the organization.

(FM) Fully Meets Requirements/Expectations - Individual demonstrates full performance of requirements of the job. The individual displays consistently high quality and quantity of work. It is expected most employees will attain this rating.

(PM) Partially Meets Requirements/Expectations - Individual falls short or some major performance expectations in one of the three categories, but fulfills other expectations. May include new employees in the process of learning the job. This is a rating that indicates improvement needed. An improvement plan should be developed.

(DN) Does Not Meet Minimum Requirements/Expectations - Individual is not performing acceptably. This rating is warning to be followed by demotion, or termination unless significant improvement occurs. Few individuals should receive this rating.

Figure 5.3   Ford performance management process.

PARTICIPATIVE MANAGEMENT BEHAVIOR QUESTIONNAIRE: CATERPILLAR COMPANY.

Employee Involvement

How do you rate your boss's effectiveness at:	Scores
1. Involving you in discussions about work-related issues?	____
2. Utilizing your skills and individual strengths?	____
3. Soliciting ideas from you?	____
4. Involving your work group in group decision making?	____
5. Involving you in decision making?	____
6. Helping your workgroup develop a common understanding of the problems and objectives that you face?	____
7. Encouraging your cooperation rather than competition?	____
8. Establishing a climate of openness and trust?	____
Group Totals	____

Task Management

How do you rate your boss's effectiveness at:	
9. Discussing with you the objectives and problems of your work assignments?	____
10. Seeking your input about how resource limitations will affect the accomplishment of your work assignments?	____
11. Reviewing your goals, timetables, and accomplishments as needed?	____
12. Clarifying with you the results expected from your assignments?	____
13. Indicating your highest-priority assignments?	____
14. Seeking your input about your current workload before giving you additional assignments?	____
15. Providing constructive feedback about those aspects of your assignments that you handled well?	____
16. Providing constructive feedback about how you could have improved your performance?	____
17. Emphasizing quality work in everything you do?	____
Group Totals	____

Communications

Listening-How do you rate your boss's effectiveness at:	
18. Listening to your ideas and concerns without interrupting?	____
19. Encouraging you to express your ideas and concerns?	____
20. Asking you to clarify areas about which he or she is unsure or concerned?	____
21. Summarizing what you have said to verify his or her understanding?	____
Sub-Group Totals	____

Two-way information sharing-how do you rate your boss's effectiveness at:	
22. Seeking input from you about decisions that affect you?	____
23. Providing you with information that affects your ability to perform assignments?	____
24. Sharing general business information with you?	____
25. Initiating two-way communication with you?	____
Sub-Group Totals	____

Personal Rapport—How do you rate your boss's effectiveness at:	
26. Getting to know you as an individual?	____
27. Maintaining individual contact on a regular basis?	____
Sub-Group Totals	____
Group Totals	____

People development

How do you rate your boss's effectiveness at:	Scores
27. Discussing your strength and areas for improvement?	_____
28. Working with you to improve your performance?	_____
29. Providing opportunities for you to develop beyond your present capabilities?	_____
30. Helping you to understand how your work contributes to your organization?	_____
31. Recognizing your job related self development efforts?	_____
Group Totals	_____

Leadership

Situational management style-how do you rate your boss's effectiveness at:

32.   Adjusting his/her management approach to meet your individual needs?

33.   Managing you differently from task to task based on your experiences, job knowledge and ability?

Sub-Group Totals   _____

Managing change-how do you rate your boss's effectiveness at:

35   Involving you in planning and implementing changes which affect you?

36   Seeking your commitment and support for changes which affect you?

37   Implementing changes with a minimum of confusion or disruption?

38   Allowing you to take risks?

Sub-Group Totals   _____

Delegation-how do you rate your Boss' effectiveness at:

39   Delegating meaningful work to you?

40   Giving you the necessary authority to accomplish your work assignments?

41   Giving assignments to you according to your strengths, weaknesses, and potential for personal growth?

42   Allowing you to complete assignments with support, but not interference?

Sub-Group Totals   _____

Decision making-how do you rate your Boss' effectiveness at:

43   Allowing decisions to be made at the lowest appropriate level of his/her organization?

44   Considering your opinions in making decisions which affect your work?

45   Encouraging you to participate in the decision making process in areas in which you believe you can contribute?

46   Explaining his/her rationale for decisions with which you disagree?

47   Managing decision making in a way that increases you commitment?

Sub-Group Totals   _____

Group Totals   _____

Overall Management Effectiveness

How do you rate your boss's effectiveness at:

48.   Managing you in general?

Summary Comments

49.   My boss could be considered a role model for other managers.

50.   I would work for my boss again.

Group Totals   _____

Grand Group Totals   _____

The following scale is to be used in scoring questions 1 through 48: 1 extremely ineffective; 2 mostly ineffective; 3 generally effective; 4 mostly effective; 5 extremely effective. The following scale is to be used in scoring comments 49 and 50: 1 disagree strongly; 2 disagree; 3 neutral; 4 agree; 5 agree strongly.

APPENDIX 7.2   ARE YOU A TEAM PLAYER?[†]

Yes	No	
_____	_____	Do you compliment your co-workers when you observe them doing a good job?
_____	_____	Are you enthusiastic about helping your teammates in any way you can?
_____	_____	Do you always do your job thoroughly and completely?
_____	_____	Do you take advantage of every opportunity to support the team effort?
_____	_____	Do you have a professional respect for everyone on your team?
_____	_____	Can you follow through and support policies and rules with which you personally disagree?
_____	_____	Do you attempt to avoid undermining those around you for personal gain?
_____	_____	Are you enthusiastic about your company and the direction in which it is headed?
_____	_____	Do you show appreciation for the efforts of others and acknowledge their contributions to the big picture?
_____	_____	Do you seek new relationships and acquaintances throughout Northwest?
_____	_____	Do you take responsibility for your mistakes and easily admit when you are wrong?
_____	_____	Does your attitude have a positive effect on those around you?
_____	_____	Are you personally dedicated to making Northwest the best in the industry?

If you find that you have a lot of answers in the right-hand column, it might indicate the need for reevaluation of your team concept. Honest self-evaluation can be difficult. It should be reassuring to remember that you, not other people or events, are in control of your attitude. If through this simple self-evaluation you see some areas that need change, go to work on them! Set your personal goals and standards high for your own personal success and Northwest's.

[†]Source: Northwest Airlines (1991:10).

Appendix A7.2 Are You a Team Player?

Yes	No	
		Do you compliment your co-workers when you observe them doing a good job?
		Are you enthusiastic about helping your teammates in anyway you can?
		Do you always do your job thoroughly and completely?
		Do you take advantage of every opportunity to support the team effort?
		Do you have a professional respect for everyone on your team?
		Can you follow through and support policies and rules which you personally disagree with?
		Do you attempt to avoid undermining those around you for personal gain?
		Are you enthusiastic about your company and the direction it is headed?
		Do you show appreciation for the efforts of others and acknowledge their contributions to the big picture?
		Do you seek new relationships and acquaintances throughout Northwest?
		Do you take responsibility for your mistakes and easily admit when you are wrong?
		Does your attitude have a positive effect on those around you?
		Are you personally dedic making Northwest the best in the industry?

If you find that you have a lot of answers in the right hand column, it might indicate the need for re-evaluation of your team concept. Honest, self-evaluation can be difficult. It should be reassuring to remember you, not other people or events, are in control of your attitude. If through this simple self-evaluation you see some areas that need change, go to work on them! Set your personal goals and standards high for your own personal success and Northwest's.

8
Leadership

> The wicked leader is he who the people despise. The
> good leader is he who the people revere The great
> leader is he who the people say, "We did it ourselves."
> Lao Tsu

Organizations, like individuals, possess personalities. Some are paranoid. Some are altruistic. Some are very informal. Some are highly communicative. For a new enterprise, the personality of the founder often determines the personality of the company. For more established companies, its personality will constantly change as the people who occupy key positions change and as external events affect the company's perspectives. Whatever the sources of influence, the organization's personality is influenced by and in turn determines the approach that the leader takes in exercising effective leadership. The characteristics that make leadership effective in one company for a particular situation might be completely ineffective in another company or for a different situation. An effective leader must be sensitive to these nuances and be sufficiently flexible to respond in an appropriate manner.

Managing an enterprise is first and foremost a matter of providing leadership. David McGregor (1960) summarizes the factors that are involved in leadership in the following description:

> There are at least four major variables now known to be involved in leadership: (1) the characteristics of the leader; (2) the attitudes, needs, and other personal characteristics of the followers; (3) the characteristics of the organization, such as its purpose, its structure, the nature of the task to be performed; and (4) the social, economic, and political milieu. The personal characteristics required for effective performance as a leader vary, depending on the other factors.
>
> This is an important research finding. It means that leadership is not a property of the individual, but a complex relationship among these variables.

This chapter deals with some of the important issues that leaders must address if they are to be effective.

8.1 CREATING THE PROPER TONE
FOR THE ORGANIZATION

In previous chapters we have emphasized the importance of adopting a system perspective of the manufacturing enterprise, establishing organizational goals and objectives, understanding the operating system, creating an organization that is responsive and integrates suppliers and vendors effectively into its operations, and viewing the employee as the enterprise's most valuable asset. No single individual–not the chairman, the president, or the plant manager–can hope to achieve all of this alone. Many people will be involved in identifying and implementing the actions that support these objectives. It is the responsibility of the leader to create the vision of where the organization is going and to set the tone for guiding the organization's actions in reaching its goals.

There is an old saying that "actions speak louder than words." An effective leader has learned this lesson and fully accepts its importance. What the leader does strongly influences those who occupy the lower positions. What the leader emphasizes becomes a signal of what is important to him or her. A leader who is sloppy in appearance invites others to be sloppy. A leader who noticeably tolerates discriminatory actions indicates that it is permissible for others to engage in such actions. A leader who is impeccably honest in all activities encourages honesty among all employees. A leader who is consistent in communicating to all employees encourages others to be open and communicative. A leader who is optimistic encourages optimism. A leader who seeks to improve the efficiency of his or her operation continuously will encourage others to do the same. As has often been stated, perception is reality. The perception of the values that the leader has and that he or she demonstrates will establish the reality of the environment in which the employees work.

Drucker (1989) describes this as the *spirit of the organization*:

> Management by objectives tells a manager what he ought to do. The proper organization of his job enables him to do it. But it is the spirit of the organization that determines whether he will do it. It is the spirit that motivates, that calls upon a man's reserves of dedication and effort, that decides whether he will give his best or do just enough to get by....[I]t is the test of an organization that it make ordinary human beings perform better than they are capable of, that it bring out whatever strength there is in its members and use it to make all the other members perform more and better.

While establishing the proper tone for the organization, the leader must remain concerned about the many topics that demand his or her constant attention. The effective leader reminds everyone continually of the organization's goals and objectives and ensures that decisions are made consistent with those goals or that a full explanation is given for any deviation. The appearance of inconsistent actions will undermine any attempt to seek long-term goals systematically. It is the responsibility of the leader to ensure that routine internal procedures allow and encourage the organization in achieving its goals. If team activities are desired, the appraisal and reward procedures must encourage team building. It is the responsibility of the leader to ensure that a proper balance is maintained between the long-term needs of the organization and the short-term requirements of productivity, profits, and service. If bonuses are based only on short-term performance, long-term needs will not be given high prominence. It is the responsibility of the leader to anticipate opportunities *and* problems and to help the organization prepare for them. These, and others, are the responsibility of the leader. They demand the leader's attention.

8.2   CHARACTERISTICS OF EFFECTIVE LEADERS

Cheves (1992) has created a list of some of the most important characteristics of an effective leader, shown in Figure 8.1. He makes the following observation about this list: "No advice on effective leadership can be final. The search for improvement in leadership will continue as organizations continue to exist. The advice represented here is derived from the extensive, diverse experience of people who have seen what is practical and effective in terms of leadership." This list is worthy of serious examination. It captures the essence of what good leaders do and think. In addition to the emphasis given to some of these items in the early paragraphs of this chapter, it is worth dwelling on several of the other characteristics that he has identified.

Communicate Well
Use clarity in instructions and information

Are Good Listeners
There is much to be learned from followers
Listen for good ideas and feedback

Are Approachable
Are seen as receptive and courteous
Give people undivided attention
"Hear" people, no matter what is on their mind
Know that being approachable leads to being informed

Delegate
To broaden responsibilities and authority of people
To develop and motivate people
To promote more commitment
To free up time for other matters

Lead by Example
They "walk what they talk."
Their actions symbolize and exemplify
   what they want and expect from people
They realize that respect is invaluable and must
   be earned through actions demonstrating high
   expectations for self and others

Read Situations and People Well
Recognize what style works in what situations
Know when to ask and when to tell
Be aware of differing needs and levels
   of ability

Are Good Teachers
Challenge followers to learn
Show people the rewards in learning
Maintain an open mind to new ideas

Care About People—And Show It
Express sincere interest
Greet people warmly, personally
Take time to chat about off-the-job matters
Publicly recognize and praise accomplishments
Maintain a sense of humor

Are Fair, Honest, and Consistent
In expectations and treatment of people
Derive respect and appreciation from treating
   people with dignity and consideration

Know How to Criticize
Do it constructively
Phrase comments descriptively
Avoid evaluative, judgmental points of view
Convey respect for the other person's worth
Be aware of nonverbal indicators
Choose words that convey equality,
   acceptance, and support

Know How to Accept Criticism
It's a fact of life
Leadership involves taking chances
Effective leaders take criticism and learn from it
Admit when they are wrong without keeping
   score on those who are critical

Figure 8.1   Characteristics of effective leaders, (Adapted from Cheves, 1992)

8.2.1   The Importance of Good Communications

The willingness and ability to communicate effectively is the hallmark of a successful leader. Communication occurs in many ways: through written documents, verbal exchange, listening, displays of interest, and through the priorities that are displayed in the daily process of solving problems. The successful leader must be able to communicate in all of these ways.

Being a good communicator is a lot like participating in a contact sport. It is not possible to be an effective communicator by isolating oneself and resorting only to the written memo or the video, although both are important for achieving good communications. One must be actively involved. People need to see their leader. They need to sense *what* the leader is conveying, not just read or hear his or her words. The good leader finds ways to interact with people at all levels of the organization. They are always prepared to listen and, when appropriate, to respond.

An organization that is actively seeking to involve and empower its employees must be willing to communicate with them effectively. Leaders at every level must know what is expected of them, the context within which they are making decisions, and the need to share information with other members of the system. An organization whose employees are strongly involved will find few people complaining that "someone did not tell us what to do." They will more often hear the complaint that "we do not know enough about the total system to do our job properly." It is the responsibility of the leaders to ensure that the necessary information is made available in a timely fashion to the people who are in need of it.

In the busy environment that characterizes most operations, it is easy to misunderstand or to be misunderstood. In the rush to attack the next problem, it is easy to forget that making a few comments about the current issue does not ensure that it is understood. Words mean different things to different people. The context within which something is said may leave an ambiguity in its meaning. A good leader will ensure that everyone understands. The extra time that will be spent is frequently far less than the time spent in correcting a misconception that may have been created.

Although Cheves separates the categories of communication and listening, the latter should be considered as a part of communication. It is not possible to communicate effectively without having an awareness of the interests, needs, and aspirations of the audience. An experienced lecturer will quickly ascertain the receptivity of an audience to the topic and the style of presentation-by listening to the audience-and will adjust to increase his or her effectiveness. An effective leader will listen to both verbal and nonverbal communication from the employees and will react accordingly. The leader must be particularly sensitive to the implications of unspoken innuendos. Accomplishing this requires special efforts. In large organizations it is frequently necessary to create a means by which interactions can occur with people at various levels of the organization without creating a sense that the leader is "going around the system" and ignoring the formal structure. Off-site group retreats, informal luncheons or breakfasts with representatives from a number of groups, chance meetings while "walking the halls," or informal conversation following a formal meeting are good ways of creating opportunities for listening without generating undue concern on the part of people. It is imperative that intermediate managers understand that superiors will not take actions that will affect them based on such informal conversations without first consulting them. To maintain the trust of these subordinates, the leader must not violate this tenant. It is usually not necessary, however, to divulge the source of the information that is learned during these informal conversations. To do so would quickly negate the tendency of people to offer comments and suggestions.

8.2.2 The Role of Criticism

Cheves' belief that giving and accepting criticism are important characteristics of effective leaders may seem inconsistent with the positive image that good leaders attempt to portay to their employees. This is because the word *criticism* is often used in the sense of fault finding or the offering of negative suggestions. This is not the sense that Cheves intended. Individuals and organizations improve when they recognize that different actions or approaches will yield improved results. Offering suggestions for new or different approaches is often necessary to accomplish these improvements. Nevertheless, criticism in any form is often accepted grudgingly and often elicits negative reactions on the part of the recipient. An effective leader will offer criticism in such a way that it will be viewed as a positive learning experience rather than a negative confrontation.

As Cheves notes, the effective leader must be capable of receiving as well as giving. No leader, irrespective of level and responsibility, can afford to adopt the attitude that they cannot improve. A good leader will retain the potential to appreciate their limitations, maintain the willingness to accept criticism, and project a desire to improve continuously. These traits might be summarized as the "retention of humbleness."

8.2.3 Handling the Uncommitted Employee

Not every employee will be willing to follow wholeheartedly the direction that an organization has chosen. Some may find that directions pose a threat to their employment, some may feel that their prestige is being undermined, and some may feel that they do not possess the requisite skills. Whatever the reasons, the manager must be prepared to encourage the organization to proceed. A single person cannot be permitted to jeopordize the success of the organization. A natural reluctance to follow cannot, however, be used to disregard the fact that some people may have a legitimate concern about new programs. These concerns must be addressed and genuine efforts made to alleviate fears and to encourage changes in attitudes. Despite all efforts, however, some employees may continue to resist. Senge (1990:222) addresses this issue in his discussion of the distinction between being committed to a program and complying with it.

> Traditional organizations did not care about enrollment and commitment. The command and control hierarchy required only compliance. Still, today, many managers are justifiably wary of whether the energy released through commitment can be controlled and directed. So, we settle for compliance and content ourselves with moving people up the compliance ladder.
>
> Enrollment is a natural process that springs from your genuine enthusiasm for a vision and your willingness to let others come to their own choice. Following are some guidelines for enrollment and commitment:

- *Be enrolled yourself.* There is no point attempting to encourage another to be enrolled when you are not. That is "selling," not enrolling and will, at best, produce a form of superficial agreement and compliance. Worse, it will sow the seeds for future resentment.

- *Be on the level.* Don't inflate benefits or sweep problems under the rug. Describe the vision as simply and honestly as you can.

- *Let the other person choose.* You don't have to "convince" another of the benefits of a vision. In fact, efforts you might make to persuade him to "become enrolled" will be seen as manipulative and actually preclude enrollment. The more willing you are for him to make a free choice, the freer he will feel. This can be especially difficult with subordinates, who are often conditioned to feel as though they must go along. But you can still help by creating the time and safety for them to develop their own sense of vision.

There are many times when managers need compliance. They may want enrollment or commitment, but cannot accept anything below formal compliance. If that is the case, I recommend that you be on the level about it: "I know you may not agree wholeheartedly with the new direction, but at this juncture it is where the management team is committed to heading. I need your support to help it happen." Being open about the need for compliance removes hypocrisy. It also makes it easier for people to come to their choices, which may, over time, include enrollment.

The hardest lesson for many managers to face is that, ultimately, *there is nothing you can do to get another person to enroll or commit*. Enrollment and commitment require freedom of choice. The guidelines above simply establish conditions most favorable to enrollment, but they do not *cause* enrollment. Commitment likewise is very personal; efforts to force it will, at best, foster compliance.

8.3   THE IMPORTANCE OF INTUITION

One additional characteristic might be added to the Cheves list. An effective leader must possess a good balance between making judgments based on intuition and judgments based on facts. In leading an organization, it is often necessary to rely on intuition about directions and possible opportunities. Effective leaders possess the capability to make timely intuitive judgments. Directions established in these manners are often responsible for major developments and new opportunities. An effective leader with an intuitive grasp of the potential and the risks that are involved can guide an organization into unchartered areas with a sense of confidence that can lead to great accomplishments. Inasmuch as intuition benefits from experience, it is difficult for people to move into an all-new environment and be able to exercise the same level of sensitivity that they may have displayed in an earlier situation.

8.4   LEADING AT ALL LEVELS IN THE ORGANIZATION

The term *leader* is often associated with the topmost elements of management. This is, however, very misleading. Leaders exist at every level. The team leader on the production floor has the same responsibility and opportunity for establishing the tone for that group as does the general plant manager for the totality of employees in the plant. It is critical that everyone recognize that he or she has a responsibility for leading, albeit some may have a more limited domain of influence than others. Each person represents to someone else what he or she considers important through the actions that the person takes or the responses made to a variety of situations. They are each leading through their actions. Pfeiffer (1989) offers the following observation concerning this.

> Most organizations consist of people in separate categories: big wheels, cogs, and specialists like accountants or chemical engineers. But the high-reliability version is a hybrid, a mix of these roles played by the same individuals under different circumstances. The big wheels are there, but use their power rarely. The chain of command is much in evidence, orders may be barked out, and subordinates behave appropriately as spit-and-polish yes-men. But when tension is running high, all work together as specialists among specialists on an equal footing in a more collegial atmosphere.
>
> The most striking and surprising role change occurs in the white heat of danger, when the entire system threatens to collapse. Then cogs can become big wheels.... The system works. It comes under greatest strain when top executives remote from frontline action issue unrealistic orders to people down the line....

Pfeiffer's description fits well within the context of employee involvement and employee empowerment. As an organization empowers its workers, it is asking each person to take responsibility for his or her actions. Each person becomes a leader in attacking problems. Employees are not to wait for someone else to tell them what to do or how to do it. Empowerment signifies delegation, a transfer of responsibility. Leadership becomes much more diffuse through this process.

8.5 LEADERSHIP ROLES

The role that the leader must assume depends on the nature of the organization. It will be different for a small company in which all employees know each other than for a large multinational that has employees in locations around the world. Cavanagh and Clifford (1983) have examined the role of managers in midsized companies.

> The CEO of a midsized growth company does double duty. First, as the strategic leader of an organization of still manageable size, he personally provides the drive behind the innovative, niche-seeking behavior on which its future growth depends; to a far greater degree than in larger companies, he is often the hands-on technical leader, product inventor, or innovator in business methods. At the same time, as the organizational leader, he must build the strongly focused and evolving organization needed for sustained success. More often than in larger, more elaborately structured enterprises, or in smaller companies where life is comparatively simple, he must create an organization that institutionalizes the entrepreneurial and operating skills of the founders. At the same time, he needs to adapt his own managerial role to the changing requirements of the business and to provide for the right kind of succession.
>
> In their personal leadership styles, the CEOs of the winner companies are notable for three common traits. Typically, they display a near-obsessional degree of perseverance, a builder's mentality, and a strong propensity to take calculated risks.

In a large organization the role that a leader must perform varies with the level of the position. At the lower positions, the leader is generally more involved in the details of the tasks or projects. For a technical group, this means understanding the technology, making sound technical decisions, and representing the technical program to others. At the higher levels of the organization the leader will be more involved in the strategic issues that influence the decision process. It is in these positions that the goals and objectives for the organization are established, that broad business decisions are made, and that operating policies are developed. It is important that this higher level management group be composed of individuals who have a detailed knowledge of the groups that they represent, since the policies and strategies that they develop must be implemented by the lower levels.

The role of the executive in a larger company is described by William E. Butler, chairman and CEO of Eaton Corporation, a Cleveland-based firm with sales in excess of $4.0 billion and more than 100 plants worldwide. In reponse to the question, "At the senior level, are you focusing people on certain sources of productivity improvement?" he said (Butler, 1993):

> Yes. It is the responsibility of the president and the executive vice president to spearhead the effort. But our operating vice presidents are the ones who really have to execute the improvements. When we reorganized in late 1991, we felt that the span of control of the group presidents had been too broad. And so we eliminated that position. As a result, the operating presidents have a direct line to the senior management of the company.
>
> We now have a flatter organization, more decentralized than before. Ten years ago, we probably had 30 or so corporate officers. We have 16 now, and that helps communication. The operations vice presidents really try to make those productivity decisions in the business.
>
> What I keep talking about inside Eaton is this holy trinity of quality, productivity, and growth. We focus on all three. We expect that, by improving our productivity and quality, we're going to get growth of existing products. My feeling is that we have to break out of the kind of stagnation that keeps many industrial cyclical companies from providing growth.
>
> I don't pretend to be an expert on everything at Eaton. If I tried to do that, it would be the death of the way we're trying to manage our company. My job is to encourage people to take their ideas as far as they can, create the kind of environment in which they can be creative, so people will step forward and make a contribution. I have to be very supportive.

Understanding the nature of the leader's role at each level as well as the importance and limitation of the multiple role is an important characteristic of the good leader. A good leader, having delegated responsibility, must not second-guess decisions that are made by his or her subordinates. Neither can a leader who has assumed a position at a higher level continue to do the tasks that he or she did when they occupied lower levels. A leader's activities must change with the assumption of new responsibilities. Although this does not eliminate the importance of contributing ideas to the lower levels, it means that the focus of attention must be on questions other than those that consumed them in earlier positions. The giving of advice and counsel without appearing to dominate or exhibit a lack of trust in subordinates is a mark of a good leader. A good leader must understand that employee empowerment applies to them as well as to other employees. They must apply it if they expect others to do the same.

8.6   LEADING IN A CHANGING ENVIRONMENT

The circumstances within which management is practiced are constantly changing, with the advent of new markets and customers, the introduction of new technologies, the entry of new competitors, the promulgation of new laws, and the changing of governmental policies. The challenge to the effective leader is to understand the nature of these changes, to offer the organization stability in terms of management while exhorting it to change with the circumstances, and to encourage the organization to view change as an opportunity rather than a threat.

Beckhard and Harris (1987) offer the following guidance to managers who would be effective leaders of organization during times of change.

To manage these issues, organization leaders need to establish the following:

- A vision of what the institution should look like, and direction toward that vision.
- A clear sense of the organization's indentity (reason to be).
- A clear sense of the organization's interdependency with its outside environment. The organizational system consists of both the formal organization and those parts of the environment that constantly affect it, such as competitors or technology.
- Clear and reachable *scenarios* (not objectives, but descriptions of end states that also define what the organization should look like at an intermediate point.) Scenarios should be clear enough to provide the basis for developing strategic plans, including contingency choices.
- Flexible enough organizational structures to manage optimally the types of work required, production, innovation, business strategies, market intelligence, information management, people management, and creative financial planning.
- Effective use of advanced technology. Managers, not technicians, shoud determine office technology, production, and telecommunication applications.
- Reward systems that equally reflect organization priorities, values, and norms and individual needs for dignity and growth.

8.7   LEARNING TO BE A GOOD LEADER

Good leaders are not born with a particular talent. They learn how to be good leaders by listening, watching, reading, studying, and so on. An indispensable criterion for a good leader is to *want* to be a leader. Individuals who by personality or motivation do not wish to be in leadership positions should not be forced into them. For those who choose the management path, the learning process needs to begin at the earliest possible time. Having responsibility for a small program usually offers the first opportunity to demonstrate the capability of managing resources and meeting objectives. Success with small programs can lead to more responsibility. Efforts must be made to learn from others. The successes and failures of good managers should be studied and evaluated. Through constructive criticism from superiors and subordinates, performance can be improved and opportunities expanded.

Peter Senge (1990:304) offers the following observations about learning.

> Learning takes time. When an individual is managing mental models, for example, it takes considerable time to surface assumptions, examine their consistency and accuracy, and see how different models can be knit together into more systemic pespectives on important problems.
>
> The management of time and attention is an area where top management has a significant influence, not by edict but by example. For instance, [Bill] O'Brien [Hanover] simply doesn't schedule short meetings. "If it isn't a subject that is worthy of an hour, it shouldn't be on my calendar." In a well-designed organization, the *only* issues that reach a senior manager's attention should be complex, dilemma-like "divergent" issues. These are the issues that require the thought and experience of the most senior people, in addition to the input of less experienced people. If top managers are handling twenty problems in a workday, either they are spending too much time on "convergent" problems that should be dealt with more locally in the organization, or they are giving insufficient time to complex problems. Either way, it is a sign that management work is being handled poorly. "It's a big year for me," O'Brien adds, "if I make twelve decisions. I may pick someone to report directly to me. I may set a direction. But my job is not consumed with making many decisions. It is consumed with identifying important issues the organization must address in the future, helping others sort through decisions they must make, and the overarching tasks of organizational design."
>
> The principle is simple to say and understand, but it's not the way most organizations operate. Instead, people at the top continually make decisons on issues such as how to run a promotion-as opposed to why they need to run promotions at all. Or they discuss how to make a sale to a particular customer-instead of inquiring about how their products serve the customers' expressed and latent needs in general.
>
> On the other hand, as the basic learning disciplines start to become assimilated into an organization, a different view of managerial work will develop. Action will still be critical, but incisive action will not be confused with incessant activity. There will be time for reflection, conceptualizing, and examining complex issues.

The time horizon of concern to the manager cannot be restricted to either the near or the long term. It is not sufficient for the manager to focus exclusively on short-term issues such as inventory control or daily sales while neglecting the long-term needs of the enterprise for new products and processes. Neither can a successful manager concentrate only on the long-term R&D needs of a firm and expect the short-term operational issues to resolve themselves. Whether or not the individual managers enjoy all the issues that confront them, successful managers will maintain a balance among them and ensure that neither the short- nor long-term needs of the enterprise are neglected. This requires, of course, balancing the amount of attention given to the details of the operation with the attention that is given to addressing the broad strategic objectives and goals of the enterprise.

8.8 DEVELOPING GOOD LEADERS

Perhaps one of the toughest roles that a leader has is to ensure that a proper replacement has been developed-someone who can effectively take over for him or her. While management development is a stated objective of nearly all business organizations, most organizations find it difficult to do it well. A good indication of the effectiveness with which a company develops its people to assume greater responsibility is the frequency with which the company recruits high-level people from outside the organization. Although it may be necessary and desirable to go outside the company to recruit people with special talents not available in the company, it should not be necessary repeatedly to go outside to fill vacant management positions. To do so reflects either a failure in earlier recruiting efforts or an unwillingness to devote the effort needed to train and develop the younger members of the organization.

The development process demands the attention and concerns of the upper levels of the corporate management. People who exhibit the talent and the desire to be managers, must be given the necessary opportunities to develop the skill. It is the responsibility of the company leadership to ensure that these opportunities are available. If the employees are, in fact, the most important asset of the company, they must perceive that hard work and good performance will lead to greater responsibility. They must understand what opportunities will be available to them. Management development must be one of the top priorities of company leaders.

8.9 THE *Management Task* FOUNDATION OF WORLD-CLASS PRACTICE

Imaginative, creative leadership at every level of an organization is critical if it is to be capable of building on these foundations. Management creates the culture within which the organization functions. Management must exhibit concern for the health and well-being of the organization's human resources. Management must insist that the organization look beyond its borders to interact with its customers, its suppliers, and the educational systems that are training its present and future employees. It is a challenge to the organization to find the proper management for the circumstances in which it finds itself.

FOUNDATION: Management is responsible for a manufacturing organization's becoming world class and for creating a corporate culture committed to the customer, to employee involvement and empowerment, and to the objective of achieving continuous improvement. A personal commitment and involvement by management is critical to success. (Heim and Compton, 1992)

REFERENCES

BECKHARD, R. AND R. T. HARRIS. 1987. *Organizational Transitions: Managing Complex Change*, Addison-Wesley Publishing Company, Reading, MA:7-8

BUTLER, W, E. 1993. Quoted in Blueprint for Best in Class, *Enterprise,* July:14.

CAVANAGH, R.E. AND D.K. CLIFFORD, JR. 1983. Lessons from America's Midsized Growth Companies, *The McKinsey Quarterly,* Autumn.

CHEVES, G. K. 1992. "Characteristics of Effective Leaders in Systems," a Supplement to *Industrial Management,* July/Aug.:4.

DRUCKER, P.F. 1989. *The Practice of Management*, Heinemann Professional Publishing, Oxford:142 (first published in 1955).

HEIM, J. A. AND W. D. COMPTON (eds.) 1992. *Manufacturing Systems: Foundations of World-Class Practice*, National Academy Press, Washington, DC:38-42.

McGREGOR, D. 1960 *The Human Side of Enterprise*, McGraw-Hill Company, New York.

PFEIFFER, J. 1989. The Secret of Life at the Limits: Cogs Become Big Wheels, *Smithsonian,* July:40.

SENGE, P. M. 1990. *The Fifth Discipline: The Art and Practice of the Learning Organization.* Doubleday Currency, New York.

QUESTIONS

8.1 What is the role of (a) the human relations/personnel departments and (b) the supervisor/manager in identifying employees who should be promoted? What relationship should be developed between these two groups?

8.2 The communication of accurate information is critical for the success of any organization. How does the type of information that is communicated differ in the hierarchical and flat organizations? Is the role of the management information service (MIS) any different in these two types of organization?

8.3 How do the leadership roles change as a person moves from the lower to the higher levels of responsibility of a company? Do the talents required for being a successful leader of a not-for-profit organization (e.g. in a university) differ substantially from that required for a leader in industry?

8.4 You have a strong desire to become a manager in your organization. Cheves has identified 11 attributes that are essential for a good leader. What actions will you take to enhance your abilities in each of these areas? How will you go about demonstrating the skills that are needed by a good leader? What will you do if your desires to achieve this objective are not rewarded?

8.5 In the process of attempting to change the philosophy of management, alter the organizational structure, or modify the objectives of an activity, an organization can experience severe turmoil and disruption in its operations. How do you undertake any of these actions while assuring that the daily tasks that must continue to be accomplished are carried out with the least interruption?

8.6 Fill in the boxes for the matrix of options in Figure 8.6 on page 199. A structured task can be considered as one that requires careful cost control, the meeting of predetermined time deadlines, and specific performance objectives. An unstructured task can be considered as a research task that is largely open-ended and rather nonspecific in terms of timing and objectives.

8.7 As CEO, you have sought to ensure that every manager in the organization has carefully developed someone who can assume his or her responsibilities. The principal technical officer of the company has just announced that he has accepted a position at another company and will leave the following Monday. In reviewing the performance and backgrounds of those who reported to the chief technical officer, you are convinced that none has the breadth of background that you believe is needed. You know that there is one member of the group that has the potential that you desire but lacks experience. What do you do in this situation?

8.8 What kinds of training programs are available that will assist the training of good leaders? Who within the organization should be responsible for assuring that such programs are (i) available and (ii) used? How do you assess the value of such programs?

READING

LARDNER, J. F. 1992. Communication Barriers to Effective Manufacturing, *Manufacturing Systems: Foundations of World-Class Practice*, J. A. Heim and W. D. Compton (eds), National Academy Press, Washington, DC:173.

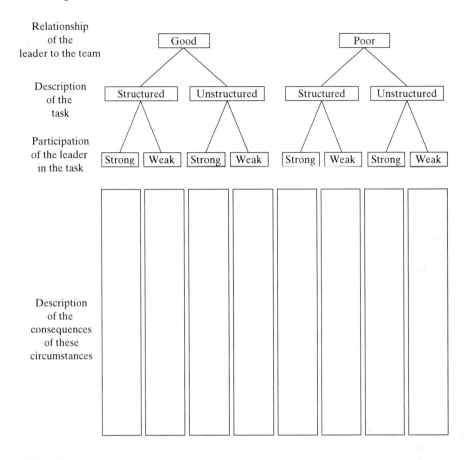

Relationship
of the
leader to the team

Description
of the
task

Participation
of the leader
in the task

Description
of the
consequences
of these
circumstances

Figure 8.6.

APPENDIX A8.1
CASE STUDY: MAGNA INTERNATIONAL, INC.

The U.S. automotive industry has repeatedly identified Magna International of Toronto, Canada, as a superior supplier of automotive components. With a strong commitment to anticipating and meeting its customers needs, Magna has sought to ensure the production of high-quality products at competitive prices. It has achieved this through an innovative philosophy of organization with a workforce that is strongly participative and creative in searching for new ways of doing business.

Magna International founder, Frank Stronach, emigrated to Canada in 1954. During his first 22 years in his native Austria he worked as a tool-and-die maker, for which he began training at age 14. His first employment in Canada was in a small tool shop. When the owner hesitated to fulfill a promise to make him a partner, Stronach left and setup his own shop in a garage. Two years later he had 20 employees and a restless foreman who wanted to leave and form his own company. Stronach's response would prove to be prophetic of his long-term management philosophy. Rather than lose Henry Koob, a good employee who was anxious for more responsibility, Stronach suggested that they open a new factory. Stronach would finance it, Koob would run it, and the two would share ownership.

So successful was the Stronach-Koob venture that it was repeated many times. By 1978 Magna International had 18 plants. The hallmark of these plants was their smallness. Stronach's early experiences as a worker on the floor had convinced him that large plants cannot be well managed. To Stronach, *large* means an employment level above 200. It is his belief that above this level the management begins to lose touch with its employees, does not have time to know the employees' families and their problems, and begins to depend on numbers rather than personal involvement. When circumstances demand that a plant with more than 200 people is needed to meet a market, Stronach believes that it is time to build another plant. Although the first plant is not allowed to grow in size, its management is not penalized by loss of markets. If a new plant is cloned, the management of the first share in the profits that are generated by the second, thus reducing the tendency to try and hold everything within one plant. This approach created a collection of many small plants, each operating as a separate business and each led by an entrepreneur.

Another distinguishing feature of Stronach's management philosophy is his intense concern about his employees. Not only does he insist on the "hands-on" involvement by his plant managers, he strongly believes that all employees must share–even have a right–to a part of the profits that they help generate. Stronach has gone so far as to create a corporate constitution that is included in the organizational bylaws of the company. Elements of the constitution include (1) distribution of 20% of annual net profits to shareholders, (2) allocation of 10% of before-tax profit to employees, (3) allocation of 7% of pretax profits for research and development, and (4) contributions of 2% of pretax profits to support the charitable, cultural, and educational needs of the communities in which they are located.

Stronach's strong committment to employees has encouraged them to commit to the company. This has created an environment in which participation in all elements of the company has flourished. Employee involvement was actively encouraged and practiced at Magna International before it became popular in North America. This involvement benefited both Magna and its customers. Chrysler's purchasing chief is quoted as saying that Magna International employees contributed well over 100 cost-improvement ideas between 1993 and 1995 that saved Chrysler nearly $100 million.

Employee loyalty and commitment was not "bought" by Stronach. Wages have often been a bit below the norm. Unions have had no success in organizing the employees. Rather, Stronach has earned the respect and loyalty of his employees through his constant attention to their needs and to an insistence that they deserve to share in the company's financial success. This sharing in the financial success of the company can be very significant for those in the management. The annual incentive for a plant manager can be 3% of gross profits, which can result in an incentive that is 10 or more times a modest base salary.

The independence that each plant manager enjoyed created an atmosphere that encouraged growth through the building of new plants and the tapping of new markets. The number of plants grew to more than 100 in the mid-1980s. Clearly, some organizational changes were needed to ensure that a broad company strategy was being maintained. "Group" offices were established that had the responsibility of oversight of 10 or so plants. These offices served to provide coordination among the plants and attempted to assist in the allocation of capital funds.

Despite the many attractive features of this organization, problems did develop. In the late 1980s, plant managers, acting somewhat autonomously, undertook an ambitious expansion progam through large increases in debt. Short-term debt increased by a factor of 10, and long-term debt nearly doubled between 1986 and 1990. With the completion of the expansion occurring just as the automotive industry plunged into depression, the company experienced serious financial problems. After losing $224 million (Canadian) in 1990 and violating some loan covenants, the value of the stock dropped from its 1986 high of over $26 per share to under $2 per share. All of this was occurring while Stronach was preoccupied with a new interest in politics as he vigorously sought a seat in the Canadian parliament.

Clearly, drastic actions were needed. The company needed redirection. Stronach renewed his commitment to the company and returned to active participation. More than 20 percent of Magna's plants were closed or consolidated. Management was reduced. Headquarters staff was cut in half. Loans were restructured and the debt eventually eliminated. The automotive industry recovered and sales expanded. Discipline was introduced into the process of approving new products. Major projects need approval of senior executives to ensure that all fit within Magna's strategic plan. The actions were sufficient and, as Figure A8.1 depicts, the company has returned to good growth and is highly profitable.

The future for Magna International is not without peril. Although highly successful as a supplier of components to the North American automotive manufacturers, the competition for business is intense. OEMs are demanding price reductions, thus squeezing supplier profits. With the number of suppliers being reduced by each OEM, Magna is seeking to build new relationships. Magna is assuming responsibility for the design, engineering, and manufacture of total vehicles–becoming a "system integrator." With a modest presence outside North America, Magna International believes that its capability to support industry efforts to produce cars that are marketed worldwide is limited. An emphasis on expanding the business in Europe has led Stronach to move to Europe and to take personal charge of those efforts, while turning over operational management of the North American segment of the company to the president and chief executive officer.

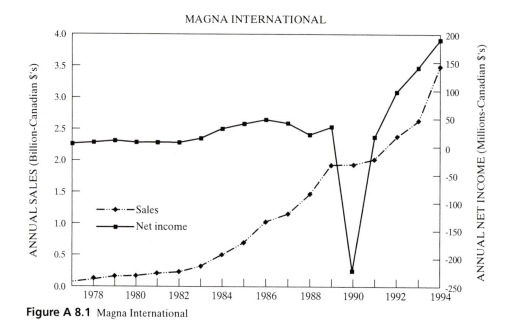

Figure A 8.1 Magna International

SOURCES:

- How Canada's Super Supplier Does It, *Automotive Industries,* Feb. 1985:55.
- How to Grow Without Getting Big. *INC.*, Dec. 1986:108.
- "Magna Carta" Helps Maintain Spirit for Growth, *Management Review,* Aug. 1987:21.
- Magna International: Northern Phoenix? *Financial World,* Oct. 1, 1990:14.
- Frank Stronach's Secret? Call it Empower Steering, *Business Week,* May 1, 1995:63.
- Magna International Profits on Big Three Outsourcing, *Wall Street Journal,* Aug. 24, 1995:B5.

ISSUES FOR DISCUSSION:

1. Although Stronach is not antiunion, he has not encouraged unionization of Magna's plants. What in his management philosophy created situations that do not make unionization probable? Could the organizational and management structure that Stronach has developed have flourished and led to such growth in an atmosphere where employees are unionized? Why?

2. When Stronach was disengaged from the company during his foray into politics, the company became more a collection of separate entities than a company with a single focus. The results were nearly disastrous. Could the company experience similar problems when Stronach retires? If so, what actions could be taken to reduce the impact of these? Will these actions likely influence the management and organization philosophy that Magna has been following under Stronach's leadership?

3. A critical component of Magna International's success is the creation of new plants, each headed by an entreprenuer who has the drive and ambition to develop a highly effective activity. It is common knowledge that the number of entrepreneurs who can successfully manage operations is very limited. How can an organization such as Magna International continually identify and attract the kind of people who will be successful managers? If you were responsible for recruiting new managers, what are the characteristics that you would use for screening people? What kind of training would you offer these people when they join the company?

4. In 1990, Magna International was forced to close and consolidate plants. If you were an employee of one of the plants that remained, how would you have viewed the importance of the company's constitution during this period of retrenchment? Would these events have influenced your participation and your willingness to search for new ways of doing business? What actions can the management take during a period of retrenchment to maintain the feasibility of management programs that have proven successful during periods of expansion?

5. In 1995 Magna International can be described as a company in transition. Many new directions are being taken. A new management structure has been put in place. New businesses are being entered. Should all of this be done simultaneously? What are some possible risks and benefits? What actions could be taken to mitigate the risks and to enhance the likely benefits?

9

Organizing the Enterprise

> Thinking about an industrial organization as an open, organic, sociotechnical system has several advantages. One of the major ones is that it can represent reality more fully and more adequately than the conventional picture of the formal organization....It brings the activities of the informal organization into the framework without excluding those of the formal organization. (McGregor, 1967).

The structure of the company—the organization—critically influences the way that problems are approached and solutions developed. Creating an organization that motivates its employees to excel, that encourages participation of its employees in teams, that encourages communications at all levels, and that supports continuous improvement is critical if the organization is to achieve world-class status. While every enterprise, however small, requires some form of organization, a study of the most successful systems suggests that smaller formal organizations are superior to those that are complex and detailed. This chapter deals with several of the aspects of organizational structure that must be considered.

9.1 THE FORMAL ORGANIZATION

At the top of the formal organization of the company is the board of directors (BOD), the legal entity that is responsible for overseeing the performance of the company. The BOD has the fiduciary responsibility for the company. It is responsible for policy and for setting general direction for the company. It is the responsibility of the BOD to assure that the company is managed properly and that its actions are taken to guarantee its survival.

In small companies and companies that are privately held, the BOD will often be small and may be restricted to the investors in the company and a limited number of members of company management. In publicly traded companies the BOD is elected by the stockholders and generally consists of nonemployees who bring a special expertise to the company, either through their management experience, their technical backgrounds, or in recognition that they are major stockholders. A limited number of the members of the upper management of the company will also be members of the BOD. The ratio of "inside" to "outside" directors has recently become a contentious issue with stockholders of many companies. To assure that company policies and actions are properly responsive to the stockholders' interest, many stockholders are demanding that the number of outside directors constitute a dominant fraction of the board. The chairman of the board of directors occupies the top position in the company.

The operating structure of the company that reports to the chairman can have a variety of forms. The president or chairman will usually be the chief operating officer (COO). While the structure below the COO is strongly influenced by the size of company, the diversity of the products that it markets, and the geographical locations in which it operates, the formal organization is often composed of two distinct types of activities: the operating units and the staffs.

The staff group often includes a finance staff, a human relations staff, and a legal staff. It may also include staffs that carry out the company's research activities, that handle all public relations, and that arrange for corporate advertising and marketing. The staffs typically focus on functions that are common to all operating groups. In most cases they are responsible for advising management and establishing common policies across all units. Although they are seldom responsible for programs or operating facilities, they may assume this role when various organizations share facilities. Inasmuch as they do not generate a product, they are considered to be cost centers, with the costs supported centrally or distributed according to some formula to the operationing divisions.

The operating groups that report to the COO can be divided in many ways. If the company has many products that are somewhat or largely independent, they will often be organized around products. For example, General Electric has a jet engine division, an electric motor division, and a plastics division. United Technologies has a jet engine division, an elevator division, an air conditioning division, and an automotive components division. If the company has one or a few closely related products but is geographically dispersed, the organization often reflects the geographical dispersion. For example, in former times Ford Motor Co. had a North American automotive organization, an European automotive organization, a Latin American automotive organization, and a Far East automotive organization. Its formal organization also differentiates among products, with its truck operation being largely independent of its automotive operation in North America. Figure 9.1 gives a representation of a hierarchical organization.

The extent to which companies give autonomy to each organizational entity is not necessarily determined by the magnitude of the sales of the group. Commonality of product and the use of similar components in various products are two reasons that are often given for centralizing some of the decision making. This can be seen in the automotive industry, where various models of cars built for a single geographic area (e.g., North America) must be designed to allow the use of several different engine designs. Recalling the earlier discussion on system optimization (Chapter 1), this centralization of decision making at a level above the individual operating unit attempts to provide a system that will assure that system issues are recognized and that decisions are made for the benefit of all units and not just a single activity.

The description above refers to what is commonly known as a *hierarchical organizational structure*. Although the number of levels involved, the titles, and

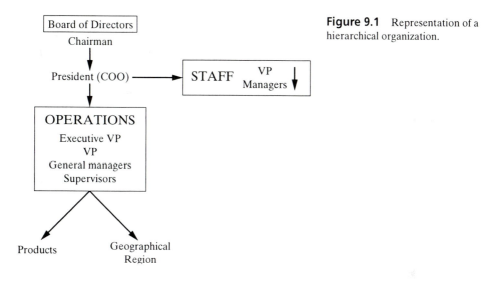

Figure 9.1 Representation of a hierarchical organization.

the number of individuals occupying each level may be different from those illustrated in Figure 9.1, the overall structure frequently follows a pattern much like this. The arrows indicate the decreasing level of authority that accompanies each level. The principal characteristic of a hierarchical structure is the need to involve two, or perhaps many, levels of management in reaching a decision. The organization is often rigid and represents, at least on paper, the structure that is to be followed in reaching decisions.

In contrast to the hierarchical organization, some companies operate with a very *dispersed structure*. In such an arrangement, decisions can be reached at the local level with a nominal amount of approval required from higher levels. This is represented in Figure 9.2, which indicates that the various operations, shown as operation A to operation F, are viewed as nearly independent companies with a minimum of structure above them. The general managers of these operations report directly to the chief operating officer and are expected to operate their businesses in ways that are essentially the same as if they were completely independent.

An example of a company that operates in this manner is the Marmon Group. This is a company that consists of more than 60 autonomous member companies that

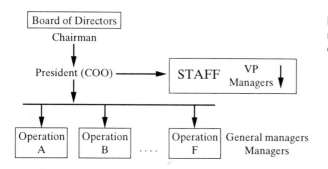

Figure 9.2 Representation of a nonhierarchical (i.e., dispersed) organization.

provide products and services worldwide. Its profits in 1991 were $127 million on revenues of $3.87 billion. Sales exceeded $5 billion in 1994.

> The Marmon Group comprises the divisions and subsidiaries of several multi-unit corporations. Each of the member companies operates on its own, under its own officers and executives.
>
> Through a consulting arrangement, a small executive-professional service organization in Chicago, Illinois-The Marmon Group, Inc.-manages and invests the financial resources of member companies and advises them on accounting, tax, finance, legal and related matters. Membership in the Marmon Group provides companies with cost-efficient management consulting services, a central source for access to operating capital and a disciplined yet entrepreneurial identity.
>
> Autonomous operation of independent companies fosters the development of industrial executives who are also skilled businessmen. Member companies are thereby assured of qualified management at all levels. (Marmon Group, 1992)

The size of these companies varies, with some having only about 20 employees while others have more than 2000 employees. While the Marmon Group has almost no hierarchical structure, this does not imply that the small organization in Chicago is inactive in the management of the individual companies. The concern of the central group is with establishing strategies, assisting in developing goals, and ensuring that the individual companies are performing adequately. Their approach to managing the individual companies is not unlike that adopted by the board of directors of independently operating companies-they are active in establishing overall policies while leaving the day-to-day operations to the individual company managers.

While the Marmon Group would be considered to be a conglomerate-an assembly of many other companies-it has not suffered the fate of many of the conglomerates created in the 1970s. In these early examples of conglomerates, companies were acquired that had little commonality, either in terms of markets, products, or technology. The motivation for acquiring a company was usually based on the interest of the conglomerate to grow, with little regard for creating a coherent business from the many parts. The overall management of the conglomerates frequently offered little guidance to their member companies except to establish financial goals and to consolidate the profits and losses of the member companies. As far as the individual member company was concerned, there was no great advantage in being a member of the conglomerate. In fact, many of them suffered because of the additional burden of supporting the structure and costs of the conglomerate. In the past few years, many conglomerates have altered their strategies and have begun to develop a more coherent business strategy. This has led many of them to divest the activities that cannot contribute to their overall business strategy. They found that the lack of a common focus was a disadvantage.

Perhaps of most interest is the organization that is intermediate between that depicted in Figure 9.1 and the dispersed organization described in Figure 9.2. Although there is some hierarchical structure in these intermediate organizations, its magnitude is forced to be minimal. There are few layers of management. The upper levels of the organization focus on policy and strategic issues. Responsibility for making and implementing operating decisions is delegated to the lower levels. A key ingredient in making this successful is the encouragement of employee participation and employee empowerment. As Jacques (1990) has noted:

> If we are to make our hierarchies function properly, it is essential to place the emphasis on *accountability for getting work done.* This is what hierarchical systems ought to be about. Authority is a secondary issue and flows from accountability in the sense that there should be just that amount of authority needed to discharge the accountability.

The current trend in industry is to move from the hierarchical structure to this more dispersed structure, frequently described as a *flat organization*. Accomplishing this can be very difficult. The first thing that is usually done is to remove some layers of management. Although this is necessary, it is seldom sufficient. If the process by which decisions are made is not modified simultaneously, this action may actually lead to increasing the delays in obtaining decisions. A nonhierarchical organization must be developed by delegating the decision-making responsibility to others, not simply by reducing the levels of management. The number of people who will be making decisions in the dispersed organization will probably be larger than in the hierarchical organization.

A dispersed organization can be successful only if everyone understands his or her responsibility and is willing and able to accept that responsibility. This requires that people feel motivated to accept this added responsibility. Each must understand his or her role in the organization. Each must feel rewarded for accepting this added responsibility. Management must be tolerant of mistakes when they occur. In moving from a hierarchical to a dispersed organization, it is essential that a new corporate philosophy be adopted and communicated effectively to everyone. This can be a daunting task in a large organization (Badore, 1992). Appendix A6.1 gives a dramatic example of the difficulty in implementing these changes in an organization that has a strong hierarchical history.

9.2   RULES, GUIDELINES, AND PROCEDURES

The need to establish clearly defined goals and objectives for the system was discussed in Chapters 3 and 4. The implementation of these goals and objectives is the responsibility of the organization. Frequently, this leads to the creation of a formal set of rules, guidelines, and procedures that deal with the proper way to handle regularly occurring circumstances.

While the reputation and success of a company demands that everyone follow the same rules and procedures in carrying out their duties, the way that this is accomplished depends greatly on its organizational structure and how responsibility is delegated. For the strong hierarchical organization, the tendency is to have a complete set of published rules, guidelines, and procedures that delineate the way each level of the organization handles its tasks. For the disperse organization the tendency is to provide general guidelines to everyone and to work continually with people to be certain that they understand the reason for these particular guidelines, to train them in the art of making decisions, and to trust their decisions.

Examples of each approach exist in industry. Many feet of bookshelves would be needed to contain the collection of operating manuals of some large multinational corporations. Other companies, some with annual sales in the several billions of dollars, operate very successfully with their rules, guidelines, and procedures confined to a single volume. For an enterprise that aspires to encourage employee participation and employee empowerment, the latter approach appears more appropriate.

9.3   COMMUNICATIONS

Achieving good communications within an organization can be unusually difficult. The larger the organization, the more difficult it becomes. The organizational chart is often represented as a road map for the flow of information. Seldom does this capture the actual flows. The informal flow of information often precedes the formal flow and frequently enables the formal system to be successful.

In a large strongly hierarchical system, information is frequently conveyed by formal written documents that are reviewed in meetings having carefully controlled attendance. When the meeting involves the uppermost levels of management, it is generally the case that proposals will undergo several levels of preliminary review. Although this may help assure that everyone has had an opportunity to review and comment on a document, it is also likely to generate a proposal that is of the lowest risk to everyone and one that does not properly reflect the divergence in views that may exist within an organization. It can also result in documents that everyone guesses will please the boss. Henry Ford was fond of pointing out that "all sorts of people around here have been telling everyone what *my* position is on issue A, when I haven't even been told what issue A is."

While the origin of the written document containing a proposal in the hierarchical organization is usually from the lower levels of the organization, the need for the document often arises from a request from the upper levels. These organizations reflect a top-down mentality. The upper levels of management set the requirements based on the objectives that *they* have determined to be in the best interest of the company. In worst cases, the lower levels may have only a vague idea of why certain actions are being requested of them. They are expected to react to requests and to submit detailed plans of how to accomplish the objectives that are given them by the management.

In a dispersed organization, the direction for decision making is often reversed. Responsibility for decisions is delegated to lower levels. Employees at the lower levels are expected to make the decisions that are needed to support the overall corporate goals and objectives. The upper levels do not review and approve; they set policy and directions. The nature of the information that is needed by the dispersed organization is quite different from that required for the hierarchical. For a dispersed organization to be successful, everyone must have a clear understanding of company goals and objectives, the resources available, and the broad steps necessary to achieve the objectives. Emphasis is on communicating common objectives rather than on focusing on the review of program details.

In many ways, communicating is more difficult in a dispersed organization than in a hierarchical organization. Although the quantity of information that is to be communicated may seem to be less in the dispersed organization, the task of ensuring that everyone understand and appreciate the significance of the communications can be very arduous. It requires frequent repetition of the message. It requires that management listen to employees and understand what they, the employees, need to participate properly in this system. It is a process that *cannot* be accomplished by circulating a memo and assuming that everyone understands it. It is a process that *cannot* be delegated.

9.4 TEAMS

Industry, in its move toward more dispersed and flatter organizations, is turning to teams to encourage its employees and to create more of the social motivation that McGregor advocated and that was observed in the Hawthorne experiments. Teams are being given the task of designing new products and new processes. The production system is turning increasingly to teams to solve problems and to provide the cohesiveness needed to achieve high-quality production. Many companies have declared that they intend to convert to production systems that use self-directed teams with technical experts being provided to support problem solving. First-line supervisors would be eliminated in this form of production. Teams will become an increasingly important element of the organization of the world-class competitor.

Creating an effective team requires careful attention and preparation. It is important to remember that many people have never experienced a team environment. If the members of the team can be given the opportunity to associate for a period of time and in a situation sufficiently long to know each other, to trust each other's judgment, to know that each will carry their fair share of the workload, and to understand that each will be rewarded in a fair and equitable manner, the members will generate a cohesiveness that will create a good team. Reaching this level of understanding and mutual commitment can come only through working together.

Success with solving a few small problems is an effective way of building the confidence needed to tackle larger problems. It is important that the first task that the team encounters be well defined and that it is quite clear what type of result is needed. It is only after team members have worked together successfully that they can be expected to tackle less well-defined problems. A newly collected group of people finds it difficult to work together effectively when there is uncertainty about their objective. It is extremely easy for people to become frustrated and discouraged if they do not experience early successes. The first step in creating successful teams is to create an environment and identify a series of meaningful but specific problems whose solutions will generate further encouragement. The experience of IBM in creating successful management teams is described in detail in Appendix A9.1. While the procedures that they outline are intended to facilitate quickly the formation of a strong management team of rather senior people, the directions that they encourage are useful in establishing teams at all levels.

The nature of the interaction of team members and the structure of the team are dependent on the objectives that are given the team and the nature of the tasks that it is to perform, as emphasized in Appendix A9.2. If it is to be successful, these differences must be recognized in the organization of the team. Wilson (1992) illustrates these issues for a very different environment, one in which the musical teams perform.

9.4.1   Team Leaders

An effective team requires a good team leader. The leader needs experience, either through special training or from having served as a member of a successful team. It is the leader's task to ensure that all members of the team agree on the problem being considered, that each member has an opportunity to express opinions, that no member is allowed to dominate the discussions or dictate the solutions, that the talents of each person are used properly, that the workload is distributed fairly, that schedules are set and met, and that the solution or recommendations that are developed are valid responses to the problem. All of these must be accomplished in a manner that encourages cooperation among the members. The important characteristic is to lead, not to dictate. While these characteristics are daunting, they can be learned. Many firms employ a few people who have experience in teaching people to be good team leaders. They serve as coordinators of training sessions and in working with the newly created teams to assure that they will be successful.

A common pattern in creating teams is to identify the leader and let the leader participate in selecting the members of his or her team. Although this is not always possible, this approach at least assures that there will be minimal personality conflicts between the team members and the leader. Once a team has worked together for some time and has developed a good working relationship, it becomes an issue of how to replace a member who leaves or how to add someone with a needed talent. In many cases new members will be interviewed by all the members who are continuing. This tends to enhance the feeling among members that they have control over their own destinies.

9.4.2   Team Size

Experience shows that small teams are generally more effective than large teams. An obvious reason is that as the number of people increase, the difficulty in accomplishing all the things that were identified as desirable becomes increasingly difficult. With a large number of team members, it becomes very difficult to ensure that all members have equal opportunity to express opinions. It is even difficult to arrange meetings that all can attend, let alone to get everyone to talk in a large group. Each of you has experienced the problem of speaking out in a large group. Some people have a natural reticence to speak out in class. Everyone has a natural fear of asking a foolish question in class, only to learn later that several classmates had the same question in mind. A few people tend to dominate all discussions. It is easier to minimize these barriers with a small group than with a large group. Once the barriers have been reduced, the team can become more effective.

Effective communication is accomplished much more easily in a small group than in a large group. As the size of the group increases, the number of possible relationships rises rapidly (Hare, 1976). Consider a group that has n individuals. The number of possible interactions, x, between any two people is given by

$$x = \frac{n(n-1)}{2} \tag{9.1}$$

In a team activity, however, the interactions are not limited to discussions between pairs of people, but include those between subgroups and between subgroups and individuals. This, of course, expands the number of possible interactions. For a group of n individuals, the total number of interactions of all subgroups, z, is given by Eq.(9.2). While not every possible combination of individuals will be needed to assure good communications within the team, the implications of this simple mathematical analysis is certainly obvious. The greater the number of people in the team, the harder it will be to assure that good communication is achieved. The actual number of interactions for a group with n members that will occur will fall somewhere between the values of x and z given in Table 9.1.

$$z = \frac{3^n - 2^{n+1} + 1}{2} \tag{9.2}$$

9.4.3 Team Proximity

The frequency of communication among members of a group is affected by the physical proximity of the group as well as by its size. Allen (1984) has studied the relationship of physical separation and the level of communications for individuals in R&D laboratories. The frequency with which they communicated at least once each week concerning technical or scientific matters was determined as a function of the physical separation of the individuals. Figure 9.3 contains the results of Allen's study for 512 individuals from seven laboratories. The probability that two people will communicate at least once a week decreases by about a factor of 100 when the separation distance is increased from 3 meters-essentially next-door neighbors-to a distance of 90 meters-a distance about three times the width of an average classroom.

The data in Figure 9.3 were taken without regard to organizational association. It might be expected that individuals who are members of the same organizational group would have an enhanced incentive to communicate with each other. Common

Table 9.1 Number of Possible Interactions in a Group Containing n Members

Size of Group, n	Total possible pair interactions, x	Total possible interactions including all subgroups and pairs, z
2	1	1
3	3	6
4	6	25
5	10	90
6	15	301
7	21	966

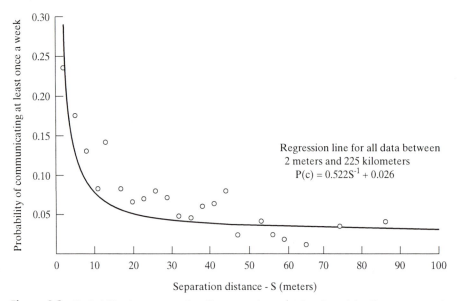

Figure 9.3 Probability that two people will communicate as a function of the distance separating them. (From Allen, 1984, 239.)

organizational objectives and common topics of interest would seem to enhance the probability of more frequent contacts among group members. Figure 9.4 presents the comparison for individuals who are members of a group with those who are in the laboratory but do not share a common organizational group.

While the probability of interaction increases with organizational association, the trend in the curves is very similar. Physical proximity remains a significant factor in the number of interactions that occur between group members. It must be noted that the interactions measured in this study are physical interactions in which one person walks to the other's location. It does not measure the interactions that take place via the telephone or computer. It is likely that the data presented in Figs. 9.3 and 9.4 understate the total number of interactions, perhaps by a large amount. The impact of telecommunications on the relationship between the frequency of interaction and physical location has not been adequately studied.

The implications of these results for the effective organization of teams is clear. Since a large part of the sociology of creating an effective team comes through personal interactions, it is important to place team members in close physical proximity. To have them come together only for meetings is not likely to create the kind of trust and understanding among its members that is needed for an effective team. Teams that are being formed with members from various parts of the organization are increasingly being co-located to a single location. Although this benefits the team, it has the disadvantage of dislocating members from their previous organization, thus making it more difficult for them to continue to function, even in a reduced role, in the home organization. If the team is to be permanent, co-location is essential. The impact of telecommunications on the cohesion of teams that are not co-located is receiving increased attention.

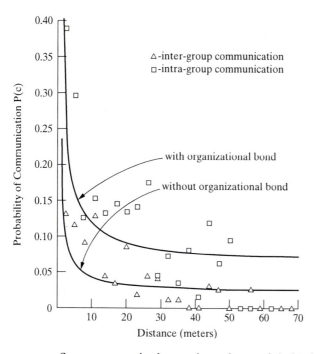

Figure 9.4 Probability of communication as a function of distance, controlling for organizational structure. (From Allen, 1984, p241.)

Some companies have adopted a special physical arrangement for their groups that recognizes the importance of close proximity and high visibility. The Mars Company has a tradition in their manufacturing plants of locating the manager in the middle of the plant with all the members of the organization located in immediate proximity in concentric circles. No high walls are permitted, thus ensuring continuous visual contact. Other companies are following the same path.

> Paul O'Neill, the chairman of the Aluminum Co. of America, is moving out of his lush executive suite into a spartan cubicle. And it's all in the name of speedier decision making.
>
> Mr. O'Neill and nine other top Alcoa executives will spend their workdays in a cluster of cockpit offices: L-shaped desks anchored with small tables for impromptu round-table meetings. The executives need only lean over a makeshift wall to confer, instead of waiting for formal meetings and wading through long-winded memos. Alcoa's existing offices were a "difficult place to have spontaneous association," Mr. O'Neill says.
>
> The changes at Alcoa are part of a growing move toward using office design as management tool. They can be seen in companies as disparate as International Business Machines Corp., the ad agency Chiat/Day Inc., and the cable channel Nickelodeon. While variations of open offices have been in and out of vogue over the years, "there is something new afoot this time, an interest in thinking of [office] design as a tool to accomplish organizational goals," says Kathleen Christensen, director of work-environment research group at the City University of New York. (*Wall Street Journal*, 1994)

As noted above, an important question that seems to have attracted little research attention is the importance of electronic communication in the creation of effective teams. While the existence of ready communication certainly assists in the transfer of information, it remains to be proven that this is sufficient to create an atmosphere that leads to a strong coherent team. Experiences with video conferences, the supplementing of computer mail with video images of operators, and voice communications all contribute to an understanding of this topic. Further study of this topic is certainly warranted.

9.4.4   Difficulties Encountered with Teams

The discussion above has emphasized the value of teams in decision making and in providing employees with a means of effective participation in the system. More will be said about this in Chapter 11 when the topic of concurrent engineering is discussed. It is worth inquiring, however, whether there are any negative issues that need to be guarded against in the use of teams. Is it to be expected that the judgment of the team will always be better than that of the individuals? Will teams find the correct answer to a problem more often than will individuals? The answer, of course, is no.

The team is expected to arrive at a solution that represents a consensus. Its opinions will be expected to reflect a common understanding of the problem. If differences of opinion exist among team members, compromise will be necessary. The risk is that the process of developing a group consensus will generate a decision that is the "lowest common denominator" of all possible options. It is often difficult for a team to generate innovative solutions, for the risk associated with the innovation will frequently be viewed by some members of the team as excessive. The leader has a strong role in helping mitigate these tendencies. Promoting open discussion and debate while encouraging the team to search for the best solution-not just a solution that can be readily accepted by all-requires great care and perseverance.

A second risk in team decisions is that members of the team will be swayed by the team member who is most vocal and who presents his or her view in the most eloquent and forceful manner. A member whose position is correct can often be led to doubt its validity if all the other team members are against it. The peer pressure that is exerted by the group on an individual can be severe. The extent to which this can happen is demonstrated by some simple experiments reported by Asch (1955).

The participants in this test were repeatedly shown two cards of the type illustrated in Figure 9.5. The question that each was asked to answer was which line on the second card is the same length as the line on the first card. Asch describes the experiment in the following way.

> The experiment opens uneventfully. The subjects announce their answers in the order in which they have been seated in the room, and on the first round every person chooses the same matching line. Then a second set of cards is exposed; again the group is unanimous. ... On the third trial there is an unexpected disturbance. One person near the end of the group disagrees with all the others in his selection of the matching line. He looks surprised, indeed incredulous, about the disagreement. On the following trial he disagrees again, while the others remain unanimous in their choice. The dissenter becomes more and more worried and hesitant as the disagreement continues in succeeding trials; he may pause before announcing his answer and speak in a low voice, or he may smile in an embarrassed way.
>
> What the dissenter does not know is that all of the other members of the group were instructed by the experimenter beforehand to give incorrect answers in unanimity at certain points. The single individual who is not a party to this prearrangement is the focal subject of our experiment. He is placed in a position in which, while he is actually giving the

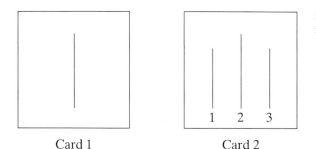

Figure 9.5   Cards used in experiment.

correct answers, he finds himself unexpectedly in a minority of one, opposed by a unanimous and arbitrary majority with respect to a clear and simple fact. Upon him we have brought to bear two opposed forces: the evidence of his senses and the unanimous opinion of a group of his peers. Also, he must declare his judgments in public, before a majority which has also stated its position publicly. (Asch, 1955:32)

The results for this test are shown in Figs. 9.6-9.8.

Figure 9.6 indicates the effect on the subject's accuracy when an increasing number of peers offered incorrect answers. As the number of people offering incorrect answers was varied, it was found that the subject was little swayed by the opinion of a single person giving the wrong answer, was less certain when the majority consisted of two people-giving the wrong answer 13.6% of the time-but gave an incorrect answer 31.8% of the time when three people in the group gave the wrong answer. After a majority of four was reached, the subject was giving an incorrect answer 35% of the time. A test of this observation was then made with 123 subjects with 18 trials for each subject (Figure 9.7). Under ordinary circumstances people were found to choose the correct line more than 99% of the time. On 12 of the 18 trials, the majority responded erroneously. In the presence of the peer pressure from the majority-maintained between 6 and 8-36.8% of the subjects chose to follow the majority and gave the wrong answer. Figure 9.8 illustrates the effect of having one member of the group always giving the correct answer. In this case the majority was less effective in leading the subject to choose the wrong answer. The rate of incorrect answers dropped to about 10% in this case.

In creating effective teams, it is important to recognize that the personalities of the individuals can be as important as the technical competence that they bring to the topic. The members must be willing and able to argue their positions. They must be capable of communicating in an atmosphere that may not always be congenial. They must be open-minded and willing to listen to other arguments. To maintain this atmosphere is a major responsibility of the leader. Increasing the sensitivity of leaders to these issues is an important part of the training that they require.

9.5   THE LEARNING ORGANIZATION

Continuous improvement through learning by the organization was discussed in Chapter 4 (Sections 4.4 to 4.6). Examples of the learning phenomenon were illustrated with the use of learning curves. Improvement in the behavior of an organization cannot take place without learning. As Lawrence and Dyer (1983) note:

Figure 9.6 Effect of the size of the majority that opposed the subject. (From Asch, 1955, :35.)

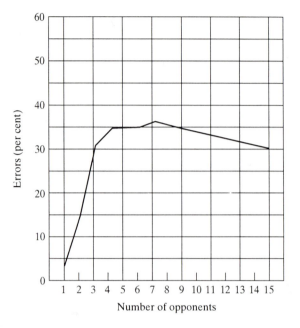

It is true...that members of an organization cannot only learn as individuals but can transmit their learning to others, can codify it and embody it in the standard procedures of the organization. In this limited sense, the organization can be said to learn. When certain organizational arrangements are in place, an organization will foster the learning of its members and take the follow-up steps that convert that learning into standard practice. Then it is functioning as a learning system, generating innovations.

Figure 9.7 Error of judgment of 123 subjects, each of whom compared lines in the presence of six to eight negative opponents (lower curve). Accuracy of judgments not under pressure (upper curve). (From Asch, 1955, :35.)

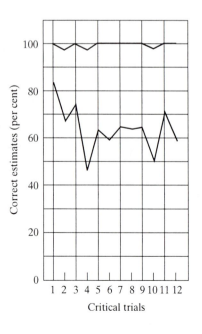

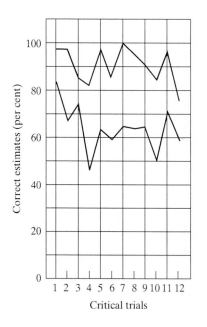

Figure 9.8   Errors in judgement when two subjects supported each other against a majority of opponents (upper curve). Lower curve same as lower curve in Figure 9.6.(From Asch, 1955, :35.)

Strata (1989) discusses organizational learning in terms of the need to nurture the organization's capability in order to accomplish this.

> What is organization learning, and how does it differ from individual learning? We tend to think of learning as a process by which individuals gain new knowledge and insights and thereby modify their behavior and actions. Similarly, organizational learning entails new insights and modified behavior. But it differs from individual learning in several respects. First, organizational learning occurs through shared insights, knowledge, and mental models. Thus organizations can learn only as fast as the slowest link learns. Change is blocked unless all of the major decision makers learn together, come to share beliefs and goals, and are committed to take the actions necessary for change. Second, learning builds on past knowledge and experience-that is, on memory. Organizational memory depends on institutional mechanisms (e.g., policies, strategies, and explicit models) used to retain knowledge. Of course, organizations also depend on the memory of individuals. But relying exclusively on individuals risks losing hard-won lessons and experiences as people migrate from one job to another.
>
> The challenge, then, is to discover new management tools and methods to accelerate organizational learning, build consensus for change, and facilitate the change process.... The values and culture of an organization have a significant impact on the learning process and on how effectively a company can adapt and change. In particular, poor communication between people and between organizations can be a major block to learning and quality improvement.218

Arie De Geus (1988), head of planning for the Royal Dutch Shell Group of companies, was intrigued by the observation that one-third of Fortune 500 industries listed in 1970 had vanished by 1983. In studying the history of 30 companies that had been in business for more than 75 years, he concluded that these companies had developed the capability to learn in ways that others had missed.

What impressed us most was their ability to live in harmony with the business environment, to switch from a survival mode when times were turbulent to a self-development mode when the pace of change was slow. ... Outcomes like these don't happen automatically...they depend on the ability of a company's managers to absorb what is going on in the business environment and to act on that information with appropriate business moves. In other words, they depend on learning. Or, more precisely, on institutional learning, which is the process whereby management teams change their shared mental models of their company, their markets, and their competitors.

Institutional learning is much more difficult than individual learning. The high level of thinking among individual managers in most companies is admirable. And yet, the level of thinking that goes on in the management teams of most companies is considerably below the individual managers' capacities. In institutional learning situations, the learning level of the team is often the lowest common denominator, especially with teams that think of themselves as machines with mechanistic, specialized parts: the production manager looks at production, the distribution manager looks at distribution, the marketing manager looks at marketing.... [H]igh-level, effective, and continuous institutional learning and ensuring corporate change are the prerequisites for corporate success....

A world-class organization recognizes that it must improve continually, that it will ultimately fail if it succumbs to "resting on its laurels." The successful organization will develop the capability to learn from its past experiences, to observe continually those who are successful, and to adapt to its environment. The organization must encourage experimentation, always recognizing that some experiments will be unsuccessful. The involvement and empowerment of employees, development of effective communications, universal acceptance of common goals, and presence of enlightened leadership are all prerequisites for an organization that aspires to learn. Without these, the organization will not possess the collective self-confidence that is required for learning and adaptation. Conveying the concepts associated with the performance of excellent managers to the concept of organizational excellence requires a special dedication on the part of all members of a management team.

9.6 ORGANIZATION INTO PROFIT CENTERS

A complex organizational issue arises when division A of a company sells its products to division B of the same company. How is the performance of division A evaluated? Division A can show a profit on its P&L statement only if it sells its product to division B at a profit. This requires that a sales price be established for the product. If the performance of division A is measured in terms of the traditional financial metrics, such as those listed in Section 4.1.1, it will be in division A's interest to raise the price as high as possible. If the performance of division B is measured similarly, it will fight to keep the price as low as possible. As you will have already concluded, it matters little to the performance of the company how this contentious issue is resolved. To the company the important issue is the cost that divisions A and B accrue in making the final product and the price that division B can charge its customer. If divisions A and B were converted into division Z, the arguments would evaporate. The consequence of creating these profit centers makes it difficult to maintain a system view.

If the company is of such a size that it is impractical to combine A and B into Z, a solution must be found to this issue. The objective must be to encourage both divisions A and B to achieve the highest quality at the lowest cost for their respective products. If the profit center concept is retained, a means must be found for establishing the transfer price, the price that A will charge B for its product. One means of determining this is to have division B purchase a fraction of its needs from a manufacturer outside the company. Assuming that this manufacturer is making a reasonable return on its investment, the same price would be reasonable for A to charge B. A similar alternative is to allow division A to sell a portion of its output outside the company. In either case, the assumption is that the free market will establish a fair price for the product. It is then up to division A to operate its business such that it derives a reasonable return for a product whose price is determined by the free market. Although neither of these alternatives is certain to produce an ideal solution, they do provide guidelines for the company to use in resolving the issues.

9.7   VERTICAL INTEGRATION OF THE FIRM

The organization of the company involves more than the structure that it creates to manage itself and the teams that it creates to design its products and operate its plants. A principal organizational issue that must be addressed is whether it will make the components or subcomponents that it uses or buy them from outside suppliers. The extent to which an enterprise makes most of its own components or subsystems is described by its level of vertical integration. A company is said to be strongly vertically integrated if it manufactures a large proportion of the parts and subsystems that are included in the final product. Conversely, a company that buys most of its parts and subsystems will be described as having a low degree of integration.

The overarching question that must be answered by the management is whether the financial benefits of having someone else supply a part or subcomponent is sufficient to offset the reduced control that would exist if the part were to be made within the company. Some companies have focused on controlling everything. They have attempted to make the maximum number of the parts that go into the final product. This was the philosophy that Henry Ford I was following when he created the Rouge manufacturing complex in Dearborn, Michigan.

The Ford philosophy was that the operation would be most efficient if it could bring raw materials into the plant-much of it by water transportation-transform those materials into components, and assemble them into complete cars, all within one complex. The Rouge complex contained blast furnaces, a steel mill, a rolling mill, a glass plant, an engine plant, a transmission plant, stamping plants, and assembly plants. Ford even sought to control the supply of raw materials that were to be used in the plant-the iron ore, the wood, and the sand-through the ownership and operation of raw material sources in Minnesota, Michigan, and Ohio. This was the earliest, and perhaps one of the most, vertically integrated manufacturing facilities. This philosophy of strong vertical integration persisted in Ford for many decades. It was only in the late 1980s that Ford sold some of these operations to other companies (e.g., the steel mill and the paint operations).

An example of an industry that operates at a very low of vertical integration is the heavy trucks industry. The manufacturers of heavy trucks generally are not strongly integrated. They buy major components (e.g., engines, transmissions, frames, wheels, etc.) from suppliers. The companies that design and sell heavy trucks concentrate on designing and assembling the truck as a system. They do not focus on the manufacture of the components.

Not only is there a variation in the level of integration among industries, there is also considerable variation among the approaches taken by companies within an industry. For example, General Motors is currently more vertically integrated than are Ford and Chrysler, Chrysler being the least so. In the telecommunications industry AT&T, was strongly vertically integrated. The aircraft industry resembles the heavy truck industry, in that they design, assemble, and service. None of the major airframe companies manufacture the engines they use. In the electronics industry, only a handful of U.S. companies have entered into semiconductor production and use the products in their final products. Most choose to acquire their semiconductors from chip "foundries."

Any simple enumeration of the principal issues that should be considered in making a decision as to whether to make or buy would include the following:

1. Financial
2. Expertise
3. Quality of the product
4. Responsiveness of the supplier
5. Control of the technology

These issues have implications for many elements of the manufacturing system. This list is therefore very abbreviated and does not attempt to recognize the impact on all of the diverse activities of the system, but rather, focuses on the areas of key concern.

9.7.1   Financial Considerations

Financial considerations weigh heavily in decisions concerning the level of vertical integration. Reaching a decision as to the desirability of making a component or buying it usually starts with a financial study. One of the principal elements in these studies is the volume of the parts that will be needed. If the volume of parts demanded by any one supplier is low and if essentially the same part will be needed by several companies, the final cost of the part or component will probably be less if someone makes it *for the industry*. This is the case for heavy trucks. If the annual sale of heavy trucks by each of six or seven companies is only 10,000 trucks per company, it is likely that one or perhaps two companies can make more economically the 60,000 to 70,000 engines that will be needed by the industry than could each of the five or six truck companies. This is the case for heavy diesel truck engines, where Cummins, Detroit Diesel, and Caterpillar are the major producers of truck engines. The same is true for aircraft engines. General Electric, Pratt & Whitney, and Rolls-Royce are the principal builders of jet aircraft engines. In both these cases, the expected annual volume for the product is a major factor in making the decision not to integrate vertically.

   The most appropriate use of capital resources becomes a second important financial concern. Financial resources are always limited. Thus the firm must decide whether it is better to invest its resources in making components or in expanding other elements of its business. The decision on this will probably involve an examination of two issues. First, which will yield the higher rate of return on the investment? That is, will the investment in the component business return a greater profit per dollar invested than the same investment in what might be called the core business? Second, is the level of competition that the firm is facing in its core business so keen that funds are needed in this part of the business just to protect or expand its market share?

9.7.2   Expertise

The technical content of the part or component that is being made often becomes a critical concern in making a decision concerning the make or buy of the component. If the technical content is high, if the technology is changing rapidly because of competition or because of new innovations, or if the processes that are used to manufacture the product are unique and critical to its success, a firm may find that it is better to support a supplier to become the expert in this area than to attempt to create that expertise within the company. This is the reason that the vast majority of firms that use integrated circuits in their products choose to purchase them from companies whose principal, or only, product is ICs. Even a company that needs large quantities of ICs may decide to purchase them rather than attempt to make them. Maintaining a competitive level of expertise in this rapidly changing field becomes ever more difficult for a company that is not a primary force in the field.

9.7.3   Quality of the Product

As has been emphasized in nearly every chapter of this book, the quality of the final product cannot be compromised by the source of the parts or subcomponents. Whether to make or buy a component will often be determined by the capability of the supplier to achieve and maintain the desired level of quality. For critical components, this may encourage a firm to make the part, even though the cost may appear to be more than a supplier would charge. In some cases it may be necessary to incur a cost penalty to ensure that quality standards are met.

9.7.4   Responsiveness of the Supplier

The capability of a supplier to respond quickly and efficiently to the needs of the purchaser is a critical concern in the make-buy decision. A supplier must be prepared to respond to its customers' needs to meet changing market demands. These market changes lead to fluctuation in the size of orders. The supplier must also be prepared to respond to changes in the design of a product. For an efficient manufacturing system, the manufacturer must develop a relationship with its suppliers, whether they are internal or external, that makes them a part of the engineering/manufacturing team. As a member of the team, the supplier must make its contribution to solving problems as they arise. It is the responsibility of the suppliers to create an environment within their own organizations that allows them to be responsive to the other members of the team. The operating philosophy and performance of the supplier often become important elements in make-buy decisions.

9.7.5 Control of the Technology

Whenever the technology contained in the product or used to make the product is considered proprietary, the firm will need to consider whether a supplier will be willing to prevent that technology from becoming known to competitors. If the answer is no or is doubtful, the firm will need to weigh the risk associated with the technology falling into the hands of a competitor. The firm may conclude that it can retain control only if it maintains the technology in-house. In an effort to control the dissemination of information, the supplier may be asked not to sell a product or use a process for a competitor. Although some suppliers may be willing to do this, such limitations often defeat the purpose of going to an outside supplier, for it effectively reduces the size of the market that they can hope to serve. If the supplier can only sell to one firm, the advantage relative to the firm making it in-house may be small. Thus the need or desire to control technologies may inhibit the purchase of certain parts from suppliers or the use of certain manufacturing techniques by suppliers.

9.8 SUPPLIER-VENDOR RELATIONSHIPS

The relationships between suppliers and assemblers have been undergoing a substantial change in recent years. In the past a supplier of a component was determined by bid, with the lowest bidder being rewarded. Each time the part changed, a new bidding process was initiated and a past supplier was at risk in terms of future business. There was no long-term commitment on the part of the assembler to the health and well-being of the supplier. The recognition that this was a wasteful and inefficient system has led many U.S. manufacturers to attempt to create a working relationship with their suppliers that benefits both. In so doing, they have begun to create a system that is patterned somewhat after that used by the Japanese.

Instead of many suppliers, any of which may be invited to submit a bid to supply a component or part to a major manufacturer, the number of suppliers in Japan is relatively few. While the following paragraph was written specifically to describe the automotive industry, it applies equally well to most Japanese industry.

> [T]he Japanese industry has long developed a closer working relationship with its supplier base than has the U.S. industry. This has created a feeling of belonging to the "family" that has contributed greatly to the capability of the Japanese industry to implement just-in-time systems, improve quality, and introduce new technology in components and subsystems. Although U.S. manufacturers are making progress in achieving some of the same relationships with suppliers, the Japanese industry continues to benefit greatly from a longstanding tradition in such relationships. One should note that this represents a form of vertical integration without the actual legal or direct financial commitment that would be required of "true" integration. (Compton, 1991)

Womack et al. (1990) describe the relationships between supplier and assembler as follows:

> [T]he system works only because a rational framework exists for determining costs, price, and profits. This framework makes the two parties want to work together for mutual benefit, rather than look upon one another with mutual suspicion.

Almost all the relationships between supplier and assembler are conducted within the context of a so-called basic contract. The contract is, on the one hand, simply an expression of the assemblers' and suppliers' long-term commitment to work together. However, it also establishes ground rules for determining prices as well as quality assurance, ordering and delivery, proprietary rights, and materials supply.

In short, the contract lays the basis for a cooperative relationship, one that is fundamentally different from the relatively adversarial relationships between supplier and assembler in the West. Similar contracts have also been commonplace between first- and second-tier suppliers in Japan since the 1960s.

No longer do assemblers consider their suppliers to be expendable. The number of suppliers is being reduced and a long-term relationship is being developed between assembler and supplier. Each views the success and financial stability of the other as important to its long-term success. As a result, the supplier is being brought into the discussions of long-term strategic plans at an early stage. The supplier is increasingly being given responsibility for the design of a component or subassembly that meets functional objectives as opposed to being asked to design a part to a strict specification, thus allowing the supplier to exercise initiative to reduce cost, improve quality, and enhance performance. To an increasing extent, U.S. manufacturers are attempting to create a family with their suppliers. It has not been easy and it has not been without turmoil in both the supplier's and assembler's organizations. A supplier once remarked that "when things are going well, we are a supplier. When things are going badly, we are a vendor." A fundamental change in philosophy of working together is under way. The impact that this can have on the organization of both groups is still not fully appreciated.

9.9   THE *Organization* FOUNDATION OF WORLD-CLASS PRACTICE

The complexity of the manufacturing system arises from many directions: the *interdependence* of the elements of the system, the influence of external forces on it, the impact that it can have on its environment, and the lack of predictability in the consequences of actions. The complexity and the difficulty in assessing the directions that should be followed can create a sense of frustration and futility for the management.

A proper coupling among the diverse units demands that each have an awareness and understanding of the objectives and capabilities of the other.

FOUNDATION: A world-class manufacturer integrates all elements of the manufacturing system to satisfy the needs and wants of its customers in a timely and effective manner. It eliminates organizational barriers to permit improved communication and to provide high-quality products and services. (Heim and Compton, 1992:31-35)

9.10   THE *Supplier* OR *Vendor* FOUNDATION OF WORLD-CLASS PRACTICE

It is essential that the barriers that have existed between supplier and purchaser be attacked as actively as are the barriers between the elements in the manufacturing organization, for example, product and process design. The sharing of goals, the exchange of information, the interchange of people, and the making of long-term commitments are some of the ways in which these barriers are being overcome.

Essentially everything that has been said in [this book] is as applicable to the supplier as it is to the manufacturer. The successful manufacturing system will create an environment that encourages, recognizes, and rewards the integrated involvement of all elements of the system.

FOUNDATION: A world-class manufacturer encourages and motivates its suppliers and vendors to become coequals with the other elements of the manufacturing system. This demands a commitment and an expenditure of effort by all elements of the system to ensure their proper integration.(Heim and Compton, 1992:37-38)

REFERENCES

ALLEN, T. J. 1984. *Managing the Flow of Technology,* MIT Press, Cambridge, MA.:235-249.

ASCH, S. E. 1955. Opinions and Social Pressure, *Scientific American*, 193 (Nov.):31.

BADORE, N. L. 1992. Involvement and Empowerment: The Modern Paradigm for Management Success, *Manufacturing Systems: Foundations of World-Class Practice,* J. A. HEIM and W.D. COMPTON (eds). National Academy Press, Washington, DC .

COMPTON, W. D. 1991. Automotive Industry, in *National Interests in an Age of Global Technology*, National Academy Press, Washington, DC:98-102.

DE GUES, A. P. 1988. Planning as Learning, *Harvard Business Review*, Mar.-Apr.:70.

HARE, A. P. 1976. *Handbook of Small Group Research*, Free Press, New York:217-219.

HEIM, J. A. and W. D. COMPTON (eds.) 1992. *Manufacturing Systems: Foundations of World-Class Practice*, National Academy Press, Washington, DC:31-35 and 37-38.

JACQUES, E. 1990. In Praise of Hierarchy, *Harvard Business Review*, Jan-Feb.:128.

LAWRENCE, P. R. and D. DYER. 1983. *Renewing American Industry*, Free Press, New York:263.

MARMON GROUP. 1992. *Annual Report*, Chicago.

MCGREGOR, D. 1967. *The Professional Manager,* W.G. Bennis and D. McGregor (eds.) McGraw-Hill Book Company, New York:41.

STRATA, R. 1989. Organizational Learning-The Key to Management Innovation, *Sloan Management Review*, Spring:63.

Wall Street Journal. 1994. Executive Suites' Walls Come Tumbling Down, June 29:B1.

WILSON, R. C. 1992. Jazz: A Metaphor for High-Performance Teams, in *Manufacturing Systems: Foundations of World-Class Practice*, J.A. Heim and W.D. Compton (eds.) National Academy Press, Washington, DC:238.

WOMACK, J. P., D. T. JONES, and D. ROOS. 1990. *The Machine That Changed the World,* Rawson Associates, New York:148.

QUESTIONS

9.1 The quality of communications among people has sometimes been found to be reduced in a hierarchical organization. To counter this tendency, it is often necessary to repeat a communication frequently to ensure that everyone understands it. Is this counterproductive? How can the correctness of communications be enhanced in a dispersed organization?

9.2 How can it be assured that rules, guidelines, and procedures are interpreted properly by each person in a dispersed organization?

9.3 The board of directors of a company can consist of executives from within the organization or people who come from outside the organization. What are the advantages and disadvantages of each of these? How should the choice be made?

9.4   What actions can management take to help people adjust to group environments and thereby increase the effectiveness of groups in the workplace?

9.5   Can multinational companies overcome the need for co-location of team members when the teams must be created with people who come from greatly separated geographical locations? If your answer if yes, how can they accomplish this?

9.6   What would be the composition of a team concerned with quality in the automobile service area of a large car dealership? What kind of interaction could they have with customers?

9.7   Company XYZ is a large manufacturing firm with six divisions organized as separate profit centers. All of the divisions use some common products that are purchased outside the company (e.g., steel sheet). Several years ago the company decided to centralize the purchase of these commonly used items in a central activity-a purchasing staff-as a means of enhancing the efficiency and reduce the cost of procurement. At a recent off-site strategy session the division general managers strongly recommended that the centralized function be abandoned and the task reassigned to each division. As president of the company, what questions will you ask, and what basis will you use to decide which approach is best able to serve company needs?

9.8   You are manager of a large activity that requires frequent exchange of information among several key managers. The success of the organization requires that all groups reporting to these managers promptly utilize this information in taking subsequent actions. Unfortunately, it is not uncommon for the people who first receive this information to be attending meetings elsewhere when the information arrives. The consequence is delays that reduce the efficiency of the overall activity. How would you go about solving this problem? How could these managers maintain control without creating a bottleneck in the operation?

9.9   You are president and principal shareholder of a small manufacturing company in Indianapolis, Indiana, that produces and services pumps that are used primarily for agricultural irrigation and oil pipeline systems. The former are marketed throughout the United States. You are currently considering a business arrangement that would require a large investment in a new plant. The feasibility of such an investment depends on both a short-term expansion in the market for agricultural systems and a long-term expansion into overseas markets, primarily in Eastern Europe and the Middle East. In Question 2.6 you identified the issues that must be addressed as you consider the advisability of making this investment. You are satisfied that the investment is sound and you are preparing to recommend to the board of directors that they support your plan. In private conversations with members of the board, you discover that there is great concern on the part of the board regarding this proposal. What will be your position if your board of directors declines to approve your proposal?

READINGS

COOK, H. E. 1992. Organizing Manufacturing Enterprises for Customer Satisfaction, in *Manufacturing Systems: Foundations of World-Class Practice*, J. A. HEIM AND W.D. COMPTON (eds.) National Academy Press, Washington, DC:128.

WILSON, R. C. 1992. Jazz: A Metaphor for High-Performance Teams, *Manufacturing Systems: Foundations of World-Class Practice*, National Academy Press, Washington, DC:238.

RAO, R. M. 1995. The Struggle to Create an Organization for the 21st Century, *Fortune*, Apr. 3:90.

APPENDIX A9.1   HOW TO MAKE A TEAM WORK*

Anyone who has ever run a business or organized a project has discovered how hard it can be to get the whole team on board to ensure that everyone knows where the enterprise is heading and agrees on what it will take to succeed.

At IBM we've used a method for some years that helps managers do just this. The technique, which we call PQM or Process Quality Management, grew out of many studies with customers to determine their needs and from internal studies as part of IBM's business quality program. PQM has been used successfully by service companies, government agencies, and nonprofit organizations, as well as manufacturers. ...

PQM does not differ radically from other planning processes: we identify goals and the activities critical to their attainment, and we provide a way to measure success. But PQM demands an intensive one- or two-day session at which *all* the key managers concerned agree on what must be done and accept specific responsibility...

PQM begins with a person who is the leader of the management team-the boss, the one whose job depends on getting the team's mission accomplished. He or she should then involve everyone on the immediate management team and no one else-nobody missing and no hitchhikers. At most there should be 12 people, since more than that is just too unwieldy. And if even one member of the team cannot attend the study, wait. PQM requires a buy-in from everyone not only to identify what is needed but also to commit to the process.

By management team we usually mean a formal group of managers, a board of directors, say, or a divisional vice president and his or her top managers. But the team can also be a collection of individuals drawn from various sectors of the company for a specific project. ...PQM demands spontaneity, so even though the boss convokes the team, a neutral outsider should lead the discussions. The leader could be a consultant or a manager or an officer from elsewhere in the company. What's important is that the leaders not be the bosses' subordinates and that their livelihood should not depend on achieving the mission. Furthermore, the discussions are best held off premises; at the office, secretaries can fight their way through steel doors to deliver "urgent" messages.

Finally, and perhaps this goes without saying, the boss had better be ready to accept challenges to the status quo. We have presided at a few disasters where, despite assurances of open-mindedness, the boss turned the study into a self-justifying monologue.... It [was] a terrible waste of time.

The first step in the PQM effort is to develop a clear understanding of the team's mission, what its members collectively are paid to do. *Collectively* is important. A marketing vice president and a finance vice president will have different ideas about their separate functional missions. But when they meet together as part of the management team, they should know their jobs as members of that team.

If the mission statement is wrong, everything that follows will be wrong too, so getting a clear understanding is crucial. And agreeing on a mission may not be as easy as it may at first seem..... A PQM study makes them [the management team] stand back and ask fundamental questions like, "Do we really understand our business well enough to form a mission statement?"

---

*Source: Hardaker, M. and B.K, Ward. 1987 "How to Make a Team Work," *Harvard Business Review*, Nov.-Dec.:112

Our advice is to make the mission statement explicit-nail it to the wall. It shouldn't be more than three or four short sentences.... The unit's mission statement [will] define the boundaries of the business and the customer population. It says what has to be done and says that achievement will be measured by the unit's demonstrable impact on...business success, customer satisfaction, and company image....The mission should be clear enough to let you know when you have succeeded and are entitled to a reward. "Increased profits" is not a rewardable mission. How much of an increase [is expected]? 0.5%? 5%? 50%? But "generate positive cash flow" might well be a rewardable mission for a management team nursing a sick company.

Once a team has defined its goal or mission, it could go straight to identifying its critical success factors (CSFs), the things it will have to do to succeed. But in our experience that's premature. At this point, few teams are relaxed enough to do the free associating needed to pinpoint their real CSFs. They are fixed on what they know and on today's problems, not on new possibilities.

To break out of old ways of thinking, we suggest a 10-minute brainstorming session in which team members list one-word descriptions of everything they believe could have an impact on achieving their mission. The usual brainstorming rules should apply: everyone should contribute; everything is fair game, no matter how crazy or outrageous; nobody is permitted to challenge any suggestion; the facilitator should write everything down so the team can see the whole list.

While thinking about these dominant influences, each member should focus intently on the team's mission. Members should look inside and outside their bailiwicks, sometimes far outside to factors like national characteristics or public policy issues....Typically a team's list will contain 30 to 50 diverse items ranging from things like costs and supplier capabilities to jogging and the weather.

Now the team should be ready to identify the critical success factors, a term used for many years in corporate planning to mean the most important subgoals of a business, business unit, or project. Here we define CSFs as what the team must accomplish to achieve its mission....Consensus on these aims is vital....Like the mission, CSFs are not the how-to of an enterprise, and they are not directly manageable. Often they are statements of hope or fear....In a sense, every CSF should be viewed as beginning with the words "We need...." or "We must...."to express buy-in by all ("We") and agreed-on criticality ("need" or "must").

In naming its CSFs, a team should be guided by the necessary-and-sufficient rule. That is, the group must agree that each CSF listed is *necessary* to the mission and that together they are *sufficient* to achieve the mission. This is a stringent requirement. The CSF list must reflect the absolute minimum number of subgoals that have to be achieved for the team to accomplish its mission....In addition, each CSF must be devoted to a single issue-pure in the elemental sense, like hydrogen or gold. The word *and* is verboten. The team has to struggle to reduce its list honestly; it can't succumb when some creative manager says, "Why don't we combine numbers three and seven so we reduce product cost *and* improve morale?"

The list should be a mix of tactical and strategic factors. If the factors are all strategic (increase market share to 15% by 199X, for example), the business might founder while everybody concentrates on the blue skies ahead. Equally, if all are tactical (reduce the delivered cost of product ABC to $20.50 by year end), the business could kill itself on short-term success. The ratio depends on several considerations, of course, including the nature of the business unit doing the study. A regional sales office would likely have more tactical CSFs, while a corporate headquarters would have an almost entirely strategic list.

The maximum number of CSFs is eight. And if the mission is survival, four is the limit-you don't worry about whether your tie is straight when you are drowning. There is no magic about eight. It just seems to be the largest number of truly critical goals that a management team can focus on continuously.

Our rules on number and absolute consensus may be tough, but they work, and it's essential to follow them. Whenever we have been persuaded to relax either rule, we have ended up with a mess, a list of moans rather than the truly visceral issues affecting the business. If someone cries, "We can't agree, let's vote," don't do it. Insist on consensus; highly paid, experienced, businesswise people should be able to agree on what's vital to their business, after all.

Reaching agreement on CSFs usually takes from one to three hours. The longest time we've seen was a day....

The third step in PQM is to identify and list what has to be done so that a company can meet its critical success factors. This might mean being more responsive to the market, exploiting new technologies, or whatever else is essential to accomplish the CSFs.... As with the CSFs' relation to the mission, each process necessary for a given CSF must be indicated, and together all those processes must be sufficient to accomplish it. Other rules we find useful are: each business process description should follow a verb-plus-object sequence; every business process should have an owner, the person responsible for carrying out the process; the owner should be a member of the management team that agreed to the CSFs; no owner should have more that three or four business processes to manage....

Now suppose we have a complete list of important business processes, each of which has an owner. The list is exclusive, since a process has to be important to be there. But it still needs ranking to identify the most critical processes, those whose performance or quality will have the biggest impact on the mission. This is the penultimate stage of our PQM.

First place the processes and the CSFs in random order on a matrix...and ask this question: Which business processes must be performed especially well for us to be confident of achieving this CSF? The object is to single out the processes that have a primary impact on this particular CSF. Many business activities will touch on it, of course; what you're after are the essential ones.... Then the list must pass the sufficiency test. If all these activities are performed well, will the team achieve its first critical goal? If the team answers no, then it must identify what else is needed.

This is usually the stage at which teams begin to be really creative, looking beyond what is already being done and breaking new ground. There's a check, though, because each new process added for sufficiency must also have an owner within the management team. So it has to be important enough to feature on the matrix.

The team then repeats this process for each CSF in turn, being careful to apply the necessary-and-sufficient test before moving on the next CSF. Then the number of CSFs that each process affects is totaled [and ranked]....

By now the chart is a valuable document. The management team has agreed on its mission, on the sub-goals, or CSFs, required to accomplish the mission, and finally, on the things that must be done to achieve those goals. Moreover, while each CSF is owned collectively by the entire team, each business process is owned by an individual member. Only one more step remains-identifying the most critical processes.

If companies had unlimited resources, each process could have equal attention for resources and management focus. But in practice, of course, managers' time and resources are always limited. So next pinpoint those activities that warrant the most attention.

Clearly, the most important processes are those that affect the most CSFs....But to get a meaningful ranking for management's attention, we also need to know how well each process is being performed.... In our PQM studies, we use a subjective ranking [of process quality].... The quality of each process is plotted horizontally and the number of CSFs the process impacts is plotted vertically. Then the team divides the graph into zones to create groups of processes...zone 1 contains the most critical processes. All the processes are important, by definition. But the higher risk [or higher opportunity] processes are found in zone 1. These activities need the team's closest attention if the company is to improve....

That's the PQM process–one way to conduct what is, in truth, a never-ending journey to zero defects.

But as we said up front, PQM requires follow-through. Decide the nature of the improvement needed, and establish relevant process measurements. Then apply the needed resources for the appropriate improvements.

We cannot stress follow-through enough. The decisions reached by the management team must cascade throughout the organization. And always there are surprises.... We recommend revisiting the CSF list about once a year or whenever a significant change has taken place in a team's mission, its makeup, or the marketplace. In a year's time, the mission usually stays the same, but the critical success factors and the most critical processes usually don't. Some of the processes will have moved from zone 1 to zone 2; others will be newly critical.... [W]hen you change the CSFs, you necessarily change the grid.... Eventually, you may even find that all of your business activities are clustered in [the top performance or quality category]. Then the only changes a new CSF list will provoke are those responding to a changing environment. Such adaptability is the ultimate goal of PQM....

APPENDIX A9.2 THERE'S MORE THAN ONE KIND OF TEAM[†]

"Team Building" has become a buzzword in American business. The results are not overly impressive.

Ford Motor Co. began more than 10 years ago to build teams to design its new models. It now reports "serious problems," and the gap in development time between Ford and its Japanese competitors has hardly narrowed. General Motors' Saturn Division was going to replace the traditional assembly line with team work in its "factory of the future." But the plant has been steadily moving back toward the Detroit-style assembly line. Procter & Gamble launched a team-building campaign with great fanfare several years ago. Now P&G is moving back to individual accountability for developing and marketing new products.

One reason–perhaps the major one–for these near-failures is the all-but-universal belief among executives that there is just one kind of team. There actually are three–each different in its structure. In the behavior it demands from its members, in its strengths, its vulnerabilities, its limitations, its requirements, but above all, in what it can do and should be used for.

The first kind of team is the baseball team. The surgical team that performs an open-heart operation and Henry Ford's assembly line are both "baseball teams." So is the team Detroit traditionally sets up to design a new car.

The players play *on* the team; they do not play *as* a team. They have fixed positions they never leave. The second baseman never runs to assist the pitcher; the anesthesiologist never comes to the aid of the surgical nurse. "Up at bat, you are totally alone," is an old baseball saying. In the traditional Detroit design team, marketing people rarely saw designers and were never consulted by them. Designers did their work and passed it on to the development engineers, who in turn did their work and passed it on to manufacturing, which in turn did its work and passed it on to marketing.

The second kind of team is the football team. The symphony orchestra and the hospital unit that rallies round a patient who goes into shock at 3 a.m. are "football teams," as are Japanese auto makers' design teams. The players on the football team or in the symphony orchestra, like those on the baseball team, have fixed positions. The oboe never comes to the aid of the violas, however badly they might flounder. But on these teams players play as a team. The Japanese auto makers' design teams, which Detroit and P&G rushed to imitate, are football-type teams. To use an engineering term, the designers, engineers, manufacturing people and marketing people work "in parallel." The traditional Detroit team worked "in series."

Third, there is the tennis doubles team–the kind Saturn management hoped would replace the traditional assembly line. It is also the sort of team that plays in a jazz combo, the team of senior executives who form the "president's office" in big companies, or the team that is most likely to produce a genuine innovation like the personal computer 15 years ago.

On the doubles team, players have a primary rather than a fixed position. They are supposed to "cover" their teammates, adjusting to their teammates' strengths and weaknesses and to the changing demands of the "game."

[†]*Source:* Drucker, P.F. 1992. *Wall Street Journal,* Feb. 11.

Business executives and the management literature have little good to say these days about the baseball-style team, whether in the office or on the factory floor. There is even a failure to recognize such teams as teams at all. But this kind of team has enormous strengths. Each member can be evaluated separately, can have clear and specific goals, can be held accountable, can be measured–as witness the statistics a true aficionado reels off about every major-leaguer in baseball history. Each member can be trained and developed to the fullest extent of the individual's strengths. And because the members do not have to adjust to anybody else on the team, every position can be staffed with a "star," no matter how temperamental, jealous or limelight-hogging each of them might be.

But the baseball team is inflexible. It works well when the game has been played many times and when the sequence of its actions is thoroughly understood by everyone. That is what made this kind of team right for Detroit in the past.

As recently as 20 years ago, to be fast and flexible in automotive design was the last thing Detroit needed or wanted. Traditional mass production required long runs with minimum changes. And since the resale value of the "good used car"–one less than three years old–was a key factor for the new-car buyer, it was a serious mistake to bring out a new design (which would depreciate the old car) more than every five years. Sales and market share took a dip on several occasions when Chrysler prematurely introduced a new, brilliant design.

The Japanese did not invent "flexible mass production;" IBM was probably the first to use it, around 1960. But when the Japanese auto industry adopted it, it made possible the introduction of a new car model in parallel with a successful old one. And then the baseball team did indeed become the wrong team for Detroit, and for mass-production industry as a whole. The design process then had to be restructured as a football team.

The football team does have the flexibility Detroit now needs. But it has far more stringent requirements than the baseball team. It needs a "score"–whether it's the play the coach signals to the huddle on the field or the Mozart symphony everyone in the orchestra puts on his music stand. The specifications with which the Japanese begin their design of a new car model–or a new consumer-electronics product–are far more stringent and detailed than anything Detroit is used to in respect to style, technology, performance, weight, price and so on. And they are far more closely adhered to.

In the traditional "baseball" design team every position–engineering, manufacturing, marketing–does its job its own way. In the football team or the symphony orchestra, there is no such permissiveness. The word of the coach or the conductor is law. Players are beholden to this one boss alone for their orders, their rewards, their appraisals, their promotions.

The individual engineer on the Japanese design team is a member of his company's engineering department. But he is on the design team because the team's leader has asked for him–not because the chief engineer sent him there. He can consult engineering and get advice. But his orders come from the design-team chief, who also appraises his performance. If there are stars on these teams, they are featured only if the score calls for a solo. Otherwise they subordinate themselves to the team.

Even more stringent are the requirements of the doubles team–the kind that GM's Saturn Division hoped to develop in its "flexible-manufacturing" plant, and that any such plant does indeed need. This team must be quite small, with five to seven members at most. The members have to be trained together and must work together for quite some time before they fully function as a team. There must be one clear goal for the entire team and yet considerable flexibility with respect to the individual member's work and performance. And in this kind of team only the team "performs;" individual members "contribute."

All three of these kinds of teams are true teams. But they are so different–in the behavior they require, in what they do best, and in what they cannot do at all–that they cannot be hybrids. One kind of team can play only one way. And it is very difficult to change from one kind of team to another.

Gradual change cannot work. There has to be a total break with the past, however traumatic it may be. This means that people cannot report to both their old boss and to the new coach, conductor or team leader. And their rewards, their compensation, their appraisals and their promotions must be totally dependent on their performance in their new roles on their new teams. But this is so unpopular that the temptation to compromise is always great.

At Ford, for instance, the financial people have been left under the control of the financial staff and report to it rather than to the new design teams. GM's Saturn Division has tried to maintain the authority of the traditional bosses–the first-line supervisors and the shop stewards–rather than hand decision-making power over to the work teams. This, however, is like playing baseball and a tennis doubles match with the same people, on the same field, and at the same time. It can only result in frustration and non-performance. And a similar confusion seems to have prevailed at P&G.

Teams, in other words, are tools. As such, each team design has its own uses, its own characteristics, its own requirements, its own limitations. Team work is neither "good" nor "desirable"–it is a fact. Wherever people work together or play together they do so as a team. Which team to use for what purpose is a crucial, difficult, and risky decision that is even harder to unmake. Managements have yet to learn how to make it.

PART 5

The Importance of the Customer

The customer is the final arbiter in the free market. Products that do not have the necessary attributes, that are too costly, and that do not have an adequate lifetime will be shunned and the manufacture of such products will exact a terrible price. The marketplace can be brutal in its assessment and in its reaction. It is critical, therefore, for a company to pay close attention to the needs and wants of its customers. Timeliness in reaching the market with a high-quality properly priced product, whether it is a revision of an existing product or an entirely new product, is increasingly becoming the mark of a successful manufacturer. To be a world-class manufacturer requires the development of these capabilties. In the following two chapters we discuss the needs and wants of the customer and a variety of techniques that have proven to be successful of translating those needs and wants into high-quality products in the shortest time.

10
Customers:
Their Wants and Needs

> Only those who become attached to their customers, figuratively and literally, and who move most aggressively to create new markets—for fast-growing and mature products alike—will survive. (Peters, 1988)

As indicated in Figure 1.1, an indispensable element of the manufacturing system is the customer. As strange as it might seem, between the end of World War II and the early 1980s, much of U.S. industry lost sight of the importance of the customer to its success. With the competition generally determined by enterprises that operated within our national boundaries, the competitive focus was on meeting or slightly exceeding the performance of national competitors rather than focusing on what the customer needed or wanted. Perhaps the classic example of this is the deterioration in quality of the cars being produced by U.S. manufacturers throughout this period. When higher-quality products became available from Japanese producers, the U.S. buying public quickly signaled their displeasure with the quality of the product available from U.S. companies by turning to the Japanese manufacturer in large numbers.

Determining customers' needs and wants is a critical determinant of the success of modern manufacturing systems. As mentioned earlier, the world marketplace has created a situation that enables customers to choose from among a wide variety of product offerings. The system that loses sight of customer needs or wants will be unable to provide for customer needs in a timely fashion. The consequence will be a loss of customer loyalty and market share. To exclude customers, either intentionally or through benign neglect, from the manufacturing system will lead to a loss of competitiveness. The successful corporation makes a fetish of determining what its current and future customers want and what they may need that is new and unique. It is this drive to satisfy the customer that is a hallmark of the successful corporation.

237

It is critical in the discussion that follows that the term *customer* be understood to refer to everyone, both *inside* and *outside* the corporation, who benefits from the actions taken by *someone* in the corporation. The *customer* is not confined to the person or firm that buys the final product. The customers of the custodial group are the occupants of the facilities. The customers of the research group are the product and manufacturing development activities. The customers of the procurement activities are the development activities and the manufacturing groups. The customers of the sales, marketing, engineering, development, and manufacturing groups are the public or another firm that purchases the product or service. This list of connections is nearly endless, for it includes all the interconnections implied in Figure 1.1.

The admonition that must be made to all members of the organization is to *know your customer* and *listen to your customer*. Should that imply that the organization pursue only those products or services that customers have said they want or need? The answer, of course, is no. Individual initiative and the search for new undiscovered opportunities must be pursued aggressively. These searches, however, need to proceed within a context that begins with listening to what customers are saying about their perceived needs and wants.

10.1 WHO IS THE CUSTOMER?

A useful first step in answering this question is to create a list of your customers. A few examples, in addition to those given above, may be helpful. Let's begin with an example that involves an organization that is familiar to all students-the university. Who are the customers of the faculty member in this class? The list must begin with the students in the classroom. It also includes the graduate students who are carrying on research under the direction of the faculty member. For the research programs, the customer is the sponsor of the research-the federal agency that provides the resources or the industry that provides the contract support. Does it end there? Certainly not. In the case of a public institution, there are the taxpayers who help defray the cost of the university and the parents who support the educational costs of their children. Other customers include the alumni, businesses that come to the university for advice and help, and state and federal agencies that use the faculty in advisory roles.

This list has dealt only with a limited part of the university. If all the customers of the internal organizations were listed, it would be many pages long. The libraries have as their customers students and faculty internal to the university and other libraries external to the boundaries of the institution. The athletic department, computer center, counseling services, health services, development offices, fraternity/sorority houses, food services, and so on, all have customers.

The list of customers, both internal and external, for a manufacturing system is no simpler. Its external customers will include some or all of the following: the distributors who serve the retailers, the retailers who sell to the public, the general populace who purchase the product, the service personnel who provide field repair; and the scrap dealers who must dispose of the obsolete product. There will be internal customers who receive support in their efforts to design, develop, and manufacture the product. Supplies of materials, services, and components will be arranged by people whose customers are also employees of the organization.

The first reaction to identifying your customer may be that this exercise is straightforward, perhaps even trivial. As Edmondson (1992:129) points out, pitfalls may be encountered in accomplishing this.

> The question of defining who your customers are seems fairly easy, particularly if you have segmented your market properly and understand who you are trying to satisfy. However, a subtlety that frequently goes undetected by many firms is that the customer set can be divided into two parts-the apparent customer and the user. The apparent customer is the person or group of people who decide what product to buy and basically have control over the purse strings. The user is the person or group who physically uses the product or is the direct recipient of a service.
>
> One way to illustrate this significant difference is to relate a new product story from the history of a well-known dog food manufacturer. After surveying their customers carefully, the manufacturer decided that the greatest unfulfilled need their customers had was to somehow provide a food that overcame the greatest disadvantage of man's best friend, specifically, the dog's bad breath. It did not take the research department long to come up with a chlorophyll additive to their regular product, and they went into production quickly. The first several weeks after the product introduction were quite successful for the firm in that they sold lots of their product. They smilingly congratulated themselves for having accurately ascertained their customers' needs and then satisfying that need. Unfortunately, the story doesn't have a happy ending in that the repeat sales of this product were virtually zero and rather shortly the product turned into a dismal failure. The simple reason; the dogs would not eat the product. Remarkably, the firm put all their bets on understanding the customer and spent no time understanding the user. The point of all this is that, to truly capture the fancy of your customer you must recognize that both the apparent customer and the user must be satisfied. Forgetting the apparent customer will probably preclude your making even the first sale, and forgetting the user will undoubtedly, as with the dog food company, preclude your getting any follow-up sales.

Other examples of an apparent customer that is distinct from the user is the car rental agency that buys vehicles from the manufacturer for rental to the public, the architect who designs the condominium for the developer to sell to the owner, and the book publisher who buys the manuscript for a book from the author for sale to the reader. To be successful, it is critical for the car manufacturer, the architect, and the author to fully understand the users' needs and not to focus solely on the apparent needs of their customers.

In addition to the need to separate the apparent from the real customer, the producer of a product may not know the uses to which the product will be placed. Consider the string of white lights that were first created as Christmas tree lights. They are now being used by designers, advertisers, and homeowners for very different purposes. The semiconductor manufacturer will sell integrated circuits for use in products that they will never see or know about. A truck manufacturer would find it hard to identify the many customers that their product serves.

Knowing your customer also requires that attention be given to the evolving patterns for use of the product. As markets evolve, new customers may develop for the product. As competing products appear, the customers for your product may be restricted or disappear. A continuing update of the customer list is critical to an understanding of the market.

It is the task of management to ensure that everyone in the organization identifies the customers for their products and/or services and to ensure that they are meeting the needs and wants of those customers. It is not sufficient for some segments of the activity (e.g., the manufacturing group) to expect someone else to provide the entire customer interface on their behalf. It is the obligations of everyone to know their customers and the needs of those customers, whether the customers are external or internal to the organization.

10.2 CUSTOMER DEMOGRAPHICS

In attempting to anticipate the future needs and wants of company customers, it is important to understand how the customer base will change over time and what the characteristics will be of those new customers who will be entering the market. Peter Drucker (1985:35) notes that "systematic innovation ...consists in the purposeful and organized search for changes, and in the systematic analysis of the opportunities such changes might offer for economic or social innovation." While he identifies several sources for innovative opportunity, only one will be included here-the demographics of the population and its impact on the population that the company seeks to serve (Drucker, 1985:88).

> Of all the external changes, demographics-defined as changes in population, its size, age structure, composition, employment, educational status, and income-are the clearest. They are unambiguous. They have the most predictable consequences.
>
> They also have known and almost certain lead times. Anyone in the American labor force in the year 2000 is alive by now (though not necessarily living in the United States...). All people reaching retirement age in 2030 in the developed countries are already in the labor force, and in most cases in the occupational group in which they will stay until they retire or die. And the educational attainment of the people now in their early or mid-twenties will largely determine their career paths for another forty years.
>
> Demographics have major impact on what will be bought, by whom, and in what quantities....
>
> Demographics alone, the combined effects of the sharp drop in birth rates and of the "educational explosion"—makes it near-certain that traditional manual blue-collar employment in manufacturing in developed countries, by the year 2010 cannot be more than one-third or less than what it was in 1970.

The following example of the importance of proper assessment of the demographics of the workforce is given by Drucker (1985:70):

> In 1909 or thereabouts a statistician at the Bell Telephone System projected two curves fifteen years ahead: the curve for American population growth and the curve for the number of people required as central-station operators to handle the growing volume of telephone calls. These projections showed that every American woman between age seventeen and sixty would have to work as a switchboard operator by the year 1925 or 1930 if the manual system of handling calls were to be continued. Two years later, Bell engineers had designed and put into service the first automatic switchboard.

Not every analysis will give a result that is this striking. It is, however, important to understand trends in a population. Demographic trends can be critical to success when considering a new product that is targeted for the group over age 60 or for those with disposable income over a certain amount or for those who are in college. It is also important to understand how the interests of a given age group are likely to change with age. As Drucker observes correctly, the numbers of potential customers having a given characteristic are probably known with great certainty, for the cohort already exists that will become that population.

10.3   LISTENING TO THE CUSTOMER

The marketplace is never static: the needs and wants of customers change constantly. Incremental improvements are regularly being introduced for existing products. New technical developments create the opportunity for all-new or vastly improved products. New competitors enter the marketplace and attract customers with entirely new offerings. The successful company must anticipate the changing environment, understand the level of changes that the customer is demanding in the performance of its products, and be prepared to take advantage of any revolutionary technologies that can make feasible a new and unusual product . The approach that the enterprise takes in understanding the needs or wants of the customer will be critical to its success.

Listening to your customers requires that they articulate their needs and wants. Edmondson (1992:130) notes the difficulty in obtaining this information from the customer.

> [N]eeds usually [are] the real requirement to which the customer should be putting the product, or, the true requirement of the customer. Wants, on the other hand, are the perceived needs the customer feels should be met. It you are lucky, the customer's needs and wants are synonymous. But this is seldom true in real life.
>
> There are a number of ways in which this complicated wants-versus-needs syndrome can cause problems. Three of the most troublesome of these illustrate the difficulty involved in sorting out this part of your product definition.
>
> The first is the customer who knows what he wants but does not know what he needs....The second...is the customer who has a hidden want but purposefully masks it as a need....The third...is finding a customer who is all too willing to tell you what your next-generation product should be in terms of your traditional product.

Not only are the customers' wants likely to be different from their needs, but different customers may have needs and wants that conflict. In the example of the university, one need only mention that students and research sponsors may view classroom teaching and research activities to be incommensurate-but for different reasons. In the sense that they compete for the time of the faculty, there is a potential conflict. In the sense that they each contribute to the intellectual activities that are a part of a modern research university-at both the undergraduate and graduate levels-it is incorrect to view them as being in conflict.

For the manufacturer of a product, the first signal that is received from the customer is the fact that the product has been chosen or rejected in either an absolute or a relative sense. It may be absolute, in that customers choose not to buy any product of this type. It may be relative, with the customer favoring a competitor's product. In either case the important question is: Why did the customer choose one product over another, or why did they choose to reject all products?

If the consumer-whoever it is-has rejected all products, this surely signals that the products do not meet the perceived need of the customer or do not offer a value commensurate with their cost. On the other hand, if a competitor's product with attributes similar to those of your product is attracting customers, it becomes a matter of determining why the customer is choosing one product over the other. Although the reasons may be many, the following product attributes frequently appear on the list of desirable products: quality, reliability, serviceability, cost, and unique attributes. A detailed comparison of all of these, and other, characteristics of the product is necessary to obtain a clear understanding of the customer's preferences. As described in Chapter 4, benchmarking your product against the products of competitors becomes an important element in "listening to the customer."

All of these factors contributed to the changes that have taken place in recent times. These can be illustrated by the changes that have occured in automotive markets since the mid-1970s. While U.S. manufacturers continued to emphasize cost and styling, Japanese manufacturers emphasized quality. U.S. consumers demonstrated that they placed a higher premium on quality, even at a higher cost, than they did on styling. Once they defected from U.S. manufacturers to become customers of Japanese manufacturers, it has proven extremely difficult to get them to return to a U.S. product. The perception by many customers of the relative value of Japanese products is highlighted by the experience of General Motors with the Nova, which was built in the same plant by the same workers who were building the Toyota Corolla FX-the NUMMI plant in San Jose, California. The products were identical except for some details of trim and the nameplate. The NUMMI plant is a joint venture of Toyota and GM, with Toyota providing management of the factory. Throughout the time that this joint venture has existed, the Toyota product has outsold the comparable GM product. Since this was a pattern that persisted even with dealers who offered both the GM and Toyota products, it seems clear that it did not result from marketing. Rather, it appears to result from the perception that customers have developed of the relative values of the two products and their loyalty to a particular manufacturer, in this case Toyota.

10.3.1   Improving an Existing Product or Process

Each year the customer is offered an array of products whose characteristics are slightly modified, improved, or expanded from those available in previous years. Examples can be seen in consumer electronics, automobiles, telephones, clothing, and foods. A successful enterprise must be sensitive to two principal concerns: anticipating those new product characteristics that will be attractive to the customer, and responding to opportunities in a timely and cost-effective manner. Developing the responses necessary to accomplish this is an important task for management.

Using market research to assess customer needs and wants. Market research is used successfully to identify the attractiveness of new product features to current and prospective customers. Customers of the current product can be asked to rank the perceived value of certain features. The new characteristics may be described in verbal form, represented pictorially, or presented in a prototype that has the proposed features. In some cases the modified product may be test marketed in a limited number of geographical areas to determine customer response. Studies of this type are regularly done by automotive companies, which use "new car clinics" to evaluate the response of customers to proposed features for the next model of a vehicle; by food suppliers who want to evaluate different tastes, packages, or contents; and by personal product manufacturers for products that are planned for marketing to particular segments of the populace.

Market research often involves the design of questionnaires that invite current users of a product to indicate the features that they would like in an improved product. Customers may also be invited to identify the most desirable features of competitors' products so that these can be considered for inclusion in the next generation of product. Ford used the results of extensive interviews with customers and the identification of the best features in competitive products to arrive at the feature content of the first successful Taurus/Sable. Analysis of negative reactions to features of existing products also serves to guide new developments.

For marketing research to be meaningful, it is essential that the participants be representative of the populace who are potential purchasers, and the number involved must be large enough to provide statistically meaningful information. Each of the above may involve considerable expense and require a substantial amount of time, so listening to the customer in these ways may not always be possible.

For out-of-date products that are replaced by new but similar products, an important source of consumer information is the fraction of owners who purchase their replacement from the same manufacturer. Known as *customer loyalty*, a consumer's return to the products made by the same manufacturer suggests that they value and trust the manufacturer just as they value the product. A manufacturer who has strong customer loyalty will probably be able to compete in a market that is more-or-less fixed in size. A manufacturer whose principal sales are to first owners will find that they will not survive in a market that is approaching saturation.

These techniques and others are employed regularly by market research groups. They are frequently very useful in identifying a series of improvements that will appeal to customers. Since the customer has experience with an existing product, he or she can therefore offer sensible responses concerning the desirability or value of proposed modifications. Experience with current products is valuable in assessing the directions that future products should take.

Responding to changing customer needs and wants. Gomory and Schmitt (1988) have described the importance of minimizing the time spent in bringing a modification of a product to market.

> In the world of computers, printers, jet engines, or medical imaging devices, when the current version of the product is in the manufacturing phase, a development team is working on the next product generation. And when that next generation goes into production, the following generation is started through the development process.
>
> The speed of this development and manufacturing cycle is vital. If one company has a 3-year cycle and another has a 2-year cycle, the company with the shorter cycle will have its process and design into production and the product in the market one year before the other. The firm with the shorter cycle will appear to have newer products with newer technologies. In fact, both companies will be working from the same storehouse of technology. It is the speed of the development and manufacturing cycles that appears as technical innovation or leadership. It takes only a few turns of that cycle to build a significant product lead. This consequence is as true for long-cycle businesses such as aircraft engines as it is for short-cycle ones such as microelectronics.

Clark and Fujimoto (1991) have compared the time spent by some selective U.S., European, and Japanese companies on various stages of development of an automobile (see Figure 10.1).[*] The total average time spent by the Japanese automotive firm-from the start of concept generation to the beginning of production-was about 19 months less than required by either the U.S. or European companies. As Gomory and Schmitt point out, it is impossible to remain competitive over a long period with this difference in time taken to accomplish the process. The importance of achieving a fast turnaround in products is the stimulus for the effort currently being expended by the U.S. automotive industry to reduce the time that it takes to introduce a new product into the marketplace. Although the advantage enjoyed by the Japanese companies has narrowed substantially since 1991, the Japanese still enjoy a timing advantage in terms of bringing new products to the marketplace.

Smith and Reinertsen (1991:3) make the following comments about the benefits of short cycle times for bringing a new product to the market.

> Many competitive advantages accrue from a fast development capability. Perhaps most obvious but least important is that the product's sales life is extended. If a product is introduced earlier, it seldom becomes obsolete any sooner. Consequently, for each month cut from a product's development cycle a month is added to its sales life, for an extra month of revenue and profit....
>
> For some products that have high switching costs the benefit is even greater, because the early introducer gains on both ends of the cycle. If a product is introduced early it gains more customers, who maintain their loyalty due to the cost of switching to another product. Their loyalty creates a sales tail that is roughly proportional to the prior sales of the product. Consequently, an early introduction develops momentum that not only carries the product's sales higher but also further into the future.
>
> As a second benefit, early product introduction can increase market share. The first product to market has a 100 percent share of the market in the beginning. The earlier a product appears, the better are its prospects for obtaining and retaining a large share of the market. In some products like software and certain types of industrial machinery there is an especially high premium on offering the first product of a given type, because buyers get locked into the first operating system or computer language they acquire. It then becomes difficult for them to switch.
>
> A third benefit is higher profit margins. If a new product appears before there is competition, the company will enjoy more pricing freedom, making higher margins possible. The price may later decrease as competing products appear. By then, however, the company will be moving down the manufacturing learning curve ahead of the competition, so there is a continuing advantage to being first....
>
> An adaptable development process can be used to exploit changes other than technical ones. Many consumer markets change quickly, for unexpected reasons. Some products are sensitive to the vagaries of color, style, or current folk heroes. Political factors such as shifts in East-West relations often have an effect on many parts of the economy. The quick can turn these events into opportunities.
>
> To gain advantage from rapid product development over the long haul a company must make swift, effective product development a way of life. In the end, the companies that can consistently develop new products quickly and effectively will step by step

---

[*] Although an entirely new car may contain many innovations, the bulk of the vehicle will use either existing or slightly modified components, thus making it appropriate to consider it as a modified product.

A. U.S. average   (Months before start of sales)

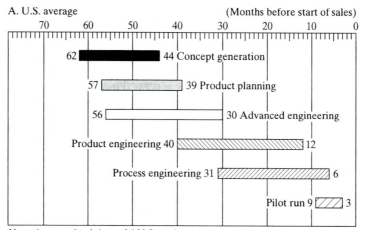

Note: Average lead time of 6 U.S. projects.

B. European average   (Months before start of sales)

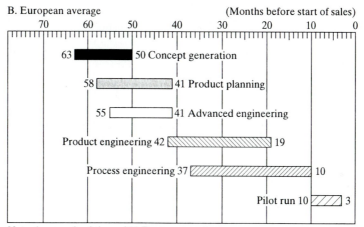

Note: Average lead time of 11 European projects.

C. Japanese average   (Months before start of sales)

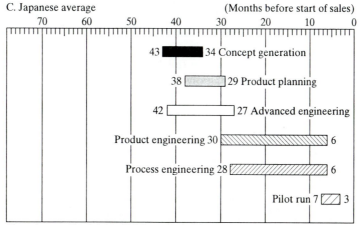

Note: Average lead time of 12 Japanese projects.

Figure 10.1   Average project schedule by stages. (From Clark and Fujimoto, 1991, p78.)

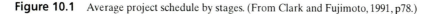

outpace those that cannot. Bringing one good product to market quickly is nice, but it will not ensure the viability of the company. Only the companies that have internalized this new way of operating will survive....

The benefits of rapid product development go beyond improving profitability of specific products and enhancing the company's image as an innovation leader. In fact, the attitudes and techniques that speed up the development process diffuse throughout the entire organization to improve the speed with which all resources, particularly the human one, can be applied....

All of the operations involved in the product development process-sometimes referred to as the *product realization process*-must be closely coordinated and work together. As Clark and Fujimoto (1991) note:

[W]hen we examine the relationship between lead time (a measure of how quickly a firm can perform the many different activities that must be accomplished to advance from concept to market introduction) and engineering hours, we find not a negative trade-off but rather a positive correlation between these dimensions, particularly for Japanese firms. Firms that are fast tend to be highly efficient; firms that are slow have low productivity. Two factors appear to be behind this relationship. One has to do with how engineers work, the other with the tight coupling of engineering activities.

The process of accomplishing this cooperative endeavor-described by such terms as *simultaneous engineering* or *concurrent engineering*-are discussed more fully in Chapter 11. Suffice it here to emphasize the essential importance of developing the capability to respond quickly to changes and opportunities. Possession of this capability is clearly one of the distinguishing features of a firm that is prepared to maintain its competitive position in the world marketplace.

10.3.2 Creating a Revolutionarily New Product

As difficult as it is to obtain a clear unambiguous message from customers regarding a product that is already in the marketplace, it is substantially more difficult to obtain useful information about an anticipated product, one with attributes that are largely unfamiliar to the customer. Such a product is described as revolutionary, not to suggest that it must utilize a revolutionary technology, but to indicate that it is unique in the marketplace at the time of its introduction. Customers frequently identify something as being of interest to them without being able to associate a value to its purchase. It is the value to the customer, either in terms of features, prestige, or investment, that ultimately leads them to purchase it. For a new product, not yet tested in the marketplace, it is extremely difficult for the customer to perceive of what its value will be to them. It is also difficult for the developer or manufacturer to assess this accurately while developing the product. It is only necessary to recall the projections in the early 1950s that a half dozen or so computers would satisfy all the needs of the marketplace. It would have been equally impossible in the 1950s to foresee the development, let alone the value, of the present-day PC. Similarly, there is no evidence that the early developers of facsimile machines foresaw the current potential for FAX machines.

Foster (1986) describes the problem as follows:

> Do companies really understand what their customers want? Sometimes certainly, but often not accurately enough to know what those customers will do when approached by a discontinuity in products available to them. Customers are notorious about wrongly predicting what they will want. They did not want plain-paper copies-carbon paper was plenty good enough they said-until they could get plain-paper copies reliably and cheaply. They did not want "tummy" TVs until Sony opened up the market for them. They would not trust paper diapers, until P&G made them strong, absorbent and cheap. So if the customer doesn't know what he wants until he gets his hands on it, it is hard to see how a company can put much faith in the thought that they will tell you that a change is coming.

It would be incorrect, however, to believe that the error of judgment is only in underestimating the market for a new product. Recall the lack of enthusiasm that the marketplace displayed for the picture phone that Bell Laboratories developed in the early 1970s at a cost of hundreds of million of dollars, the lack of enthusiasm for Ford's Edsel, or the problems that DuPont faced in convincing the consumer that Corfam was as good as leather for shoes. The translation of technical opportunities into products that will attract a large following in the marketplace can be a daunting challenge.

Planning an entirely new product or process presents a genuine dilemma in that an established base does not exist of customers who have experience with the product and who can, therefore, be expected to offer good advice about the value of the product proposed. This does not imply that those who are asked will be void of opinions. It suggests, however, that the credence that is to be placed in their answers should be carefully assessed. It is particularly difficult for a consumer to offer guidance on a new technology. They frequently can state the performance that they would expect in a product, but they often care little about the technology used to achieve that performance. For example, a customer identifying fuel economy as one of the desirable characteristics for a new automobile may care little what combination of smaller engine, lighter-weight vehicle, or improved aerodynamic styling is used to achieve the objective. Similarly, a consumer asked about a product that provides a previously unexperienced use may not be able to project the final value of the product. The mistaken projections for the growth of the computer industry, the prefabricated home construction industry, or the earlier attempts to introduce picture phones are examples at the extremes. Edmondson (1992:132) argues that the description of a new product or process must "be detailed and based on a thorough understanding of the customer's requirements of a feature-by-feature basis.... [T]his description must reflect complete understanding of what your competitors are offering today and are likely to be offering in the future-once again, on a feature-by-feature basis." This need for understanding leads Edmondson to offer the following advice on how this information should be collected:

> Many firms feel that marketing is the function that should be responsible for product definition, while others feel that the engineering department is responsible. It is my very strong view that both departments must act in partnership to gather the information necessary to form a clear, concise product definition. There is an appropriate division of duties between R&D engineering and marketing within the task of product definition. I suggest that the marketing people be held responsible for understanding

the marketplace, that is, knowing who are the dominant customers and users of products similar to the one you are trying to define. Further, marketeers should have a good idea of which customers are the innovative leaders and best suited to describe future needs for the industry. Additionally, the marketing people are more likely to have insights into what competitors are about to do.

On the other hand, the engineering department is best qualified to bring technical expertise to bear on customer problems and potential solutions. Engineers also probably have a better idea about the technical suitability of the products that are currently being offered by competitors. Stated more succinctly, the marketing people should be responsible for developing the market research plan and managing the places where information is sought, while the engineers should be responsible for providing the technical creativity and working with customers' technical people to develop a set of features for the new product.

How, then, should the problem of identifying the characteristics of an all-new product be approached? The best information available must be collected and analyzed. If time allows, test marketing of a limited production of the new product has been shown to be a good means of determining market acceptance. This provides an opportunity to determine the most desirable attributes of the final product and to adjust those attributes based on limited production. This is, in short, a very effective way to listen to what the potential customers say about the new product. If the final product has a short life cycle, is appealing because of its fad appeal, or is very seasonal, test marketing may not be possible. Experience, intuition, and sense of the marketplace must then be utilized to reduce the risk of failure. The manager who understands the marketplace and can sense the direction in which it is evolving will have the best chance of making the correct decision.

10.3.3   Niche Markets

In listening to the customer, an opportunity may be found to create a unique product that fills a particular niche in the marketplace. As Foster (1986) notes:

> Even if companies do know what the customer wants in general, it's not clear that these wants can be accuarately and speedily converted into the specifications for products that economically meet the customers' needs.... Further, the discontinuity, when it comes, may not come in the main market sector, the sector where a company knows the most about its customers' needs. It will probably come in a niche.

With increased competition from manufacturers located in all parts of the world, the market is tending to fragment into smaller and smaller discrete specialized segments. The creation of specialized products or services that serve unique markets offers one form of protection from encroachment by other manufacturers. If the market is modest in size, it may not be cost-effective for a competitor to develop a product for it. If the processes used to create the product are unique, it may not be cost-effective for a competitor to create comparable facilities. If the product appeals to a particular group of people who can be reached in a unique way by one manufacturer, this may offer some protection from competition. The laptop computer started as a niche market but has since grown into a more general market. The very expensive, very exclusive prestige car market is a niche market in which the large-volume manufacturers do not find it economical to compete. Their focus on mass production has led them to create facilities that cannot be employed economically in producing products for small markets. The mini-steel mills that specialize in one product have revolutionized the U.S. steel industry. Small local vineyards and breweries are good examples of industries that are capitalizing on a local product. Niche markets are common for many products, and the creation of a product that creates a new niche market can offer an attractive opportunity for a manufacturer. Clearly, the customer's needs and wants must be well understood to be successful in satisfying or creating a niche market.

If the protection offered by a niche market is so attractive, it might be asked why this is not the dominant pattern of manufacturing and marketing. Many issues contribute to the limited applicability of this approach, only a few of which are mentioned here. First, since niche markets are, by definition, small segments of a larger market, the manufacturer who concentrates on the niche will not enjoy the benefits of lower unit manufacturing costs that generally accompany large-scale manufacturing. Second, a niche market may suddenly disappear. A specialty product can be made obsolete by an improvement in a product being offered to the general market or by a new product that appeals to a broader market but with attributes that destroy the uniqueness of the current niche product. Third, the consumer who was targeted for the product may disappear because of changes in interests, needs, age, or boredom with the product. Despite these risks, many manufacturers have found niche markets to be very profitable.

10.4 LISTENING TO THE MARKET

As Drucker (1985:88) has emphasized so forcefully, the search for opportunities for new products should begin with a search for situations in which some significant change is occurring. It is not necessary, however, to remain passive in this search for opportunity. One need not wait until a change has been demonstrated so dramatically that all can recognize it. The opportunity will go to those who have the capability to anticipate that change will soon occur and to prepare for its occurrence. The successful anticipation of changes will provide a powerful advantage to the manufacturer that can have a new product or process ready for the emerging opportunity.

This is described in *Enterprise* magazine (1993) as follows:

> For many corporations, hearing the voice of the market involves going beyond traditional marketing research. Overseen by senior management and directed by cross-functional teams, the research explores customer needs early on to ensure product development gets everything right the first time. It stresses customers' total experience with a product or service, and it often makes use of sophisticated databases and modeling tools to share information companywide and to produce high-quality goods.

The following sections offer specific examples of tools that can be used to understand the message that the market is giving.

10.4.1   Anticipating and Understanding Market Trends

Markets for products characteristically develop slowly, accelerate as the consumer recognizes that a new product has desirable features, and slow as the market begins to saturate. This characteristic pattern is shown in Figure 10.2 and is described, appropriately, as the *S-curve growth pattern*. The qualitative basis for a curve of this form is as follows. The opportunity for a new market often becomes apparent to a number of companies at about the same time. Each company will seek to capture the market with a unique offering. Eventually, one of these products will be demonstrated to have the most desirable attributes and will become the market leader. From this early experimentation with many products, the market will have established a rather clear direction for the products that can be expected to satisfy a growing market. This early stage can be characterized as a *gestation period*.

As a product with the most desirable attributes emerges as being the clear choice of the consumer, some competitors will drop out of the competition or will modify their products such that their products resemble the product of choice. With a demonstrable product having the desired attributes, the market will attract more customers and will grow rapidly. This period can be characterized as the *high-growth period* of the market. During this period of rapid growth, investment in plants and facilities must be made quickly and at substantial cost, with only a few companies having the financial capability to continue. Usually, a few manufacturers will dominate the market. A company considering entry into the market at this time will be successful only if it can build the necessary capability *quickly* and *efficiently*.

In the later stages of the market-the *saturation period*-the market will have reached a large fraction of its potential customers and the rate of growth slows. Surviving manufacturers will be attempting to maintain market share through competitive pricing and incremental improvements in product attributes. As the saturation level for this product is approached, entry into the market will require that the new producer immediately become a low-cost, high-quality producer. It will not be possible to spend time learning, as established competitors will be relentless in preserving their market share.

Figure 10.2   Time Performance Potential.

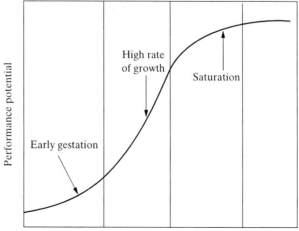

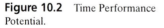

This description is as appropriate to processes that provide unique capabilities for the manufacture of products as it is for products. The introduction of NC or CNC machine tools to replace manually controlled machines and the introduction of electronic controls on unit processes such as welders, heat-treat ovens, and so on, are examples of processes that followed the same trends.

Assessment of the opportunities for a new product or process must begin by listening to the market. By understanding the current market and noting how far it has evolved toward saturation-its progress along the S-curve-gives a good indication of the strategy that must be followed. A plot of the cumulative sales of a product, using all the products that serve that market, as a function of time will immediately indicate the stage of development that characterizes the market. While a visualization of the S-curve offers some valuable insight into the status of the market, it is often desirable to have a somewhat more quantitative description of the market and to be able to estimate such quantities as the time at which the market will be 50 or 90% saturated, or what volume of products will have been sold when the market is 80% saturated. Knowing these values somewhat more precisely than is possible by simple visualization can be accomplished using the techniques discussed in Appendix A10.1.

10.4.2   Markets and Strategies

The opportunities for achieving success are different at each stage of the development of the market. Entry into the market at each stage also carries certain risks and requires a different strategy to be successful. A particularly interesting opportunity exists when the market is approaching the saturation stage. A large fraction of the potential customers will have already acquired the product. The market continues to exist because of replacement sales or because a few new customers are entering the appropriate demographic group for which the product is appealing.

In a stagnant market the opportunity exists to upset the status quo, to introduce a new product or a new process that has very different attributes. A successful new product will initiate the development of an all-new market segment and can be represented by a new S-curve. In effect, the new product or process will be sufficiently unique to destroy the old market. This may involve utilization of a technology that has recently emerged, as was the case with the replacement of piston aircraft engines by jet engines, or the use of a new material that allows the design of a product with all new attributes, as occurred with the replacement of vacuum tubes with transistors and integrated circuits, or the combination of attributes that existed in several products into a single new product, as with cellular phones. Saturation and stagnation are properly used to describe a market that has largely exhausted the capabilities that continuous improvement of an existing product or process can provide.

Market stagnation or saturation present the opportunity to find or develop the means by which the complete replacement of the existing products can occur, perhaps with a revolutionary product or process. The strategy is therefore to search for the opportunities that are determined by the status of the market.

Throughout this search for improved or revolutionary products or processes, it is essential that the decisions be made within the context of the overall goals and objectives of the enterprise. It is far too easy to follow a path that is determined by expediency. Imagine that you are a publisher of physical science textbooks who has just been given the opportunity to publish an attractive reference series for social sciences. Do you accept? The answer will be very different if the goals and objectives of the publisher are to market to all disciplines in contrast to a publisher who desires to specialize and be the leading publisher of physical science textbooks. The choice of whether to do it or not must be made carefully within the context of the long-term strategies and goals of the corporation.

Although making such decisions seems straightforward, experience suggests that it is often difficult to forgo what appears to be a good immediate opportunity, even if it is on the fringe of the company's normal product offerings. It is tempting to move from one opportunity to another, without regularly examining whether each opportunity fits within the corporate goals and objectives. This can eventually lead one outside the mainstream of the business. Although this may occasionally be good, for it may offer an opportunity to diversify and gain access to potential new markets, the consequence can be a loss of focus and a dispersion of effort. The newspapers and trade journals regularly carry accounts of companies "that have lost their direction" and are divesting of certain activities to concentrate on their core business.

Customers and markets send many messages to the manufacturer. A good filter must be developed to remove the signal from the noise. Although a deliberate strategy is needed in choosing among the many possible product offerings, it is essential to remember that strategies and goals are not static but must be reexamined with regard to their currency and value. The successful manager will constantly search for opportunities but only choose those that fit within the overall goals of the organization. In the final analysis, good experience and intuition are indispensable for the successful manager.

10.5   CONCLUDING OBSERVATIONS

As difficult as it is to know what the customer wants or needs, the foreign manufacturer who seeks to export a product to an overseas market faces a still more difficult problem. Turning again to the automobile industry for an example, the German consumer is much more concerned with the performance of the vehicle, its capability to respond at high speeds on the Autobahn, and the technological content of the product than is the average U.S. purchaser. Because of the much higher taxes imposed on automotive fuels in Europe than in the United States, the European market has placed a higher value on good fuel economy than has the U.S. market. The result is that high-performance cars with good fuel economy dominate European car markets.

Knowing, understanding, and incorporating the priorities of the local market into products that are being designed and manufactured in other countries places an unusual burden on the system. Historically, the answer to this problem has been to locate the design, development, and manufacturing facilities in the country in which the product will be marketed. Since the size of the market in many countries may not warrant the cost and duplication of facilities in all countries, the burden on "listening to customers" throughout the world becomes increasingly heavy and costly. It may even be desirable to restrict some of these functions to one location rather than to diversify.

The message that must be learned is that one must talk to the customer. It makes no sense to attempt to delegate the responsibility of talking with the customer to a single group in the company (e.g., the marketing group). The president, the product engineers, the manufacturing engineers, and the service personnel must all feel an urgency in listening to their customer and in translating the concerns and interests of the customer into actions that reflect their particular responsibility for the product. Personnel in each of these functions must be prepared to translate the comments that they hear into improved products or processes. Satisfying the customer must become the common objective of everyone in the organization.

Recalling again the broad definition of the term *customer*, the same admonition applies to the people who are serving another unit of the company rather than the ultimate customer. It is well for the internal groups to continue to ask:

- Are we doing things that help *you* do *your* job better?
- Are we able to do things for you that you cannot do for yourself?
- Does organizational separation of the function contribute to its success or hinder its success?
- Am I better able to serve others in the company because of the efforts that I am expending on *your* behalf?

10.6   THE *CUSTOMER* FOUNDATION OF WORLD CLASS PRACTICE

A key element in the effective management of the manufacturing system is the determination of your customers' needs and wants, whether those customers are internal to the organization or are purchasers of the final product or service. Three important steps have been identified in developing this understanding:

- Identify who your customer is and why they chose your product.
- Listen to what your customer is telling you about your product and its performance.
- Assess the opportunities that exist in the marketplace for new and improved products.

Accomplishing this is never straightforward and seldom easy. It requires the attention of all members of the system. The benefit to the manufacturer of accomplishing this successfully cannot, however, be overestimated. It is, after all, the principal objective of the manufacturing system to produce a product or a service that fulfills a need for their customers.

A manufacturing organization serves a variety of customers. In addition to the customers who expect to purchase high-quality products and services, the owners or stockholders may also be thought of as customers in that they expect a reasonable return on the investment that they have made in the company. The employees are customers in that they expect an employer to recognize their contribution to the success of the company and to provide them with a reasonable reward for their efforts. These are the stakeholders in the organization in that each has made a personal commitment to its success. The stakeholders have special expectation and needs that must be met.

FOUNDATION: World-class manufacturers instill and constantly reinforce within the organization the principle that the system and everyone in it must know their customers and must seek to satisfy the needs and wants of customers and other stakeholders. (Heim and Compton, 1992)

REFERENCES

CLARK, K. B. AND T. FUJIMOTO. 1991. *Product Development Performance: Strategy, Organization, and Management in the World Auto Industry*, Harvard Business School Press, Boston.

DRUCKER, P. 1985. *Innovation and Entrepreneurship: Practice and Principles*, Harper & Row, New York.

EDMONDSON, H. E. 1992. Customer Satisfaction, *Manufacturing Systems: Foundations of World Class Practice*, J. A. Heim and W.D. Compton (eds.) National Academy Press, Washington, DC:128-136.

Enterprise. 1993. Voice of the Market, July:22.

FOSTER, R. N. 1986. *Innovation: The Attacker's Advantage*, Summit Books, New York.

GOMORY, R. E. AND R. W. SCHMITT. 1988. Science and Product, *Science,* 240, May 27:1131.

HEIM, J.A. AND W. D. COMPTON (eds.) *Manufacturing Systems: Foundations of World-Class Practice*, National Academy Press, Washington, DC: 30-31.

PETERS, T. 1988. *Thriving on Chaos: Handbook for a Management Revolution*, Alfred A. Knopf, New York:46.

SMITH, P. G. AND D. G. REINERTSEN. 1991. *Developing Products in Half the Time*, Van Nostrand Reinhold, New York.

QUESTIONS

10.1 What are some of the bases that can be used to make trade-offs between conflicting wants and needs that have been identified for various customers of a product?

10.2 What are the trade-offs between providing a new product that will immediately "do the job" compared to a more ideal product that requires longer and more expense to bring to the market? Is it better to incorporate design changes into the product at some later date-as part of the product learning curve-or to delay the product's introduction date and include the changes before it appears in the marketplace?

10.3 For a very unique and new product, how high a priority should be placed on achieving high quality at the time of introduction of the product? What are the trade-offs?

10.4 What are the penalties for entering a market with a new product prematurely? To what extent can research and development costs be avoided by entering the market late? What are the risks in a late entry?

10.5 What are the legal ramifications of being a supplier of components and/or subassemblies to your competitor? How do you keep problems from developing?

10.6   Since the end use of many products is not well defined and will probably vary for each customer, how should the product specifications be set for the team that will be designing the product? If the end uses are not known precisely, how can the customers be clearly defined?

10.7   You are manager of one of several plants that manufacture components used in pumps. The pumps are designed and developed at the company's principal manufacturing center in Roanoke, Virginia. Since your products, rubber seals, are a critical component, you have created a technical capability for compounding rubber that can create final seals that have certain very desirable properties, such as being impervious to oils, having good frictional properties, and good temperature characteristics. Your plant, located in Davenport, Iowa, is currently operating at approximately 60% capacity. A manufacturer of pumps who is a competitor of your company, located in Cedar Rapids, Iowa, has come to you and asked you to supply seals for a new product that they expect to introduce in the next 18 months. The volume that they will need would bring your plant to approximately 90% capacity. To respond, what information will you need? How do you respond? Why?

10.8   As director of the corporate research function, you are expected to support the company efforts in product and process throughout the world. The corporation has three very distinct divisions. One division supplies the military, one the building industry, and the other the consumer. Who are your customers? How do you listen? How do you allocate your resources?

READINGS

EDMONDSON, H. E. 1992. Customer Satisfaction, *Manufacturing Systems: Foundations of World-Class Practice*, J. A. Heim and W.D. Compton (eds.) National Academy Press, Washington, DC:128-136.

GOMORY, R. E. AND R. W. SCHMITT. 1988. Science and Product, *Science,* 240, May 27:1131.

APPENDIX A10.1 THE LOGISTICAL GROWTH FUNCTION

In its simplest form, the logistical growth function describes the maturation of a system as it moves from the early development phase through rapid growth to the eventual level at which it becomes saturated. The characteristic shape of this function-looking a bit like a distorted S-was displayed in Figure 10-4. The concept of the logistical growth function was first used to describe biological systems (Lotka, 1956). The concept is most easily visualized by considering the number of bacteria that will exist at any time t in a Petri dish that contains a fixed amount of nutrient. With the introduction of one cell into the dish, growth begins and the cells divide and multiply. At a somewhat later time, the number of cells will have grown large, and as each divides, the total number will increase rapidly. Eventually, the population will become so large that the remaining supply of nutrient will be small and the rate of growth of the cells will slow. When the food supply has been exhausted, further cell growth will cease and the number of cells that exist in the dish will become fixed. Saturation in the population will have been achieved.

This formalism has been found to describe many physical systems, including the expansion of air travel, the length of surfaced roads, the length of oil and gas pipelines, the growth of plants, the introduction of new technological devices (e.g. ATMs), and the substitution of one form of energy for another. *The logistical growth function will describe the time behavior of any system that is moving toward saturation. An essential characteristic is the existence of a final state that is fixed in size.*

Consider a system that at time t has $N(t)$ units occupied. The saturation level at very large times is designated as N_∞. The fraction of units occupied at any time t is given by $f(t)$, where

$$f(t) = \frac{N(t)}{N_\infty} \tag{A10.1}$$

The fraction of empty sites that remain to be filled is $(1 - f)$. The rate at which f changes with time can be seen to be

$$\frac{df}{dt} = \alpha f(1 - f) \tag{A10.2}$$

Equation A10.2 can easily be integrated to give

$$1n \frac{f}{1 - f} = \alpha (t - t_0) \tag{A10.3}$$

where α and t_0 are constants of integration. This equation has the following characteristics. At $t = t_0, f = \frac{1}{2}$. For $t = 0, f$ will be very small with a value determined by the constants α and t_0. At very large values of t, f will approach unity. It is easily shown by taking the derivative of Equation A10.2 with respect to t and setting it equal to zero that the maximum rate of change of f occurs at $f = \frac{1}{2}$.

The number of elements occupied at any time t is found by substituting Equation (A10.1) into (A10.3), giving

$$N(t) = \frac{N_x}{1 + e^{-\alpha(t-t_0)}} \qquad (A10.4)$$

Equation A10.3 has a very simple form and can be plotted easily. If a series of values for f are known at times t, a plot of the quantity $f/(1 - f)$ versus time on a log-linear graph will yield a straight line. This can be seen using an example from plant growth. The observed height of sunflower plants was recorded as a function of time

Table A10.1

Days	Height (cm)
7	17.93
14	36.36
21	67.76
28	98.10
35	131.00
42	169.50
49	205.50
56	228.30
63	247.10
70	250.50
77	253.80
84	254.50

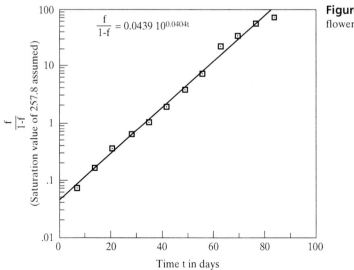

$$\frac{f}{1-f} = 0.0439 \; 10^{0.0404t}$$

Figure A10.1   Growth of sunflower seeds versus time.

and is given in Table A10.1 (Lotka, 1956:74). The graph, Figure A10.1, is the plot of $f/(1 - f)$ using an assumed value for the height at saturation of 257.8 cm. The goodness of fit is demonstrated by the correlation factor of 1.00, as shown on the graph.

For the case above, the determination of the saturation value was reasonably straightforward. It was easy to estimate since the total growth history was available for examination. For this formalism to be of value as a tool in estimating the ultimate size of a market or the extent to which a market is saturated, a means must be found to estimate the value of N_∞. This is accomplished by estimation and curve fitting.

To illustrate how limited market data can be used to accomplish this, suppose that the data available for the height of the sunflower existed only for the first 35 days. How could this be used to estimate the value of N_∞?

1. Plot the height versus days of growth. (see Figure A10.2)
2. Estimate the value of N at which the curve will saturate. Choose this for the initial estimate of N_∞. In this example, a reasonable choice might be $N_\infty = 400$ cm since saturation is not evident and the value of N_∞ must be substantially greater than 120 cm.
3. Calculate $f/(1 - f)$ using the assumed value of 400 cm and plot, as in Figure A10.3
4. The poor fit to a straight line indicates that an improper choice was made for the quantity N_∞.
5. Repeat the process with a smaller value of N_∞. Choose the largest value that is possible for N_∞ that gives a straight line and a good fit. This is the best approximation that can be found for N_∞ with the limited amount of data available. A curve for an assumed saturation value of 200 cm is shown in Figure A10.3. While a slightly larger value of N_∞ also gives a good fit, in this case 250 cm, the information that can be obtained using the more approximate value of $N_\infty = 200$ cm is quite good. A comparison of the values of $t_{10\%}$, $t_{50\%}$, and $t_{90\%}$

Figure A10.2   Sunflower plant growth.

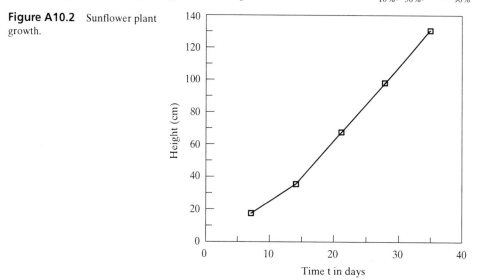

Height (cm) vs. Time t in days

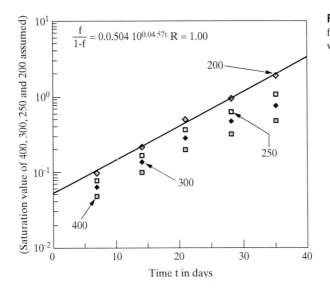

$$\frac{f}{1-f} = 0.0.504 \; 10^{0.04.57t} \; R = 1.00$$

(Saturation value of 400, 300, 250 and 200 assumed)

Time t in days

Figure A10.3   Logistic curves for sunflower plant growth using various assumed values of N_∞.

obtained using $N_\infty = 200$ cm with those obtained using the correct value of N_∞ = 257.8 cm is given below. As more data are obtained of the total population, refinements can be made in the estimate of N_∞ that is used, thus giving better approximations to the final description.

An identical procedure can be used to estimate the level of the total market available and the times after introduction at which various levels of the market will be achieved. The availability of early but good data are sufficient to obtain a reasonable early estimate of the extent of development of a market. This can be of immense value in the planning of actions. As was shown above, an approximate value of N∞ is

Figure A10.4   Logistic plot of the growth of the automatic teller machine market. (From Krolik, 1992.)

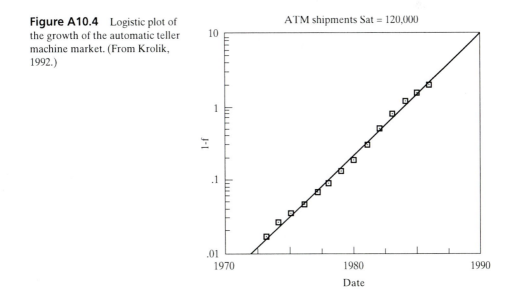

ATM shipments Sat = 120,000

1-f

Date

sufficient to give early indications of how the market will develop. The use of this technique is illustrated in Figure A10.4 for the recently developed market for automatic bank teller machines. This demonstrates very rapid saturation of the market, which had the strength to go from $t_{10\%}$ to $t_{90\%}$ in just over 10 years.

Modis and Debecker (1988) has used the technique of logistic curves to examine the development of various segments of the computer market. Figure A10.5 is a plot for companies A and B. Company B has produced two types of computers, occupying different market niches.

Figure A10.6 is a linear plot with a superimposed logistics fit for company E. Modis and Debecker also consider the market for PCs as a whole and provide the logistics analysis shown in Figure A10.7 If the logistics analysis is correct, Figure A10.7 would suggest that between 90 and 98% of all new models of PCs had been created by the year 1994.

It must be remembered that total saturation never occurs, in a mathematical sense, and that the last 10% of a large market can be very large. In the early stages

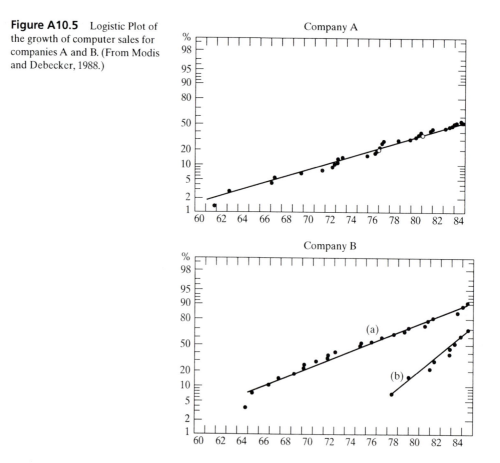

Figure A10.5   Logistic Plot of the growth of computer sales for companies A and B. (From Modis and Debecker, 1988.)

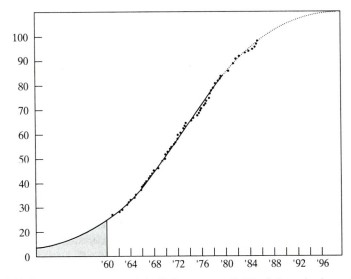

Figure A10.6   Data points and logistics fit for company E. The fit is on the data points to the end of 1979. (From Modis and Debecker, 1988.)

of the development of a market, it is often helpful to be able to bracket the value that will represent saturation. This can give a good indication of the importance that should be placed on entering the market, waiting for a later opportunity, or beginning the search for a revolutionary alternative.

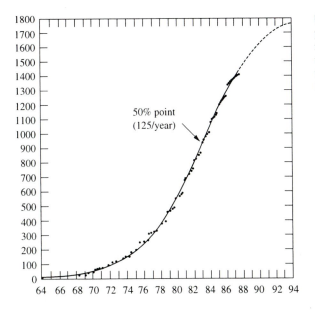

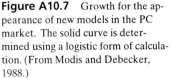

Figure A10.7   Growth for the appearance of new models in the PC market. The solid curve is determined using a logistic form of calculation. (From Modis and Debecker, 1988.)

REFERENCES

KROLIK, K. 1992. Personal communication.

LOTKA, A. J. 1956. *Elements of Mathematical Biology*, Dover Publications, Inc., New York:74.

MODIS, T. AND A. DEBECKER. 1988. "Innovation in the Computer Industry," *Technological Forecasting and Social Change*, 33:267.

APPENDIX A10.2 CASE STUDY: HARLEY-DAVIDSON, INC.

The name Harley-Davidson has always carried a certain mystique. It meant something that was adventurous, something that was flashy, even something that was raw. To its avid followers, it was an example of an American institution that had conceived of a unique product for those who were committed adventurers. Appearing first in 1903, the Harley motorcycle soon became a tradition. The early bikes were not flashy, but they were tough, one of the first going over 100,000 miles. By 1907 police were using them for patrols. They served as scout and dispatch vehicles in both world wars. When the Indian Motorcycle Manufacturing Co. ceased production of its motorcycle in 1953, Harley was the sole U.S. producer of motorcycles. It enjoyed the advantage of a cadre of devoted customers who contended that the characteristics of the Harley bike were so unique that they would never think of owning one made by any other manufacturer. It is hardly surprising that complacency would evolve when, as the only domestic manufacturer, you had 100 percent of the market.

It is this kind of circumstance that generated an opportunity for someone else and created a crisis for Harley. All of this started slowly, without much fanfare. In 1959, Honda Motor Company of Japan announced the introduction of a small motorcycle to the U.S. market. Not surprisingly, Harley was not only not worried, they were unimpressed. No one could expect a small motorcycle to compete with the large heavy bike that characterized the Harley, particularly a bike made by a Japanese company. The principal competitors in the late 1950s was not the Japanese but the British-made middle-weight motorcycles.

In response to the European imports, Harley introduced the Sportster in 1957, and in the early 1960s it introduced a midsize range with the help of an Italian subsidiary. The Sportster sold well and became one of Harley's most popular product-for a time. In this time frame Harley also added smaller bikes as they attempted to attract customers that were not wedded to the big bike. Most of these efforts failed.

Not only did the Harley mystique not carry over to the other segments of the bike market, they were having troubles in the market that they had "owned" since the beginning. The bikes they were making were low quality, they required frequent maintenance, they leaked oil, and their technological sophistication had stagnated. Dealers became disgruntled. The service network was slow and unresponsive. To make matters worse, manufacturing operations were inefficient and costly. All of this was occurring during a time in which ownership of the company had been assumed in 1969 by American Machine & Foundry.

Although the new ownership provided the capital that was necessary to double production by 1972, little attention was paid to solving more basic problems. The competition, realizing the opportunity that existed, blitzed the market with new products. Honda introduced the Gold Wing bike in 1975 as a direct competitor to Harley's big V-twin touring bike. It was the biggest, heaviest, most sophisticated bike that the Japanese had ever offered.

Harley-Davidson knew that it had to respond. They initiated two long-range programs. The first was a program to upgrade existing engines. The second was to develop a family of high-performance, technically superior engines that would compete directly with the Japanese engines. Recognizing that the company could not survive if it had to wait for these new products, the head of the styling department and grandson of one of the company founders, William G. Davidson, undertook to make the current product more acceptable to the new buyers. Here was a designer that mixed with his customers. Here was someone who loved motorcycles, loved to ride them, and loved to associate with other bike riders. He even looked the part, with beard, jeans and leathers. What Willie G., as he was known, knew is that bikers were ready to tell people what they liked about their product, if only someone would listen. And Willie G. listened.

As with many product trends, Willie discovered that bikers in California were sending some interesting messages. Bikers were regularly customizing their bikes. Taking the Harley's big touring bike, the Electra Glide, they were removing the saddle bags, windshield, and heavy front forks and replacing them with lighter, extended forks that moved the front wheel farther in front. By combining elements of two production bikes, the Sportster and the Electra Glide, Willie created a new bike that offered customers a factory-built bike that had all the features that the California bikers had been achieving by customizing. The Super Glide was the first of the customized factory-built bikes. It was followed by the Low Rider, basically a Super Glide with special paint, special trim, and a lower seat; the Wide Glide, a bobbed rear fender, stepped seat, extended forks, and high bars; the Fat Bob, a Low Rider bike with spoked wheels and higher bars; and the Wide Glide. The inspiration for the Wide Glide again came from observing the California customizations. This was followed by the Softail, which has been described as the ultimate custom bike. Willie G. had created genuine excitement in the marketplace with designs that changed production products cosmetically to satisfy customer wants.

While all of these new products were important, other fundamental problems were not being solved. The competition from the Japanese had become severe. By 1980 the company's share of the super-heavyweight market had dropped to 31% from almost 80% in 1973. Honda, Suzuki, and Nissan were all offering products that competed directly with the large Harley bike. Moreover, the small Harley bikes were not being well received. Complicating the situation was the decision by American Machine & Foundry to divest Harley. Harley-Davidson became a privately owned company when 13 Harley executives arranged a leverage buyout in 1981, almost coincident with the large market drop in motorcycle sales that occurred in 1981.

The situation was grim. The new management recognized that many things must change if they were to survive. Quality was at the top of the list. Through attention to detail, adoption of employee involvement, and acceptance that change must begin with changed attitudes by the upper management, the manufacturing system was able to provide high-quality products at an acceptable production level at a competitive cost. The competitive challenge was the next priority. With assistance from the government, Harley was able to obtain a temporary reprieve from the Japanese. Supported by owners, dealers, and friends, Harley was given a favorable ruling in 1983 by the International Trade Commission, which imposed heavy import tariffs on Japanese bikes having engine displacements greater than 700 cubic centimeters. This offered a breather to the company.

Next, the company had to focus on those products that had the greatest appeal to customers. The time was past when customers would buy anything offered. They had alternatives and they exercised them by buying competitors' products. How could it listen better to its customers? The answer was truly innovative. Harley created the Harley Owners Group–the H.O.G.–strikingly similar to the nickname for the large bike-the HOG. H.O.G. is created to develop a long-term bond between the company and owners. Organized rides, motorcycle events, and a bimonthly publication, *Hog Tales*, provides customers with ways to know what is going on and to respond. It promotes a feeling of belonging to a family. With approximately 200,000 members, the H.O.G. provides an informal atmosphere for owners and dealers to meet company executives at rides and to tell them what they like about the product. That is why you regularly find Richard F. Teerlink, president and CEO, and his wife Ann, riding into a rally on a new Ultra Glide to hold a "town meeting" and hear what customers are saying. And this is not an isolated event. All Harley executives actively participate in rallies around the country.

Harley was also successful in broadening its customer base. As a company often associated with a fringe-the Hell's Angels and hippies-it has managed to retain the mystique of its product while appealing to a much broader population. About half are supervisors, machine operators, and other skilled workers. The other half are mostly lawyers, doctors, bankers, entertainers, engineers, and scientists. Harley has even established a 'Ladies of Harley' club to encourage more women to ride. Harley now competes in only the big-bike market segment-doing what it does best.

Harley's story is one of successful recreation. Having flirted with bankruptcy in December 1985, the company returned to profitability. In 1994 it reported net income of $104 million on sales of $1.54 billion. Only once in the previous 10 years has it failed to report a profit. It has captured 60% of the big-bike market, up from 23% in 1983. Its current market share is limited by its capacity. Harley also believes that customers deserve fair treatment in pricing. Even though Harley cannot supply all of the demand for its products, Teerlink makes no apology to the financial community when it comes to pricing policy. He says: "We will not compromise quality for quantity. Nor will we abuse our customers by limiting demand and charging excessive prices, because when the company was down our customers were there. The family was there. You don't abuse the family."

SOURCES

MOSKAL, BRIAN S. 1993. Born to Be Real, *Industry Week,* Aug.2,14.

REID, PETER C. 1990. *Well Made in America: Lessons from Harley-Davidson on Being the Best,* McGraw-Hill Book Company, New York.

ROSE, ROBERT L. 1990. Vrooming Back: After Nearly Stalling, Harley-Davidson Finds New Crowd of Riders; It Sheds Hell's Angels Image, Markets to Professionals and Aging Baby Boomers Biker Night at Bloomingdales, *Wall Street Journal,* Aug. 31 Sec. A:1.

ITEMS FOR DISCUSSION

1. The fraction of the people who own large motorcycles for their utilitarian value, as a source of transportation, is very small. People ride them for the joy of belonging and to experience the excitement of the ride. Can confusion exist between the needs and wants of customers for a product such as this? Why?

2. As the population ages, will the marketing strategy for the large bike need to change? If you were the president and CEO of Harley-Davidson, what actions might you take to protect against the demographic changes of the population? How would you approach these options?

3. What unique features make the H.O.G. concept successful? Is this concept applicable to other companies? Why? If so, would it need modification? How? If not, could it be changed in ways that would still make it a useful tool for communication with customers?

4. Even though Harley-Davidson had a near monopoly on the market in the late 1950s, it was nearly forced into bankruptcy in the middle 1980s. How could it have so misread the events that were taking place in the marketplace? What are the principal ways by which a company can prevent missing the importance of seemingly small changes in the marketplace that turn out to have been the early development of monumental changes?

5. What events are occurring today that might affect the dominance of the markets enjoyed today by Federal Express and Boeing? If you were CEO of one of these companies, what would you do about it?

11

Translating Customer Wants and Needs into a Product or Process

Having obtained an indication from the customer of their wants and needs, the challenge to the *system* is to translate these into a product, process, or service that has the requisite attributes. In addition, it must accomplish this in an expeditious manner if the product is to be brought to market promptly and successfully. The first challenge is to interpret the customers' identification of needs. What are the true needs and wants of customers in terms of the reliability of the product? How important is it that the product never fail versus the ease with which it can be repaired? How important is the "image" of the product? How does this product relate to other products being offered by the manufacturer? What is the importance of the product retaining a high resale value when it is replaced by a new product? Answers to these questions lead to the creation of general specifications for the product. From these, the engineering activity will create a design that culminates in a prototype of the product. This must be tested to determine its compliance with the stated performance criteria. If modifications are needed, as is often the case, a partial repeat in the development cycle will be required. Having accomplished the development of a product that will meet the needs and wants of customers, it must now be produced in a cost-effective manner and at the highest possible quality level. Effective implementation of the manufacturing processes now becomes a high priority.

Clearly, there is no clear demarcation among the various steps in this total chain of events. Each depends on the other. The discussion that follows explores many of the issues more fully and focuses on the tools that have been shown to be successful in enhancing the efficiency with which the product realization process can be accomplished. The intent is to develop the best product possible, in the shortest time, that will have the lowest cost.

11.1   THE PRODUCT REALIZATION PROCESS

The product realization process encompasses the entire spectrum of activities beginning with the initial idea and concluding with the delivery of the product to the
customer. Each step of the process contains risks, the magnitude of which depend on
the level of uncertainty that is involved and the level of experience that people have
in solving the problems.

The first and, arguably, one of the most important decisions that must be made
is the clear identification of project objectives. Smith and Reinertsen (1991:43) emphasize the importance of establishing clear objectives.

> We have repeatedly found rich opportunities to save time at the beginning of the de-
> velopment cycle. These opportunities deserve attention because, of all the actions de-
> scribed in this book, actions taken at the "fuzzy front end" give the greatest time savings
> for the least expense.
>
> The true cost of this phase is usually many times higher than managers suspect. In this
> phase the most important influence on cost is the cost of delay, not of the manpower as-
> signed to the project. The calculated cost of delay is often 500 to 5,000 times higher than
> the visible costs of assigned personnel. Managers unaware of these costs will tend to ig-
> nore the "fuzzy front end." Those who understand these costs will instead focus a great
> deal of attention on this phase.

Bowen (1992:94,96) has also the emphasized the importance of the early phases of the product design process and the impact that early decisions have on the
total cost of the project.

> The starting point, $t=0$, of a commercialization project occurs after initial research and
> development have determined a significant value to, and acceptable risk for, moving the
> development into an implementation or commercialization phase. The technical feasi-
> bility of the product or process has been verified, but there are still uncertainties that re-
> quire refinements, integration functions, and pilot developments. The product gets a title,
> resources are allocated, and someone is assigned to track the expenditures, most often
> there is a carefully constructed schedule with gates, hurdles, or phase reviews.
>
> A slow ramping of expended dollars or engineering hours occurs during the initial 40-
> 70% of the elapsed time. The early period is mostly expenditures of people on design and
> primitive prototypes. The later rapid rise is associated with purchases and installation of
> equipment and facilities and the training of operators. The rates are quite different in
> these two periods. The teams involved in these projects perceive the actual time of the
> decisions that triggered the expenditures as occurring very early in the process.... I have
> found that within 15-20% of the total elapsed time, 80-85% of the key decisions about
> future expenditures have occurred. Complex processes involving numerous variables
> and elements of subsystems (such as information, technology, human, financial, and mar-
> keting) result in longer than anticipated execution consequences and, thus, strongly in-
> fluence feedback loops in the manufacturing system. We will call this observed
> relationship between essentially sunk costs and decisions the 15/85 rule.
>
> The 15/85 rule poses many questions about how projects are managed:
>
> 1.   When is the appropriate time for senior management to become involved in new
>   product and process projects?

2.   What should the leadership of such teams be, and are there opportunities to change leaders midstream?

3.   How much of the loss due to engineering change orders, redesigns, and the like is due to the long cycle times (decisions to implementation), and how much is a result of uncertainty about the component part and system that is being designed, built, or implemented?

4.   What is the value of knowledge (know-how and know-why) at the early stages, that is, before 15 percent of the time has elapsed?

5.   How should project teams be staffed, organized, and managed to promote organization learning as well as to accomplish project objectives?

Bowen illustrates the importance of these factors with the representation shown in Figure 11.1.

The issues addressed by Smith and Reinertsen and by Bowen must be addressed each time a new product or new process is to be developed. The answers to these questions depend on the extent to which the new product or process is revolutionary as opposed to evolving from a current product. The more revolutionary the product or process, the more difficult it is to assure that all the key questions have been asked, let alone all answered. In such circumstances, the value of a team approach to solving these problems is greatly enhanced. For a modest improvement in an existing product or process, it may be possible to proceed with a more limited effort and with the responsibility for implementation being kept within a local group.

As an example, consider the processes that are to be used in the manufacture of the product. If the processes are all new, adequate opportunity must have been provided the manufacturing group to experiment with them and to convince themselves that the processes are workable and are ready for full-scale production. If only modest changes are needed in existing practice, a small number of tests will often suffice to show feasibility. If the process that is expected to be used for production is already in use somewhere–but is new to the enterprise–it is important to seek assistance from those who have experience with it. This would probably start with vendors who have provided the facility or the machines for implementing it. If the new process is owned by a competitor, it may be appropriate to approach them and inquire whether they would be willing to license–if it is patented–or to sell rights to the development. A natural approach is to offer to trade some of your technology for theirs. These discussions must be initiated early enough to ensure that the information will be available and understood in time to incorporate it into the planning for the new facility.

Many of the questions that arise can be answered if sufficient time and resources are available. It must be recognized, however, that the window of opportunity for the introduction of new technology may be quite small (see Section 10.3.1). When the date for the release of purchase orders for the manufacturing equipment arrives, it is important that answers to most, if not all, of the questions above be available. It is critical, therefore, that the prove-out and validation effort be initiated sufficiently early that the manufacturing activity can be confident that the risks associated with the introduction of the new technology will be limited. As Bowen (1992:98) has observed:

Figure 11.1 Schematic relationship between project expenditures and the decisions about the product or process. (From Bowen, 1992, p.95.)

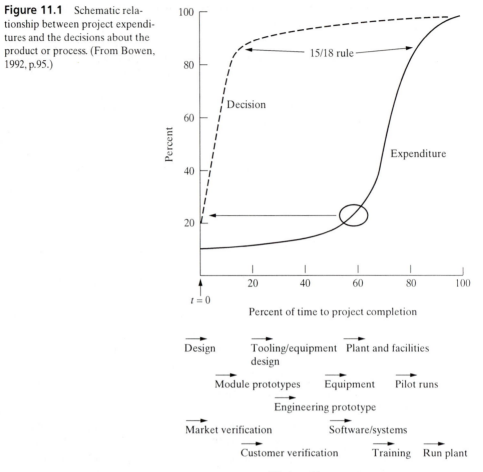

What is the value of the *right knowledge* early in the commercialization project? Extensive research will be required to quantify the answer to this question, but the nature of the empirical observations represented in the figures suggests multiples of 10, and perhaps as much as 100 to 1000, in savings. The cost and consequences of early, ill-informed decisions put into place processes (expenditures) that can be measured and judged only much later.

A Dataquest, Inc. study arrives at similar assessments. In examining the cost for a change that is made at various stages in the development of major electronic products, they estimate that the cost of a change often increases by a factor of 10 at each of the following stages:

- During design
- During design testing
- During process planning
- During test production
- During final production

In other words, a change during final production can be expected to cost 10^4 times as much as a change made during design.

The traditional system of accomplishing the product realization process is shown in Figure 11.2. Traditionally, this has been accomplished by a process that involves the segmentation of the task into discrete parts. Activities (e.g. departments) were assigned responsibility for the task that fell within their area of expertise. The design was accomplished sequentially, with each group completing its task and handing the result of their work off to the next group–a process that has been described as "throwing the design over the wall." This procedure not only segmented responsibility, it also created a segmentation of understanding, in that no group had access or was expected to be familiar with the problems and constraints within which each of the other groups worked. Thus it was not uncommon to find that decisions made by one group created serious problems for other groups. Time and effort were required to resolve the resulting conflicts and discrepancies.

Heavy boxes drawn around traditional organizational activities are meant to emphasize the independence that the individual groups exercise in the process. The light outline around several of the groups (those at the top of the figure) indicates that the individual groups generally acknowledge a tenuous responsibility to the total group but that their responsibilities continue to be focused on their respective organizations. Upon completion the design that is released to the manufacturing activity is expected to be adequate for the planning of the manufacturing system, for the procurement of facilities and for the implementation of the production system. If the design placed demands on the manufacturing group that were unachievable, could not be accomplished with the planned facilities, or resulted in an unacceptable cost, the individual groups were called on to modify their earlier design.

The problems that can arise from the sequential nature of this process are obvious. Companies were prepared to live with these limitations under circumstances that the time spent in the product realization process was not a critical element of success or when all competitors were using the same procedure to bring products to the marketplace. The consequences of accepting this procedure became intolerable, however, when U.S. companies were confronted with products from Japanese companies that were more optimally designed, completed in less time, and had higher quality and lower costs. In examining the practices of Japanese companies, it became clear that they were not following a series of sequential steps in their design practice. Perhaps influenced by a cultural heritage that emphasizes cooperation and consensus, the Japanese had developed a process that involved a highly *cooperative* effort among the various *groups* who were involved in all aspects of the *design and manufacture* of the product. The working team membership came from *all* of the interested activities. This allowed design decisions to be reached by a consensus of experts. This process has come to be known as the *concurrent engineering* and/or *simultaneous engineering* process.

11.2   CONCURRENT/SIMULTANEOUS ENGINEERING

11.2.1   A Description of the Process

The task to be accomplished by the concurrent engineering process is no different than that for the traditional development process–the design of a product that meets the needs of the customer, that is functional, that is manufacturable, that is easily assembled, that can be serviced easily, that is of high quality, and that meets the cost objectives that have been set for it. While the concurrent engineering process contains many of the same elements and requires that many of the same questions be addressed and solved, there are significant differences in the approach that is used and, consequentially, in the results that are achieved.

First, the concurrent engineering process recognizes the overarching importance of the customer, thus placing their wants and needs at the top of the process, as shown in Figure 11.3. Second, while each group must perform many of the same functions, the concurrent engineering approach requires that the individual groups focus on the final product rather than on the completion of their individual tasks. Thus, the heavy box encircles the system rather than focusing on each individual organizational group. Third, planning the manufacturing facility and the processes is an integral part of the concurrent engineering process. Whereas the release of the design in the traditional

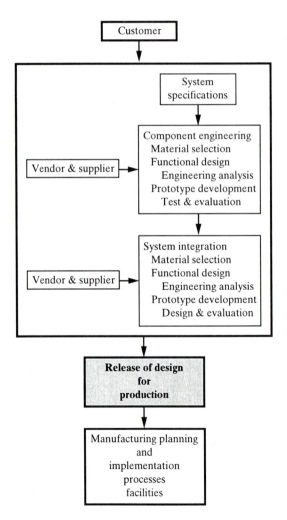

Figure 11.2 Traditional steps in the creation of a new product.

procedure is accomplished without the advice and consent of the manufacturing group, the manufacturing activity is an integral part of the design activity in the concurrent engineering approach. In this procedure, the release of a design for production cannot occur without the involvement of the manufacturing activities.

A central element of the concurrent engineering process is that the design is accomplished by a *team* that includes *members* from *all elements of the system:* marketing, sales, design engineers, manufacturing engineers, service engineers, finance, and so on. Each member brings to the process the expertise of their discipline/activity, while working cooperatively to help accomplish a design that is optimal for the *system.* Each is expected to participate in the design process in a way that assures that the final design will reflect the combined thinking of the team. To be effective, the team must be committed to working on all elements of the design, from initial concept to

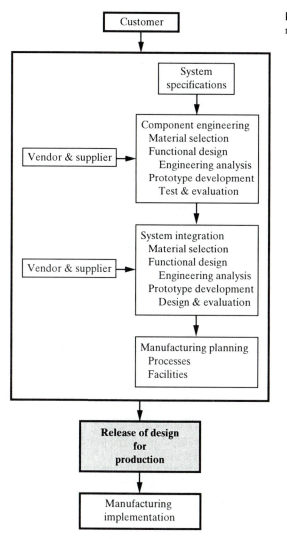

Figure 11.3 Steps in the creation of a new product using concurrent engineering.

manufacture, in a cooperative and committed fashion. The objective is to accomplish a design that will provide a product that has the desired attributes, that can easily be manufactured and serviced, and that will meet the quality and cost objectives that have been established. In short, the team seeks to optimize the *life cycle* of the product, from initial concept to final disposal.

A variety of definitions of the concurrent engineering process have been offered by various authors. The following are representative:

> In Concurrent Engineering, the key ingredient is teamwork. People from many departments collaborate over the life of a product–from idea to obsolescence–to ensure that it reflects customers' needs and desires. Marketing, engineering, and manufacturing, for example, work together from the outset to anticipate problems and bottlenecks and to eliminate them early on. In so doing, they avoid delays in bringing the product to market and costly failures in service. Accounting and purchasing have their say, too, and help to ensure low product cost and reliable supplies of parts and materials. (Rosenblatt and Watson,1991)
>
> *Concurrent Engineering* (CE) is defined as the earliest possible integration of the overall company's knowledge, resources, and experience in design, development, marketing, manufacturing, and sales into creating successful new products, with high quality and low cost, while meeting customer expectations.
>
> The most important result of applying CE is the shortening of the product concept, design, and development process from a serial to a parallel one.... (Shina, 1991a:1)
>
> ...a systematic approach to the integrated, concurrent design of products and their related processes, including manufacturing and support. This approach is intended to cause the developers, from the outset, to consider all elements of the product life cycle from conception through disposal, including quality, cost, schedule, and user requirements. (Winner et al., 1988:2)

Simultaneity and concurrency refer to the creation of the product design and the manufacturing system in a unified process, in contrast to the more traditional process, in which the design was completed and then passed along to the manufacturing group to create the manufacturing system that could make it.

11.2.2   Benefits of the Concurrent Engineering Process

Although more time and human energy are often required to complete the initial step in the simultaneous/concurrent process, experience has shown that the process will generate a design that requires fewer changes, the final product will be more economical, and overall, less time will be consumed between the start of the initial concept and introduction to the marketplace (McKnight and Jackson, 1989). Following are some of the statistics that have been published concerning these benefits:

> CE has...scored well at Mercury Computers System Inc, a Lowell, Mass.-based maker of add-on processor boards for VME-bus. The company was able to shorten appreciably the cycle for shipping a new board from design approval to a customer for testing down to just 90 days instead of 125 days it would otherwise have taken. (Shina, 1991b:23)
>
> Production ramp-up to high volumes for a line of medical electronic instruments was reduced from eight months to one month. The average number of engineering changes made within six months of release of an electronic assembly dropped from 3.7 in 1983

to 1.7 in 1985. The design iterations per circuit board design dropped from 5.8 in 1982 to 4.6 in 1986. (Shina, 1991b:23-25)

A case in point is Hewlett-Packard's 54600 oscilloscope...from idea to finished product, it took us about one-third the time to complete this project than it would have without CE. (Wheeler,1991)

Hauser and Clausing (1988) indicate that production costs at Toyota were reduced by 61% from earlier levels through application of CE. The responsiveness of the system was also found to be enhanced. As experience with concurrent engineering is achieved, the ramp-up of production to high volumes is achieved in less time, as illustrated in Figure 11.4 for a line of medical instruments. Other studies have indicated that concurrent engineering has resulted in 30 to 70% less development time, 65 to 90% fewer engineering changes, 20 to 90% less time to reach the market with a new product, and 200 to 600% better quality. All the relevant metrics that have been used to characterize the concurrent/simultaneous engineering process indicate that great value can be gained with it.

11.2.3 Implementing the Concurrent Engineering Process

It is not easy to implement the concurrent engineering process. The following observations concern the effort needed to make this an effective procedure:

> Starting and sustaining CE is not easy. It takes dedication and discipline, as well as a sweeping cultural change. Out goes the typical short-term business focus. In its place, a world-class CE culture zeros in on continuous improvement. It relies on continuous improvement. It relies heavily on teamwork among all employees connected with a product's development, plus close relations with customers and suppliers.

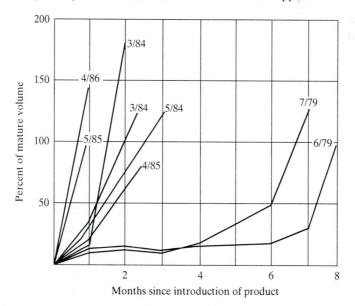

Figure 11.4 Time after introduction that high volume was achieved following experience with concurrent engineering. (From Shina, 1991b, p.23.)

Implementing concurrent engineering... takes commitment by the entire organization, from top to bottom. It takes education at all levels so that everyone speaks a common language. It takes repetition: managers and engineers must constantly impress upon their peers, and those above and below them, their commitment to CE.

Another nearly universal problem in getting CE under way is that the members of the product development team are not all in the same place and so are unable to communicate as fully as desirable. The best solution is to locate the members of the team near each other. This encourages ongoing communications–that is, not just at formal management and design reviews. It also fosters a group spirit of responsibility for seeing that the product gets to market as quickly as possible with the least amount of trouble....If the team members cannot all be in close proximity, they should at least have electronic communication with each other via an electronic network or commercial e-mail service. (Turino, 1991)

When implementing concurrent engineering...engineers should be made aware of the values of teamwork, the sharing of ideas and goals beyond their immediate assignments and departmental loyalties. These are skills that are not taught to engineers in their formal education in U.S. technical colleges and universities, and which have to be reinforced by having these characteristics valued just as highly as the traditional engineering attributes of technical competence and creativity. The successful new product interdisciplinary teams are the ones that are focused on aggressive but achievable goals for concurrent engineering and design for manufacture. The characteristics of teamwork and cooperation can be rewarded by making them an integral part of the performance evaluation process for engineers. Shina (1991a:3)

Earlier chapters of this book concerning team building, appraisal of team performance, and the management commitment to the process will, hopefully, contribute to a reduction in the educational void that Shina has identified.

11.3 COMMUNICATING CUSTOMER WANTS AND NEEDS TO THE SYSTEM

It would seem straightforward to communicate the needs and wants of customers to people throughout the manufacturing system. Unfortunately, this is not always the case, for as described earlier for quality, the language used by the consumers is often not compatible with the engineering terms that are used to provide specifications or describe attributes. The consumer will usually describe the attributes that they desire in terms of their needs (e.g., higher quality, good feel, ease of lifting, or sturdy). To the engineers who must design and manufacture the product, these can mean various things. For example, does good feel mean a fine texture or good balance? Does ease of lifting mean low weight or readily accessible handles? Does sturdiness denote a product that is not easily upset or one that is not damaged by dropping? It is obvious that the interpretation that is given to the terms that are frequently used by the customer can greatly influence the attributes contained in the product.

In section 11.4, we discuss an important procedure by which the desires of the consumer can be translated into terms that are amenable to engineering interpretation and will result in a product that meets customer expectations. Although this procedure is focused on achieving a product of high quality, it is equally applicable to other important product attributes. Other tools that are important for accomplishing these objectives–design for manufacturability and design for assembly–are also discussed.

11.4 TOOLS FOR ACCOMPLISHING CONCURRENT ENGINEERING

11.4.1 Quality Function Deployment

Quality function deployment (QFD) is a *procedure* by which the customers' wants and needs are spread throughout all elements of the organization to ensure that the final product satisfies these wants and needs. The concept of QFD was introduced in Japan in 1969 by Katsukichi Ishihari. It was later developed for U.S. manufacturers by Sullivan (1986) and by Hauser and Clausing (1988). Sullivan describes QFD as "a system to assure that customer needs drive the product design and production process." It is a process by which customer requirements are translated into the appropriate technical requirements of the product and into a manufacturing process that is capable of creating the product. Hauser and Clausing (1988) described QFD as

> A set of planning and communication routines, quality function deployment focuses and coordinates skills within an organization, first to design, then to manufacture and market goods that customers want to purchase and will continue to purchase. The foundation... is the belief that products should be designed to reflect customers' desires and tastes–so marketing people, design engineers, and manufacturing staff must work closely together from the time a product is first conceived.
> [QFD] is a kind of conceptual map that provides the means for interfunctional planning and communications. People with different problems and responsibilities can thrash out design priorities while referring to patterns of evidence ...

In discussing quality, it must be presumed that the customer wants a product of high quality–the user-based approach to quality. The enterprise seeks to design the product so that the attributes of the product satisfy the desires of the customer–the product-based approach to quality. The manufacturing activity has as its objective the production of a product that meets the design objectives–the manufacturing-based approach to quality. All of these actions are taken within a context that certain cost objectives must be achieved if the product is to be successful in the marketplace–the value-based approach to quality. As the various steps in the QFD process are addressed, the team members must remind themselves constantly of the definition of quality that is relevant to the stage of the process being considered, for the focus will change as the process proceeds from product design through process design.

QFD has been used with products and processes as diverse as a new automobile (Sullivan, 1988; Hauser and Clausing, 1988), the advance launch system cryogenic tanks for the *Challenger* spacecraft (Juarez, 1990), and wave solder processes in integrated-circuit manufacture (Shina, 1991a). The QFD procedure is equally applicable to the optimal development of a service function (e.g., the design of a library system, the providing of fast food, or the creation of an efficient traffic control system), as it is to the creation of a manufactured product. The term *quality* can be replaced with other desirable attributes and the procedure will also apply, although a focus on something other than quality may not lead to as broad an impact on the enterprise as is likely or desirable with an emphasis on quality. It is possible, for example, to use this procedure to treat costs, or reliability, or serviceability. While these characteristics might be considered as being contained within the broad rubric of "quality," a specific focus on any one of them might yield somewhat different results. In other words, QFD is very general and can easily be adjusted to many applications.

The QFD process can be visualized as follows. It begins with the creation of a list of customer wants. The second step involves a translation of these customer wants into a list of engineering requirements that must be met to achieve the wants and needs. This might be viewed as preparing a listing of "what" is wanted and a corresponding list of "how" the wants are to be achieved. It is convenient to present these in a simple matrix, as illustrated by Figure 11.5. This can be generalized into a matrix of what versus how.

For the case that the new product is a new briefcase these could include the following items:

Customer Wants–the "Wants"	Engineering Requirements–the "Hows"
Prestigious Appearance	Leather or Stylized Vinyl
Able to Hold Legal Size Documents	11" x 16" Dimensions
Storage Compartments in the Interior	Storage for Sheets of Documents/Computer and Card Storage
Security of Contents	Combination Locks
Sturdy Construction	Reinforced Handle and Body

As is frequently the case, there is not a one-to-one correspondence between the engineering requirements and the customer wants. For example, can a prestigious-looking design be accomplished using stylized vinyl and an expandable structure? Some customer requirements can be affected negatively by some of the engineering requirements: for example, what will be the impact on cost of using leather? It is easy to visualize these dependencies by the matrices shown in Figure 11.5.

The conclusion of the QFD process is identification of the manufacturing processes that are to be used. Since all the relationships that exist are difficult to capture on a single matrix, it is usually necessary to create a sequence of matrices. These can be visualized as shown in Figure 11.6, where the "how" of matrix I becomes the "what" of matrix II, and so on. The process is repeated as often as necessary to complete the design and to provide the manufacturing details. In the example above the engineering requirements for the briefcase would be the "what" of the second matrix and would indicate how items such as expandability are to be accomplished.

Sullivan identifies a series of steps that need to be taken in order to implement quality function deployment. Although these steps are simplified somewhat in this book, the essence of the procedure is retained. Sullivan (1986:40) describes this cascading of the process in the following way:

The QFD system concept is based on four key documents.

1. *Overall customer requirement planning matrix*: translates the voice of the customer into counterpart control characteristics; that is, it provides a way of turning general customer requirements–drawn from market evaluations, comparisons with competition, and marketing plans–into specified final product control characteristics.

2. *Final product characteristic deployment matrix*: translates the output of the planning matrix–that is, the final product control characteristics–into critical component characteristics.

3. *Process plan and quality control charts*: identify critical product and process parameters, as well as control or check points for each of those parameters.

Figure 11.5 Matrix relationship among customer wants and engineering requirements. This is generalized into a matrix between "what" is desired and "how" these desirable attributes are to be achieved.

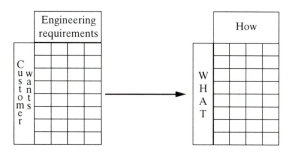

4. *Operating instructions*: are based on the critical product and process parameters; these instructions identify operations to be performed by plant personnel to assure that important parameters are achieved.

The example that is used in this book was chosen because of its probable interest to students–namely, the creation of a textbook. Much of the content of this example was developed by teams of students. This example presents an interesting issue. The person who chooses the textbook for the course–the faculty member–is not the person who purchases the product–the student. The apparent customer and the user, using Edmondson's distinction, are not the same. Although it would be hoped that the two customers would have common wants and needs, this might be as untrue for the textbook as it was for dog food (Chapter 9). For the sake of simplicity, however, we assume that the textbook is developed by the publisher with the student as the ultimate customer.

The process begins with a list of the project objectives, including such items as a targeted cost for the book, the date of publication, and the need to use existing or to create all new printing facilities. These must be provided or concurred in actively by management of the publishing house or the division that is responsible for the book. Next those who will be responsible for carrying this process from beginning to end must be identified. For the textbook, the representatives might be a student, a professor who would be using the textbook, the author, a representative of the publisher's marketing department, a person from graphics layout, the technical editor, the series editor, the production editor, compositor and the printer. The identification of the team members is a critical first step in accomplishing the process in an effective manner.

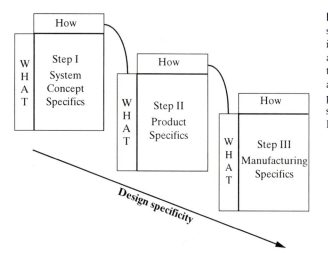

Figure 11.6 Cascading the design through its various steps, starting with the initial customer wants and concluding with the identification of the specific processes that are to be used to manufacture the product. (Adapted from material supplied by the American Supplier Institute, Detroit.)

The process requires that consensus be reached by the team. In accomplishing this the team uses a set of charts, with each completed chart serving as the basis for subsequent charts. The intent is that all segments of the organization, through their team representatives, will participate actively in the decisions that lead to the final product. The team's efforts begins by identifying the customers wants and concludes with the identification of a manufacturing system that will assure that the resulting product has the desired quality. The steps are approximately as follows:

1.   *Identify the set of customer requirements believed to be critical to the customer.* Use terms that the customer normally uses to describe the important product attributes. Market research will have been the source of these requirements. These are listed on the left side of Fig. 11.7. A numerical value–on a scale of 1 to 10–is assigned to each of the requirements based on the importance that the customer currently gives to these in the marketplace and based on the fact that these requirements appear in best-of-class competing products. The intent is to assure the team that they have identified a proper set of requirements. If, for example, it is found that the customer is buying products in the marketplace or that the leading manufacturer does not offer a product that satisfies one or more of these requirements, the team will need to reevaluate the appropriateness of including these items on the list. If the plan is to evaluate the requirements for a product that is not currently in the marketplace, the product most similar to that which is available can be used. Sum these evaluations and enter them as a numerical ranking. Using these numerical rankings, identify the most important customer requirement for this product and rank order all the others relative to it. The order is entered into the final column of Fig. 11.7.

2.   *Create an influence matrix of customer requirements.* The Customer Requirements identified in Fig. 11.7 are transferred to the left-hand side of Figure 11.8 and listed according to their rank ordering. In this example, a decision was made that the two lowest-ranking customer requirements–freedom from errors and good resale value–would not be carried forward to Fig. 11.8. It is very important that *all members of the team*

	Evaluation of requirements		Relative numerical ranking of requirements	Rank ordering
	According to customer ranking (1-10)	According to best-in class competitor (1-10)		
Easy to read	9	9	18	2
Easy to understand	10	9	19	1
Easy to carry/handle	6	7	13	5
Good aesthetics	4	8	12	5
Content relevant to course	9	6	15	3
Reasonable price	7	7	14	4
Good resale value	5	5	10	8
Freedom from errors	5	6	11	7

Customer requirements (Student)

Figure 11.7   Customer requirements. Competitive evaluations and rank ordering (example-textbook).

agree whether an item should be removed from the list or carried along to the next level. The product characteristics that are important in meeting each product requirement are listed across the top of Fig. 11.8. To the extent possible, the various product requirements should be independent of each other. Note that certain product characteristics may influence more than one customer characteristic. A value between 0 and 5, indicating the importance of the product characteristic in meeting the customer requirements, is assigned to each characteristic. The vertical columns of Fig. 11.8 are summed to provide a numerical measure of the importance of each product characteristic. It is likely that some characteristics will have a strong influence while others will have a minor influence.

At this point the team will have generated a list of important product characteristics that should ensure that a product containing these characteristics will satisfy the customer requirements that were listed. The team should, at this time, review the two lists carefully to ensure that nothing important has been omitted.

3.   *Develop the critical determinants of characteristics.* The critical product characteristics are transferred from the top of Fig. 11.8 to the left-hand side of Fig. 11.9 and are listed in order of priority–the highest value that appeared on the bottom line of Fig. 11.8 represents the highest-priority product characteristic. Those items having low importance may be omitted, again with the agreement of all team members. To the extent possible, a numerical objective is identified for each product characteristic. This should represent the value of this characteristic that the team feels must be achieved to meet the customer requirement. For example, it might be established as being 110% of the best current product. If the characteristic relates to quality, it might represent the desired failure rate in use, or the mean time to repair,

PRODUCT CHARACTERISTICS
(Assign a value between 0 and 5 indicating importance in meeting each customer requirement—5 is of highest value)

	Order		Currency of material	Organization of material	Technical content	Literary style	References	Diagrams and tables	Index	Examples	Problems	Appearance	Weight	Dimensions		
CUSTOMER REQUIREMENTS (Student)	1	Easy to understand	3	5	4	5	1	4	4	5	1	0	0	0		
	2	Easy to read	2	4	3	5	0	4	2	2	0	3	2	2		
	3	Content relevant to course	5	0	4	0	2	2	0	4	3	0	0	0		
	4	Reasonable price	3	0	0	0	1	1	2	0	0	1	2	1		
	5	Easy to carry/handle	0	0	0	0	0	0	0	0	0	0	5	5		
	6	Good aesthetics	0	2	0	0	0	2	0	0	0	3	0	2		
	7															
		SUMMATION of IMPORTANCE RATINGS	13	11	11	10	4	13	8	11	4	7	9	10		

Figure 11.8   Influence matrix of customer requirements and product characteristics (example-textbook).

or the number of "things gone wrong" in the first one, two, or five years of ownership, or for the textbook, the number of misspellings per page of text, and so on. The critical determinants for meeting all the product characteristics are identified. These are listed at the top of Fig. 11.9. One determinant may influence several product characteristics, as is the case for the frequency of revision of the text. The importance of each of these is indicated by entering a cross, a sigma, or a triangle in each block. If one or more of the critical components is highly important to several product characteristics, it will be necessary to ensure that any conflicting characteristics are treated properly. For example, frequent revisions may create a problem with layout of text and figures that will increase the price beyond that which the customer will accept. If all of these critical determinants are to be satisfied, a compromise may be necessary. A glance at Fig. 11.9 indicates immediately the determinants that must be controlled carefully , since some of these affect many of the desired product characteristics. Those determinants that appear as highly or moderately important and are grouped in the upper part of the figure must be given special attention.

4. *Create an action list.* The critical components list from Fig. 11.9 are transferred to the left side of Fig. 11.10. For each critical determinant, the actions that are needed to achieve the critical determinants of the product characteristics are identified. For example, achieving frequent revisions requires a commitment on the part of the author to provide the revised text.

Critical determinants

X--Highly important
Σ--Moderately important
Δ--Slightly important

PRODUCT CHARACTERISTICS (Listed in order of importance)	Numerical objective for each product characteristic	Frequent revisions	Section headings	Chapter summaries	Layout of text and figures	Three color printing	Type font	Binding and cover	Weight of paper
Currency of material	Revision in 1999	X	Σ	Σ	X				
Diagrams and tables		Σ			X		Δ		
Technical content		X	Δ	X					
Organization of material		Δ	X	X	Σ				
Examples	2/chapter	Σ			Δ				
Literary style		Δ		X	Σ				
Dimensions	7" X 9 1/4"					X	Σ	Σ	Δ
Weight	6 lb.				Σ		Σ	Σ	X
Index	Author & subject	Σ			X				
Appearance			Δ			X	X	X	X

Figure 11.9 Critical determinants of characteristics (example-textbook)..

For a product that involves subassemblies, each of these characteristics is likely to depend on some component or subcomponent. If the example under discussion had been an engine, these might include the material of construction, the fuel handling system (e.g., fuel injection versus carburetor), the existence of a turbocharger, or the specific design of the combustion chamber in the cylinder. For a transmission, it might have been the surface quality of the pinion gears that comprise the final gear train. For such an example the QFD process would have identified a set of subcomponents or parts whose characteristics must be critically controlled if the important components are to have the desired level of performance. The student is referred to Appendix A11.2, for an example that includes additional steps to account for this added complexity. The textbook example will not require this step.

The final step in the QFD process involves determining the actions that are needed to accomplish the actions listed in the middle of Fig. 11.10. For example, providing frequent revisions of text requires that communications between the author and editor be rapid and with a modest amount of text translation, thus indicating the need for electronic communication.

Throughout the process, the team will be working within certain constraints that are established by the organization. Cost objectives for the final product were stated at the beginning, as was the time available to bring the product to market. For example, the team might conclude that frequent revisions cannot be accommodated and

CRITICAL DETERMINANTS of CHARACTERISTICS	ACTIONS NEEDED TO ASSURE CHARACTERISTICS	PRODUCTION PROCESSES THAT ASSURE CHARACTERISTICS
Frequent revisions	Commitment of author to constant rewrites	Electronic transfer of text and graphics between editor and author
	Rapid manuscript review	Use of computer-based editing
	Good working relationship between author and editor	
	Rapid indexing	Assignment of adequate support by publisher
Layout of text and figures	Standardization of format	Use of computer-based editing
	Capability of author to provide material in proper format	

Figure 11.10   Actions to achieve critical determinants (example-textbook).

have the book meet the target objectives. If this is the case, the QFD process must be repeated with modifications being made to the product and the resulting publication process to reduce the cost. This will result in changing many of the conclusions that were reached by the team. In such cases, it is critical that all members of the team review the changes that are being proposed and that the team reach a new consensus on all the entries. In the unusual circumstance that it is found that the objectives cannot be accomplished within these constraints, the team must meet with senior management, in this case the management of the publishing house and the author, and determine whether the constraints will be relaxed or whether the product characteristics will need to be changed. Such decisions must have management concurrence.

In the introductory paragraphs that described the QFD process, emphasis was placed on the importance of creating a team that is broadly representative of the elements of the organization that is responsible for creating, designing, manufacturing, selling, and servicing the product. In reviewing the four steps identified above, it is clear that the identity of the members of the team who will be most important in arriving at decisions will change as the process moves to completion. This does not mean that the team membership will change–only that those who will have the most expertise in making decisions will change among the various members. The sales, marketing, and engineering members will probably make the most significant contributions during the initial stages of the discussions. The manufacturing and engineering representatives will probably have the principal input during the final stages. What must be emphasized, however, is that no member must allow decisions to be made without assuring the team that the consequences are proper when viewed from their area of specialty. The reasons for this are many, but perhaps the most important is to ensure that the compromises that are made in translating an idea into a final product do not result in a design that has problems that would need to be solved at a later date.

The completion of the QFD process is both time consuming and difficult. It will also be frustrating, and often unsuccessful, if the team is not able to call upon the full complement of resources of the organization. Accomplishing the QFD task for a complicated product such as a car or for a more simple product, such as a textbook, involves consideration of a vast array of details. The team members may find it necessary to consult with many people in their organizational area and to ask to have detailed studies and analyses undertaken at various stages. They will need the support of many people to accomplish the tasks that are required for them to be good team members. Furthermore, as mentioned above, the team must be allowed to iterate the many steps to ensure that the best solution is reached.

If the process is completed in a timely and thorough fashion, it will generate important benefits. The involvement of the entire organization, through the team, can be expected to greatly reduce the possibility that unpleasant surprises will arise or problems will be overlooked in either the design or the manufacture of the product. The resulting product can be expected to include the attributes that the customer considers important. The involvement of the manufacturing and engineering groups in establishing the requirements and in creating the processes to satisfy those requirements should help assure that the product will be of high quality. Finally, the process can be expected to reduce the number and frequency of the engineering change orders that often plague the introduction of a new product, thus reducing the time and costs spent in correcting unpleasant mistakes and oversights.

As mentioned earlier, the steps outlined in this discussion may need to be expanded, depending on the type of product that is under development. The example from Sullivan (1986) illustrates this with an example from automotive that includes a greatly expanded assessment of competitive products, including an analysis of the best in class for the competition, the inclusion of characteristics that will be useful for marketing the product, and consideration of the correlation among the various features that one has identified. This is summarized in Appendix A11.2. When all the items are included, along with others important to a particular product, it is common to see a final display of all the information that looks something like that shown in Fig. 11.11. This characteristic appearance has led some to refer to the QFD process as the creation of the "house of quality."

11.4.2   Design for Assembly and Design for Manufacture

In the discussion of quality function deployment, an explicit presumption was made that it was straightforward to provide a manufacturing plan that would be more or less optimal for the components that were identified to be critical to satisfying the requirements of the customer. In other words, it was assumed that completing the right half of Figure 11.10 was straightforward. Since this is not necessarily the case, it is important to utilize the tools that are available for enhancing the probability that the design will be capable of easy manufacture and assembly.

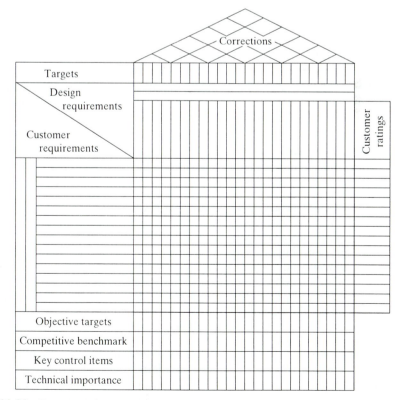

Figure 11.11   Representation of the "House of Quality". (From Sullivan, 1986.)

In the discussion that follows, design for manufacture and design for assembly are treated as elements of a common topic. This is not always the case, as *design for manufacture* (DFM) is often thought of as the process by which the best way to make a part is determined and *design for assembly* (DFA) is focused on the assembly of the final product. Since many of the principles that are important for one of these is found to be equally important for the other, they will be treated together. Whitney et al. (1988:200-201) describe the situation as follows:

> Assembly, with its close coupling to design, vendor control, quality control, etc., requires a new, more highly integrated approach to manufacturing. Assembly is inherently integrative. Parts that were separately designed, made, handled, and inspected must be joined together, handled together, and tested together. The assembly process focuses attention on pairs and groups of parts. Decisions that affect assembly also affect nearly every other aspect of production and use of a product. Assembly, therefore, is a natural forum for launching an integrated attack on all the phases of a product, from conception and fabrication to quality and life cycle. The recent interest in design for assembly is an outgrowth of the realization that assembly is an important phase in the life cycle of a product. Other phases, however, are also important. They may last longer and cost more. Thus, design for assembly cannot be done in isolation.

Designing for manufacture must begin with designing for function. This means that the designer must assure that the function of the part, after inclusion in the final assembly or in the hands of the consumer, will satisfy the requirements established by the user. Since a variety of designs may satisfy these criteria, it is important to find a way of eliminating those that are least desirable. Geoffrey Jacobs, manager of value engineering for British Aerospace's Filton Division, offers a striking example of the importance of subjecting early conceptual designs to *function*-oriented studies (Jacobs, 1980). Asking such questions as, "What is the minimum required function, and how can this be satisfied in the most effective manner?" can lead to simplified designs that can result in greatly reduced costs. He offers, as one example, the redesign of a locking mechanism on a passenger service unit of the Concorde airplane. The original assembly consisted of 12 parts and was manufactured using a die casting that required no machining. By following the function-oriented approach it was found that this mechanism could be replaced by a simple spring, saving 97% of the original cost and 37% of the weight. Since about 180 of these units are used on each Concorde, the total saving was significant. Whitney et al. (1988:210) note that

> Product function analysis is an activity in which designers and engineers seek ways of simplifying or rationalizing a product's design by starting from what the product *should* do rather than how it performs that function *now* or how that function was performed in previous designs. Decisions regarding fabrication or assembly methods can affect users as well as factory personnel and field costs as well as factory costs. These decisions are related to design, not manufacturing. Because they affect the character of the product, they are strategic in their impact. No one department should make these decisions alone, nor can the decisions be parceled out for decentralized action. These are decisions that the concurrency team must address.

While the QFD discussion emphasized the importance of designing to satisfy the requirements of the customer, the importance of meeting certain technical objectives that are established by the environment within which the part will operate was not emphasized. It is critical that the requirements that are imposed on the part as a result of its use be carefully identified (e.g., the loads, the temperature extremes within which the part must operate, the response that a system must provide to an external input, the mean time between failures that are likely to occur, etc.). These are the detailed engineering characteristics that must be developed to complete Fig. 11.10. The design will then require that a variety of variables be considered and a choice made among them to ensure that the parts will meet the requirements that were identified as being *critical* while minimizing the overall system costs. The following are examples of important variables that must be investigated:

1. Choice of materials
2. Choice of processes to be used
3. Parameters that must be considered in choosing a process
   a. Tolerances required of final parts
   b. Time required for each step
   c. Capital investment required
   d. Status of facilities
   e. Experience with the process
   f. Time required to launch
4. Estimated cost for the part

An initial choice must be made among the many variables. The task then becomes one of iteratively improving the initial design to obtain one that will result in the least cost, best quality, and shortest time to completion. A number of people have developed tools that are useful in sorting through the many possible alternatives and focusing on those that are most effective.

Boothroyd and Dewhurst (1988a) have been leaders in the development of a quantitative method of design for manufacturability. It involves two principal steps. Step 1 attempts to determine whether a part should be separate from other parts of the assembly. Representative questions that are asked are:

- When this part is in operation, does it move relative to other parts being assembled?
- Must this part be made of a material that differs from that of the other parts being assembled?
- Must this part be separate from all other parts assembled to facilitate assembly or disassembly?

If the answer to all these questions is no, the next step is to inquire whether the assembly can be created as a single piece rather than as an assembly. Boothroyd and Dewhurst (1988b) offer an example of a design that benefited from these considerations. A cylindrical base assembly was composed of several parts machined from standard shapes, including bars and plates, and then assembled with standard bolts and screws. Even though the assembly costs were small compared to the cost of the materials, an exploration of the questions listed under step 1 led to a consideration of using a casting with machined operating surfaces. The casting resulted in a material cost saving of about 42% and no assembly costs. Since machining costs were less than 10% of the original costs, the overall cost of the product was reduced by 60%.

Step 2 of design for manufacturability involves an estimate of the costs of handling and assembling each part, using the appropriate assembly process. This involves determining whether the assembly should be undertaken manually, with soft automation (e.g., with robots) or with high-speed fixed automation, and exploring whether design changes can be made in the product such that assembly costs can be reduced. The objective of step 2 is not to finalize the assembly process, but to use the results to guide the designer in further simplification of the product design.

Applications of the DFM techniques have resulted in some dramatic improvements in designs (Vogt, 1988). It has been reported that Ford reduced the number of parts in a throttle body by 43%, assembly costs by 38%, and material costs by 14% by using DFM techniques. Federal Products Corp. eliminated 25% of the parts and 20% of the costs from a column gage using DFM techniques. The number of parts in a reciprocating power saw was reduced from 41 to 28. When IBM Corp. undertook the design of its Proprinter, it set as an explicit objective the automatic assembly of the new design (Newman, 1987). In accomplishing this, it achieved a printer with features that exceeded that of its predecessor, the IBM 5152, with a design that had 60% fewer parts. The IBM 5152 used 152 parts, only half of them being functional, since 74 were fasteners. The Proprinter had 61 parts and no discrete fasteners.

Foo and Kinney (1990) illustrate with the following example some of the other benefits that result from the team approach to using DFM.

> The participation of a designer(s) on the team was vital. Not only was the designer a major participant in the investigation and analysis but the designer was the primary interface in feeding data and action items back to the design organizations. One outcome of the team was the design for manufacturability (DFM) checklist. The checklist contains a set of DFM guidelines and is now used by manufacturing and design organizations to officially review a design for manufacturability and testability prior to the release of the design to manufacturing. A key element of the DFM review is a check to ensure that all components used in the design are on the approved parts list of the factory.

Recognizing that assembly can be one of the most time-consuming steps in discrete manufacturing , it is critical to minimize the effort this requires and to ensure that the procedure results in the fewest possible defects. Andreasen et al.(1988) have identified a series of principles that systematize the design process and enhance the ease of assemble. The following are included in their DFA principles.

1. Design so that the fewest parts result. This principle is usually stated as "design simply."
2. Integrate parts to achieve the fewest number of parts and subassemblies. This is an extension of principle 1.
3. Design so that each step in the assembly process is clearly unique and distinct. This means that ambiguity about part identity or part orientation, which can confuse assembly and lead to errors, should not be permitted in the design.
4. Use a subassembly–a chassis–that allows other components to be added during the total assembly. The chassis can form a carrier for the assembly operations.
5. Design a stacked assembly to enhance ease and accuracy of assembly.

6. Design so that components do not require tight tolerances and so that surface tolerances are not critical for the mating surfaces.

To this list Crow (1989) adds:

7. Design verifiability into the product and its components.
8. Design "robustness" into products to compensate for uncertainty in the product's manufacturing, testing, and use.
9. Utilize common parts and materials to facilitate design activities, to minimize the amount of inventory in the system, and to standardize handling and assembly operations.
10. Design for ease of servicing.

In considering the methods by which the assembly processes are to be accomplished, choices must be made among a manual process, an automated process, or a combination of the two. Andreasen et al. make the following important observations. If the design of the product makes it easy to assemble manually, it is highly probable that it will be easily assembled automatically. The corollary of this statement is that products designed for automatic assembly will probably be easy to assemble manually (Andreasen et al., 1988). IBM discovered this to be the case with the design of the Proprinter. The new design that was created to allow automated assembly was so good that it could easily be assembled manually, thus eliminating the need for a large investment in automated equipment.

Boothroyd and Dewhurst (1988a) have developed a detailed classification system that allows the determination of the most economical method of assembly. The following variables are analyzed for each assembly to determine whether the process should be manual, utilize a special-purpose machine, or be accomplished with an adaptable or programmable machine.

1. Quality of parts to be assembled
2. Number of parts in the assembly
3. Annual production volume
4. Number of variations in product style
5. Frequency with which design changes are made
6. Number of different products
7. Economic factors influencing the market

The equations and criteria that Boothroyd and Dewhurst have developed for ranking the relative importance of each of the foregoing variables in terms of determining the optimal assembly system can be summarized as follows:

1. Automatic machines are more susceptible to interruptions from defective parts than are manual operations. Thus the use of automatic assembly is favored only for parts that have small numbers of defect.

2. As the number of parts in an assembly increases, the number of machines needed to accomplish automatic assembly will increase, thereby increasing the cost of the assembly system.

3. When the annual production per shift is large, the use of automatic assembly systems is favored.

4. Increased product complexity requires additional machines or additional fixtures, thus increasing the cost of automated assembly.

5. Frequent changes in the design of the products or numerous products, either of which significantly influence the assembly process, can lead to obsolete assembly facilities and thus increase the cost of automated processes.

6. Interest rates, the rate of inflation, and the length of time that one must wait for a manufacturer to supply a new tool–influenced by the level of business activity in the tool industry–each influence the cost of machinery to accomplish assembly and thus impact the decision on whether to use manual or automatic processes.

Boothroyd and Dewhurst(1988a) offer some dramatic examples of how the number of parts influences the costs of assembly. In a detailed study of a simple assembly consisting of a sheet metal base plate and two sheet metal side brackets that support a spindle, the costs were evaluated for a variety of assembly techniques and for a variety of part consolidations. For this particular product, the cost of manual assembly is always higher than that of assembly by robot and varies between a factor of 2.0 and 1.6, depending on the design. This particular example demonstrates, however, that certain designs can reduce the cost of assembly to such a point that the choice between robot and manual assembly will need to be on bases other than the cost of assembly per piece. For the assembly of the four parts with bolts, washers, and nuts–20 of which were required–the savings in utilizing automated assembly is $0.60 per assembly for a product whose combined material and machining cost is about $1.18. For a design that consists of only eight parts–four screw fasteners–the cost of assembly is reduced to about $0.57. For two different designs, one with four parts and one with two parts, the manual assembly costs have become $0.20 and $0.12, respectively. While the material and machining costs are also less for these designs, the cost of assembly has now decreased to such a level that the savings in using automated assembly is estimated to be $0.04 per assembly for the design with two parts.

An additional important concern regarding the method of assembly arises with regard to the quality of the final assembled product. If the assembly is such that a worker will become fatigued, either because of the rate of movements required to accomplish the assembly or because of the complexity of it, experience has shown that automated machinery is usually superior. Furthermore, if the objective is to maintain a very low defect rate in the final product–approaching a few defects per 100,000 or per million–it may be that automated machinery will be the only feasible method of achieving this. In these cases the economics of the situation are driven by the value that is placed on the quality of the final product rather than the explicit cost of the process itself. This, you will recall, is consistent with Garvin's description of quality in terms of a user-based approach (Chapter 3).

There are certain basic design guidelines that should be followed irrespective of the particular process that is used for assembly. To reduce the chances for error, it is critical that parts have a recognizable orientation–or alternatively, they should be completely symmetric–so that there can be no confusion concerning the relative desired orientation of the mating parts. Ambiguity of orientation, of material, or of order of assembly should be avoided. As noted above, complexity should also be avoided. For example, the use of three different fasteners, each of a unique but nearly the same size, will almost certainly lead to errors and resultant quality problems. Furthermore, parts that must move, either by hand or on a conveyor, should have a stable base to ensure that the orientation does not shift during transport.

All of these principles and guidelines appear to reflect a common sense approach to designing parts. While the general correctness of them may be obvious to all concerned, it must be acknowledged that it takes great discipline on the part of the designer to adhere to them and to accept the constraints that they impose on the design process. Furthermore, since the designer may not have a working knowledge of the manufacturing processes that are likely to be employed, the manufacturing engineer who is participating in the design process is expected to be able to provide this information to the team. This is the value of the concurrent engineering process.

11.4.3   A Word of Caution

With all tools, limitations may exist that will influence their usefulness. Boothroyd and Dewhurst (1988a) note that some people would like to embed the DFM and DFA techniques into CAD/CAM systems. They caution that this may not be advisable. The principal benefit derived from the application of DFA and DFM is to provide early guidance to the design team during the conceptual phases, to help them eliminate undesirable alternatives, and to allow them to focus their efforts on those alternatives that will reduce costs and improve quality. The problem with embedding the DFM and DFA tools into systems that are intended to create the final design drawings and the detailed process plans is that this attempts to merge tools that are intended to be used at two very different stages of the process. Experience suggests that under these circumstances one or both of the tools–those intended for early choice selection and those concerned with late detailed design–can be compromised.

Bancroft (1988) has also noted concerns relative to a strict application of DFM and DFA tools.

> Uniaxis assembly eliminates some of the advantages of hard automation. Hard tooling usually costs more than soft tooling because the hard tools have to be custom-made for each product. Adding to the cost is the fact that hard tools generally cannot be reused for a different product. This fact also adds to the risk of hard tooling. If demand for the product should stop unexpectedly, the investment in hard tooling will be lost. It might also be lost when switching to a new-generation product.
>
> Another DFM rule–reduce the number of assembly steps–also needs to be reevaluated. Sometimes it is easier to add several simple steps than one complicated one.
>
> DFM suggests that the number of parts should be reduced and that more functionality should be incorporated in one part. This idea could cause numerous problems for a low-volume product. It reduces the number of standard parts and leads toward custom parts.

Perhaps the only response that can be made regarding these concerns is that the team must always test their designs, their conclusions, and their recommended actions against commonsense principles. Experience is an important asset for the team. This emphasizes the care that must be taken in choosing the members of the concurrent engineering team.

11.5   EVALUATING THE DESIGN

Throughout the design process, the team will be seeking to assure itself that its detailed decisions are adequate to provide a product with the desired attributes. There are various means by which this can be accomplished.

Perhaps the greatest revolution in evaluation in recent years has been achieved through the use of *simulation*. The evolution of fast high-powered computers has stimulated many companies to invest in the development of programs that allow them to simulate how a product will perform in a variety of circumstances. The aircraft industry is one of the most sophisticated in this regard. The use of simulation and analysis throughout the design of an airplane is so extensive and complete that the first prototype of a new airplane is flown and ultimately, sold. In the automotive industry, it is now common to use simulation to determine the characteristics of each element of the powertrain so that the final vehicle will have the desired fuel economy, generate less than the mandated level of emissions, and offer the driver the proper level of performance in terms of acceleration and braking. Major efforts are under way to develop simulation programs that will assure the designer that the mechanical design of the vehicle will provide the necessary protection to occupants in the event of a crash while remaining free of bothersome noise and vibrations. In the electronics industry, simulation programs are used to test the design of every integrated circuit before it is built. Although the design of most modern products could not proceed without the availability of these important tools, the process does not stop with these analyses. Ultimately, a prototype of the part must be created.

The creation of one or several prototypes has often required the skills of one or more artisans and frequently consumed many weeks. In the interests of reducing the time to create prototypes, many techniques are being developed to create the first model of a new design (Kruth, 1991; Ashley, 1991). In some cases, the design team will show the prototypes to customers to assure themselves that the design meets the customers needs and wants. In other cases, the prototypes will be used for testing conditions that the simulation was not able to accomplish (e.g., performance under extremes of temperature, pressure, stresses, etc.; durability under extreme loads; etc.) These tests can be very valuable in discovering weaknesses in the design that may need to be corrected before the product is placed in the hands of the customer.

All the information that is learned from the testing of prototypes must be provided to the concurrent engineering team for evaluation. In many cases the team will find that these tests suggest that some change needs to be made in the design before it is released for full production. When this occurs, the team must repeat the design process. Obviously, the second time through should take much less time than the first.

The serial method of designing a product, described in Section 11.2.1, frequently required that the process be repeated several times, with a resultant delay in finalization of the design. The concurrent-simultaneous engineering process has been found to greatly reduce the frequency of redesigns. The team, with members from various backgrounds and disciplines, can be very effective in identifying problem areas early and in developing solutions that are more nearly optimal.

Although prototypes are important for evaluating performance, it is frequently found that detailed behavior of the prototype does not match the performance of the part when it is produced on production equipment at large volumes. Subtle differences in tolerances, joining, material variations, and so on, often combine to make the production part more sensitive than was the earlier prototype. It is frequently necessary, therefore, for a final prove-out of the design to be accomplished with parts that are made on the final production equipment. Obviously, any problem that is discovered at this late date may lead to serious delays in introduction of the product. It is the responsibility of the design team to ensure that these problems are kept to a minimum. The presence of the manufacturing engineer on the design team is critical in accomplishing this.

Whitney et al. (1988) draw a sharp distinction between DFM and value engineering that must be appreciated by the concurrent engineering team.

> It is readily apparent that design for producibility is different from value engineering, an activity aimed chiefly at reducing manufacturing cost by astute choices of materials or methods of making parts. Value engineering occurs after major product design is finished, and thus it is neither concurrent nor likely to be very thorough. The required thoroughness cannot be accomplished except through a concurrent process. More importantly, design for producibility includes and sometimes subordinates reduction of manufacturing costs within the larger goal of optimizing the entire life cycle of the product. The reason is that, although employees assemble it, customers or repair personnel may disassemble it, and the actions of these others can be made easier, safer, or more congenial to the character of the product by decisions made simultaneously with design or assembly decisions. (Whitney et al., 1988:209)

11.6   RESPONSIVENESS AND FLEXIBILITY

In considering an all-new product or process, it must be assumed that the attributes of the initial representation of the product will require some modification or "fine tuning." Just as no player can guarantee hitting a home run each time he or she comes to bat in a baseball game, a manufacturer cannot assume that it can produce a product with every detail correct the first time. Changes will be necessary. The firm that has the capability to adjust rapidly, efficiently, and effectively to changes found to be needed in a new product will have a distinct advantage in maintaining his competitiveness. The responsiveness of the system to change can be as important for a firm that is introducing an all-new product as it is for the one that is merely updating a product. Furthermore, a system that is adaptive during the product development phases will probably be adaptable to changing market demands.

Assuring that a system will be responsive to the changing needs of the customer can have a profound impact on a manufacturing operation. Facilities that have been designed such that it is very expensive or time consuming to change them can prove to be an impediment to change. In such cases the manufacturer may be faced with replacing facilities prematurely or losing markets to manufacturers who are able to meet the changing demands of the marketplace. In either case, the result can be financially disastrous.

The need for flexibility raises important issues for manufacturers. If it is desirable to have flexible facilities, how is the financial value of flexibility established? Experience with manufacturing systems suggests that the more flexible systems are the more costly. How much greater cost can be justified for a system to assure its capability to adjust to future, but currently unknown, changes in the product? Answering these questions is very difficult. For example, an important element in determining the price that will be charged for a new product is the cost of producing the part, which, in turn, depends on the cost of the facilities. If the useful life of the facilities are dependent on a future product, the costs allocated to the present product should not bear the total cost of the facilities. Uncertainty is introduced by this reasoning, since future product needs and profitability cannot be assured. Allocation of some of the cost to the future, and thus an assignment of value to flexibility of the facility, enhances the risk that the enterprise must assume in making the decision to invest in flexible facilities.

The need for flexibility and responsiveness to changes is not limited, of course, to the manufacturing activity. If the company is to be responsive, the entire *system* must be responsive. The time required to translate an idea into a marketable product involves the entire system. The time required to design the product, to develop prototypes, to test and modify the early prototypes to provide a final design, to create or modify the manufacturing facilities, to "launch" the manufacturing activities and reach full-scale production, to advertise the product, and to train the sales and service personnel must be as short as possible.

CONCLUDING COMMENTS

There can be little value in having the most efficient manufacturing plant in the world if the design of the product that is to be manufactured is not compatible with the capabilities of the plant or if it has taken so long to develop the product that it has missed its window of opportunity. The procedures discussed in this chapter are tools for ensuring that the resulting product will satisfy the requirements of the customer at the lowest cost and highest quality. Achieving this in the shortest time is critical to remaining competitive in the world marketplace.

In addition to the difficulty of accomplishing these goals, it is essential to recognize that the manufacturing system is never static. World competition has replaced dependable and predictable markets with markets that are continually fragmenting and changing. Customer have more choices and more opportunities to sample products whose attributes represent emerging trends. The consequence is that new products are appearing at a rapid rate, established products are experiencing dramatically shortened lifetimes, and specialization in product offerings are tending to fragment an otherwise stable marketplace into many "niche" markets.

For the manufacturing system, these changes in the market environment offer both challenges and opportunities. The system that can reduce the time that it takes to bring a product from concept to the market–the time for product realization–will have a distinct advantage. The system that can bring its manufacturing facility to full production with a new product in the shortest time with the fewest number of defects will have the advantage of early introduction of the product. The system that can design a new product in such a way that it can be manufactured easily and properly will have a cost and time advantage.

The environment for the manufacturer increasingly demands that it have the capability to respond to an ever-increasing rate of change in the demands of the customer. It is this demand from the customer that places such a high premium on the capability to respond quickly with high-quality, cost-effective products. The manufacturing system must become a dynamic organism that is responsive to multiple inputs, many of which are unpredictable. Achieving sufficient understanding of the system to allow it to respond to these forces in a sensible and predictable way is one of the primary challenges to management. Accomplishing this is a formidable task. The U.S. manufacturer has not, in the past, excelled in this. Those that can achieve it will have a clear advantage.

REFERENCES

ANDERSON, M.M.,S. KAHLER, and T. LUND. 1988. *Design for Assembly*, Springer-Verlag, New York:95-148.

ASHLEY, S. 1991. Rapid Prototyping Systems, *Mechanical Engineering*, April:34-43.

BANCROFT, C. E. 1988. Design for Manufacturability: Half Speed Ahead, *Manufacturing Engineering*, Sept.:67-69.

BOOTHROYD, G. and P. DEWHURST. 1988a. Product Design for Manufacture and Assembly, *Manufacturing Engineering*, April:42-46.

BOOTHROYD, G. and P. DEWHURST. 1988b. Making it Simple: Design for Assembly, *Mechanical Engineering*, Feb.:28-31.

BOWEN, H. K. 1992. Implementation Projects: Decisions and Expenditures in *Manufacturing Systems: Foundations of World Class Practice*, J. A. Heim and W.D. Compton (eds.) National Academy Press, Washington, DC:93.

CROW, K. A. 1989. Ten Steps to Competitive Design for Manufacturability, The Second International Conference on Design for Manufacturability, Management Roundtable, Boston.

FOO, G. and L. KINNEY. 1990. Integrated Pull Manufacturing–The Integration of MRP and JIT Systems, Proceedings of the First International Conference on Systems Integration (ICSI '90), IEEE:415-425.

HAUSER, J. R. and D. CLAUSING. 1988. The House of Quality, *Harvard Business Review*, May-June: 63-73.

JACOBS, G. 1980. Designing for Improved Value, *Engineering*, Feb.:178-182.

JUAREZ, J. M., E. J. GUNTHER, J. E. TOOMEY. 1990. Quality Function Deployment Applied to an ALS Cryogenic Tank, AIAA Aerospace Engineering Conference and Show, February, AIAA-90-1807:1-7.

KRUTH, J. P. 1991. Material Incress Manufacturing by Rapid Prototyping Techniques, *CIRP Annals* 40(2):603-614.

McKNIGHT, M. and J. M. JACKSON. 1989. Simultaneous Engineering Saves Manufacturers Lead Time, Costs and Frustration, *Industrial Engineering*, Aug.:25-27.

NEWMAN, J. 1987. Matrix Printer: No Pulleys, Belts, or Screws, *IEEE Spectrum*, May:50-51.

ROSENBLATT, A. and G. F. WATSON. 1991. Concurrent Engineering, *IEEE Spectrum*, July:22.

SHINA, S. G. 1991a. *Concurrent Engineering and Design for Manufacture of Electronic Products*, Van Nostrand Reinhold, New York.

SHINA, S. G. 1991b. Concurrent Engineering, *IEEE Spectrum*, July:23-26.

SMITH, P. G. AND D. G. REINERTSEN. 1991. *Developing Products in Half the Time*, Van Nostrand Reinhold, New York.

SULLIVAN, L. P. 1986. Quality Function Deployment, *Quality Progress*:39-50.

TURINO, J. 1991. Concurrent Engineering, *IEEE Spectrum,* July:30.

VOGT, C. F. JR. 1988. Beyond CAD and CAM: Design for Manufacturability, *Design News,* Mar. 7:18-19.

WHEELER, R. 1991. Concurrent Engineering, *IEEE Spectrum,* July:32.

WHITNEY, D. E., J. L. NEVINS, T. L. DE FAZIO, R. E. GUSTAVSON, R. W. METZINGER, J. M. ROURKE, and D. S. SELTZER. 1988. The Strategic Approach to Product Design, in *Design and Analysis of Integrated Manufacturing Systems*, W. D. Compton (ed.), National Academy Press, Washington, DC.

WINNER, R. I., J. P. PENNELL, H. E. BERTRAND, and M. M. G. SLUSARCZUK. 1988. The Role of Concurrent Engineering in Weapons System Acquisition, Institute of Defense Analysis, *IDA Report R-338* (Unclassified), December

QUESTIONS

11.1   How can young engineers with somewhat limited experience be utilized effectively in QFD teams?

11.2   More that one project is often under development within an activity at the same time. What are the responsibilities of the teams for communicating with each other during this period of codevelopment? How can conflicts over the allocation of resources be resolved for the various concurrent design teams?

11.3   What kind of training should be undertaken by team members before they undertake a design that will utilize the QFD process? How crucial is training for achieving project success?

11.4   Estimates of costs are needed at many stages of the design process. Since it is particularly important to have reasonable cost estimates in the very early stages of the design process, what tools are available for making these estimations early in the process when the design is still in its fuzzy state?

11.5   In addition to building and testing a prototype of the product that has been designed, what other methods are there for evaluating the validity of a design? How and to what extent should these tools be regularly used?

11.6   A set of questions raised by the 15/85 ruleare given in Section 11.1. Offer reasonable answers to each of the five questions.

11.7   Discuss possible modifications of QFD that might need to be made for application of this approach to a service organization.

11.8   You and six other members of the concurrent engineering design team have been working for the past nine months on the development of a complicated electromechanical product. You are the team leader. The team has developed a good working relationship and is nearing a period in which numerous critical decisions must be made. The team has just begun its weekly meeting. The manufacturing engineer who has been working with the team has just entered the room and has informed the team that his plant manager has told him that he must resign from the design team to help launch a new production line for a product that is to be in the marketplace late this year. How do you handle this morning's meeting? What actions will you take?

11.9 Carry out a QFD for an umbrella.

11.10 Carry out a QFD for a plastic insulated cup.

11.11 Carry out a QFD for a ball point pen.

11.12 Carry out a QFD for a backpack.

11.13 Carry out a QFD for a gourmet coffeemaker.

11.14 Carry out a QFD for a child's toothbrush.

11.15 Carry out a QFD for a pair of roller blades.

11.16 Carry out a QFD for an ice cream product.

11.17 Carry out a QFD for a library that is serving a university student body.

11.18 Carry out a QFD for a bookstore.

11.19 Carry out a QFD for a student food service in a dormitory.

11.20 Carry out a QFD for a sleeping bag.

READINGS

BOWEN, H. K. 1992. Implementation Projects: Decisions and Expenditures, in *Manufacturing Systems: Foundations of World Class Practice*, J. A. Heim and W. D. Compton (eds) National Academy Press, Washington, DC:93.

SULLIVAN, L. P. 1986. Quality Function Deployment, *Quality Progress* :39-50.

APPENDIX A11.1
CASE STUDY: FORD MOTOR COMPANY–TEAM TAURUS

It was 1979 and the economy was being hammered by the second major oil embargo by Mideastern suppliers. Inflation was high. Company after company was being forced to cut costs, reduce employment, and delay planned expansions. For the domestic automotive industry, the problems were compounded by an aggressive Japanese industry that found that their small, high-quality, good fuel economy products were nearly perfect for a market that was anxious for new alternatives. Sales of Japanese cars soared as the public turned away from domestic products.

For Ford, the alternatives were stark. Its products did not produce the excitement in the buying public that was evident for the Japanese product. Ford's presence in the pace-setting market of California had essentially disappeared. The quality of existing products was poor. Styling, design, and feature content of the product was well below that of the successful foreign competitors. Faced with financial losses that amounted to $200 million in 1979 and more than $2 billion in 1980, its resources were severely limited. But it had little choice. It had to respond aggressively or face almost certain bankruptcy.

The immediate problem was with the Thunderbird. Product plans called for a "freshening," a change in appearance with modest changes in fundamental design, in 1983. Although not the highest-volume product, recent sales had been disastrous, dropping from 297,000 in 1978 to 41,600 in 1982. It was necessary to go forward to keep its share of the market from further erosion.

The initial rendition of the style for the new Thunderbird was standard for the time–boxy, rectangular, and bulky, with little character. A simple question from the company president led to a dramatic shift in direction. When the head of the design team acknowledged that the proposed design was what the team thought management wanted instead of what they thought was right, direction was given to design it as the designers thought it should be, to follow their instincts. The resulting design was the first timid step that Ford was to take in the revolution of automotive design. The new Ford Thunderbird, with its aerodynamic design, attracted lots of attention. Although it did not employ major new technology, it demonstrated that customers in this segment of the market were attracted to the new styling theme.

While success with the Thunderbird was important, it did little to address the problems with the large-volume products that Ford was offering–namely, those that competed at the center of the market, the Ford LTD and the Mercury Marquis, which sold 350,000 units in 1984 and accounted for 20 percent of Ford's production. If Ford was to become competitive in this segment of the market, quality had to be improved drastically and product content had to be upgraded to meet the needs of customers who now had choices among Japanese and European offerings. Clearly, an all-new product was needed. But an entirely new car with a new engine and transmission could easily cost $3 billion. A commitment to invest this amount at a time that the market was weak and the company losing money was not just risky; it was a "bet your company" type of decision. If it failed, the company would probably be unable to recover.

There was no mistaking the importance of the task or the magnitude of accomplishing it. It was necessary to create a car that would compete successfully with the Japanese imports and that would attract customers who had become disenchanted with U.S. producers. As planning for this new product was starting, the company was also undertaking a shift from a hierarchical, "management knows best" philosophy to an involvement of workers in the decision making (see Appendix A6.1). It was a time of great turmoil within Ford.

The task was given to Lew Veraldi, a vice-president who had extensive experience in Europe, where he was responsible for the launch of Ford of Europe's most successful car, the Fiesta. He understood cars and he understood the importance of this project. The project was given the code name Taurus, for no reason except that the wives of Veraldi and John Risk, the leading product planner for the car, were both born in May. This was to be the first major car that Ford had done using the team concept, the concept of bringing together people with an understanding of all elements of the project at the beginning of the project and keeping them together throughout the project. Because of the complexity of an automobile, this was concurrent engineering in the extreme.

Recognizing that existing Ford products were not appealing to consumers, a major new approach to understanding the wants and needs of the customer was undertaken. Veraldi and Risk set as their goal the design of a world-class car–the "best in class." Through extensive market research, hundreds of people around the country were contacted, dozens of cars were reviewed to examine the features that they thought were important, and every effort was made to understand the reasons behind Ford's loss of appeal to the customer. Not only did they discover the many things that people like and do not like about their cars, they were treated to a constant barrage of complaints about existing Ford products. After more than two years of questioning and probing, a list of more than 400 items were collected that represented what people identified as being important. Heading the list was quality. Veraldi set out to find a way to get them all into the Taurus.

Having experienced recent success with the Thunderbird, company management was convinced that the *designers* had to be given the *responsibility* for the styling of this new vehicle. Every detail was examined and reexamined. Sitting side by side, the designers for the inside and outside of the vehicle were committed to make the interior and exterior design flow together a result that had not been achieved in the past, as the two tasks were done separately with minimal interactions. Because the U.S. public viewed front-wheel drive to be at the cutting edge of technology, Veraldi concluded that the Taurus would be perceived to be "best in class" only if it was front-wheel drive, with the result that an all-new engine and transmission were required. Market research continued to emphasize the need for an exciting product, new styles were explored and refined, new engines and transmissions were designed and tested. Always in mind was how to include the 400 or so ideas that had been collected from consumers. Not all of these ideas found their way into the final design, not all ideas that the designers liked could be afforded, and not all options could be accommodated in the powertrain. But all views were considered. Team Taurus had insisted on it.

In November 1982, the product-planning committee officially agreed on the design of the Taurus. Board of Director approval was given in December. But this did not mean that the work was finished. Market research continued to search for the best combination of comfort, features, and ergonomics, all constrained by the cost target. The advertising campaign to present the car was developed. A name for the final product was researched and after much debate it was decided that the Ford vehicle would remain Taurus and the Mercury version would be called Sable. Planning for the manufacturing now swung into gear.

Progress was being made by executives of Ford and the United Auto Workers in conveying the importance of this car to Ford's survival. Although employee participation and employee involvement were being embraced with skepticism by many plant workers, everyone knew that the quality of this car must be substantially better than in the past. Manufacturing the Taurus/Sable was going to be an enormous challenge. Many new procedures were needed, including taking such steps as removing the doors after painting to allow the interior to be introduced and mounting the engine and transmission from the bottom of the car. New welding techniques were needed to provide the stylized smooth appearance. The cars were to be assembled at two plants, in Atlanta and Chicago. Employment involvement groups were formed to provide the channel for receiving the suggestions of workers. Suggestions flowed in as to how to simplify the design to ease production.

The plan was to introduce the car to the public in October 1985. The Atlanta plant was scheduled to begin production–it's Job One–in July 1985. Just as the launch began in Atlanta, a serious problem appeared. Production sheet metal for the body was not fitting properly. The variability among the production panels was much greater than it had been with the prototype panels. Doors and trunk lids would not fit properly. Had this happened in earlier years, cars would have been shipped and then repaired, if the owner insisted. But that action would not be consistent with the image of this as best in class. Veraldi had only one choice–stop production and delay the date of introduction. This was unheard of previously. An introduction date would never be allowed to slip, not even to correct a quality problem. Not only was this important to assure that the car met the image that was being projected, but by so doing management made a dramatic statement to workers that profits would be sacrificed before a shoddy product was shipped. In the past, profits meant everything. Employee involvement and employee participation were given a gigantic boost with this decision.

Ultimately, the problems at Atlanta were solved. The launch date was reestablished as December 26, 1985. Production at Chicago began on January 20, 1986. It was relatively smoother, thanks to the learning experience of teams of Chicago employees who had spent days at Atlanta. In January 1986, the Taurus was named Car of the Year by *Motor Trend*. It was one of the Ten Best Cars of *Car and Driver* in 1986, 1987, 1988, 1989 and 1990. While the Taurus/Sable fell short of selling the 550,000 annually originally planned, it was a huge financial success. Ford profits recovered, in large part because of Taurus, to $3.3billion in 1986.

Just as Ford had responded to a genuine threat to its existence with an imaginative, gutsy new product that caught the eye of customers, competitors would seek to offset the advance that Taurus had gained. The Honda Accord presented that challenge. In 1989 it surpassed the Taurus to become the best selling-car in the U.S., a position that it held for several years. Honda sought to maintain a constant public interest in the Accord by freshening it every two years and redesigning it every six years, practices that Ford has not adopted.

Ford was not to bring out a newly designed replacement for the Taurus until the fall of 1995. Its key concern as it undertook the design of this new car, was how to create a car that would appeal to younger buyers *without* alienating current buyers. The path that it chose to follow has many similarities to the Team Taurus plan of the 1980s-concurrent engineering, colocation of teams, listening to the customer, and achieving best in class. The new product was viewed by many people as a bit avant-garde for this segment of the market . Ford was again taking large risks that the public was ready for something really new. Still the buying public can be fickle. Although this was not a "bet your company" decision, it remains to be seen whether early success of the new Taurus will be as dramatic as for the first.

SOURCES

KERWIN, K. E. HILL and K. NAUGHTON 1995.The Shape of a New Machine, *Business Week,* July 24:60.

NAUGHTON, K. 1995.Ford's Global Gladiator, *Business Week,* Dec. 11:116.

SMITH, DAVID C. 1985. Ford's $3-Billion Mid-Market Plunge, *Ward's Automotive News,* Feb.:26.

TAUB,ERIC 1991. *The Making of the Car That Saved Ford,*Dutton, New York.

ITEMS FOR DISCUSSION

1. In spite of making every effort to do exhaustive testing of prototype vehicles and components, Ford learned that the prototypes may not be fully representative of final production. What are some of the things that can done to help ensure that these problems will be minimized?

2. The Taurus lost its leadership to the Honda Accord in the late-1980's. While Ford appeared to remain content with the outstanding product that it produced in 1986, it did not aggressively seek to continuously improve it and to eliminate some of the problems that appeared its design. Honda, on the other hand, sought to make changes every two years. In a strategic sense, how does a company justify the costs of doing this? How can a company find the resources necessary to continuously change its existing products every two years while investing in the major new products that are needed but require so much more manpower and attention? Does this suggest that a company should not attempt to provide a spectrum of products that will compete in a large fraction of the market?

3. Ford is in the midst of creating a worldwide organization that assigns the design and development of centers that will be responsible for a vehicle 'platform' for the world. The Taurus was so successful because it was the Best-in-Class for that market. The market in this case was North America. How does one create a process that will lead to the design of a Best-in-Class for the world market? Can such products exist, or will there need to be modifications for each regional region? Will this destroy the gain that will be achieved by the central thrust of a design team?

4. In the 1970's, the Ford designers knew that the products that they were designing lacked broad appeal. They were constrained by what the local competition was doing, namely not reaching out with new designs, and by a management who was conservative and unwilling to take risks with all new designs. Did the designers have other options in these circumstances? Could/should the Board of Directors play a role in these decisions? Is it only in circumstances of extreme organizational stress, for example, threatened bankruptcy, that imaginative ideas will be encouraged?

5. Are large organizations doomed to experience this phenomena of excelling, followed by resting on their laurels, followed by a threat to their existence, followed by a renewed energy and revitalization? If not, how would you go about stimulating them to always excel?

READINGS

PETERSEN, D. E. AND J. HILLKIRK, 1991. *A Better Idea*, Houghton Mifflin Co., Boston.

LACEY, ROBERT, 1986. *Ford: The Men and the Machines*, Little Brown and Co., Boston.

APPENDIX A11.2 QUALITY FUNCTION DEPLOYMENT

Sullivan (1986) was an early proponent of the quality function deployment process as a means of achieving companywide quality control. The following discussion is adapted from his 1986 discussion and is presented here to illustrate some of the details that must be addressed in using the matrices that were discussed in Section 11.4.1.

There are six key terms associated with QFD.

1. *Quality Function Deployment*–an overall concept that provides a means of translating customer requirements into the appropriate technical requirements for each stage of product development and production (i.e., marketing strategies, planning, product design and engineering, prototype evaluation, production process development, production, sales). This concept is further broken down into product quality deployment and deployment of the quality function (described below).

2. *The Voice of the Customer*– customers' requirements expressed in their own terms.

3. *Counterpart Characteristics*–an expression of the voice of the customer in technical language that specifies customer-required quality; counterpart characteristics are critical final product control characteristics.

4. *Product Quality Deployment*–activities needed to translate the voice of the customer into counterpart characteristics.

5. *Deployment of the Quality Function*–activities needed to assure that customer-required quality is achieved; the assignment of specific quality responsibilities to specific departments. (The phrase quality function does not refer to the quality department, but to any activity neeeded to assure that quality is achieved, no matter which department performs the activity.)

6. *Quality Tables*–a series of matrices used to translate the voice of the customer into final product control characteristics.

The QFD system concept is based on the four key documents described in Section 11.4.1.

1. Overall customer requirement *planning matrix*
2. Final product characteristic *deployment matrix*
3. *Process plan* and *quality control charts*
4. *Operating instructions*

   The overall QFD system based on these documents traces a continuous flow of information from customer requirements to plant operating instructions; it thus provides what W. Edwards Deming calls "a clear operational definition" a common purpose, priorities, and focus of attention.

The process is initiated by listing the product requirements in customer terms, as shown by the example in Figure A11.1. The primary requirements, which are the very basic customer wants are usually expanded into secondary and tertiary requirements. The customer wants are derived from a variety of sources: marketing research data, dealer input, sales department wants, "buff" magazines, special customer opinion surveys, and so on. This is one of the most critical parts of the process and is usually the most difficult because it requires obtaining and expressing what the customer truly wants.

The customer requirements are translated to the vertical axis of the top portion of the planning matrix. The final product control characteristics that you believe should be assured to meet the customer-stated product requirements are entered across the top of the matrix in Figure A11.2. These characteristics are the product requirements that relate directly to the customer requirements and must be achieved during the design, manufacture, assembly, and service process for the final product to meet customer needs.

The product characteristics must be capable of being expressed in measurable terms since the characteristic output is to be controlled and compared to objective targets. Using the results of Figs. A11.1 and A11.2, the first step in creating the planning matrix, shown in Figure A11.3, requires identification of the relationship between the customer requirements and the characteristics of the final product. Symbols are used to identify whether the correlation between the requirements and the characteristics are strong, medium, or weak. The absence of symbols or a dominance of weak correlations suggests that some customer requirements have not been addressed or that the characteristics listed may not be sufficient to satisfy customer needs.

Using the results of Figs. A11.1 to A11.3, the planning matrix shown in Figure A11.4 is completed. The market evaluation is contained in the evaluation of four competing vehicles, with a numerical rating giving the importance of each requirement to the customer. These data are derived from evaluations of the control characteristics of competitive vehicles and are listed in the box in "Control characteristics of competitive evaluations" in Figure A11.4 A discrepancy between the market evaluation and

Figure A11.1   A statement of the primary, secondary and tertiary customer requirements. (From Sullivan, 1986, p. 40.)

Primary	Secondary	Tertiary
Product is dependable	Reliable, trouble-free	•Always starts •Absence of troubles that: —Stop the vehicle —Result in inoperative function —Result in inconvenience
	Lasts a long time	•Absence of unexpected component wear-out •Absence of unexpected appearance deterioration
	Easily and quickly serviced	•Can be serviced quickly •Parts are readily available •Service is effective

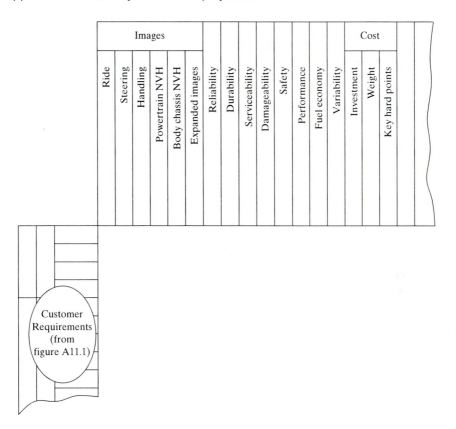

Figure A11.2   Matrix listing the final product control characteristics. (From Sullivan, 1986, p. 41.)

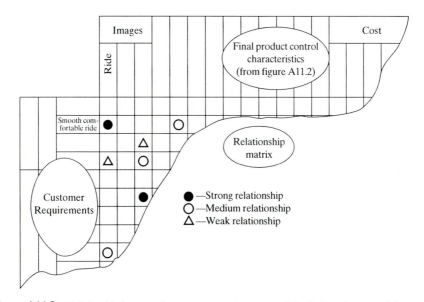

Figure A11.3   Relationship between the customer requirements and the final product control characteristics

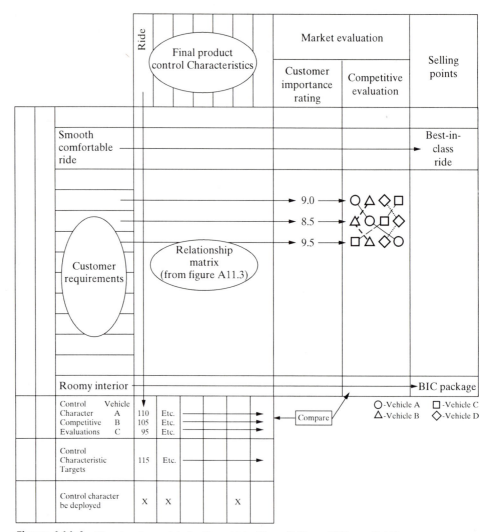

Figure A11.4 The planning matrix for the vehicle. (From Sullivan, 1986, pp. 42-46.)

the engineering evaluation of the competitive products will suggest that the customer's evaluation differs from yours. On the right-hand side of Figure A11.4, a list is given of key attributes to be used in marketing the vehicle. These are based on the degree of importance to the customer, your company's capabilities, and the cost of their incorporation.

The numerical control targets are given at the bottom of Figure A11.4 for the various product characteristics. These numerical values will be measured on the final vehicle to confirm that the objectives have been met. As shown, the ride of the new vehicle is expected to achieve a value of 115, higher than that of any of the vehicles tested, in support of the marketing strategy that the new vehicle will have a best-in-class ride. The items that must be be controlled to achieve the desired performance are identified with checks at the bottom of Figure A11.4.

The planning process is completed when Figure A11.4 is finished. The information contained in Figure A11.4 is used to create the deployment matrices shown in Figs. A11.5 and A11.6. For this example, the deployment matrix involves the translation of the characteristics of the total vehicle into characteristics of the subcomponents. In Figure A11.5, customer and control characteristics are listed in greater detail. With the identification of the front suspension as a critical subcomponent in achieving a smooth, comfortable vehicle ride, the next step is to identify the components in the front suspension that must be controlled. The importance of the springs is examined further in Figure A11.6, where finished component characteristics and critical part characteristics are identified.

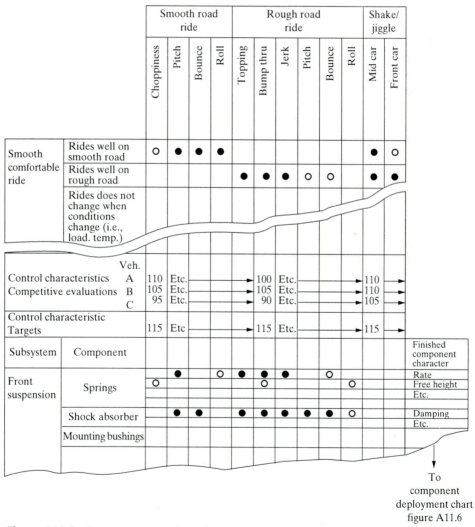

Figure A11.5   Deployment matix derived from the planning matrix. of Figure A11.4. (From Sullivan, 1986,. p. 47.)

Figure A11.6   Process plans and control chart for the spring. (From Sullivan, 1986, p. 48.)

From component deployment matrix or figure A11.5

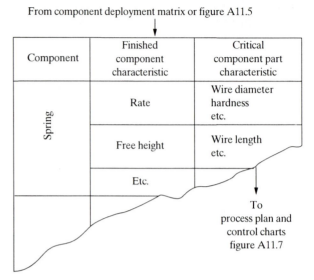

Component	Finished component characteristic	Critical component part characteristic
Spring	Rate	Wire diameter hardness etc.
	Free height	Wire length etc.
	Etc.	

To process plan and control charts figure A11.7

Figures A11.7 and A11.8 are the process plan and the control chart, respectively, for the spring. These identify the control points, the plan for monitoring these points, the control method, and the frequency with which they are to be examined. From these, the operating instructions for the manufacturing facility are derived. As Sullivan points out:

From component deployment matrix

Component: spring

No.	Process	Critical component part characteristics						Control points	Process monitoring plan			
		Wire diameter	Wire length	Hardness	Etc.	Etc.	Etc.		Check points	Monitoring method	Frequency	
1	Draw wire	●		○		○		•O.D. •Inclusions •Cracks •Etc.	•Material tempertaure •Die Temperature •Speed	•Check sheet •$\overline{X}$ & R. •Etc.	•1 hr. •3 checks hour 5 min. apart	
2	Cut wire		●				○	•Length				
3	Roll spring			○								

To quality control chart or figure A11.8

To operating instructions

Figure A11.7   Component deployment matrix. (From Sullivan, 1986, p. 48.)

From process plan chart

Component: spring

Process flow		Control point	Control method	Sample size frenquency	Check method
Raw material	Process				
▽		• Chemistry			
		• Hardenability	──────── See supplier plan ────────		
		• Etc.			
	Draw	• O.D.	X̄ & R	5 pcs. /hr.	Electronic guage
		• Inclusions • Cracks	N/A	100%	Eddy current tester
		• Etc.	Etc.	Etc.	Etc.
	Cut	• Length	X̄ & R	5 pcs. /hr.	Checking fixture
		• Etc.	Etc.	Etc.	Etc.

To operating instructions

Figure A11.8   Process plan and control chart for the Spring. (From Sullivan, 1986, p. 49.)

QFD offers many advantages. It is a customer driven system for evaluating the relationships between:

- Customer requirements and characteristics to be used to develop and control the product.
- Customer (market) and company evaluations of competitive products.

QFD also provides means of translating customer requirements into critical product control characteristics, component characteristics, process control characteristics, and operating instructions. There is a continuous flow to assure that all programmed elements are directly related to customer requirements.

Finally, QFD promotes a team approach by fostering a better understanding of requirements and what is needed to meet those requirements.

REFERENCE

SULLIVAN, L.P. 1986. Quality Function Deployment, *Quality Progress*: 39-50

Managing the Operations

The term *manufacturing system* has been used to describe the collection of activities that interact and critically influence the successful transformation of materials into products. In seeking to optimize operation of the system, it is necessary to understand and control the individual elements of the activities that take place within the manufacturing system. Some writers have described the system depicted in Chapter 1 as *Manufacturing* with a capital *M*. Those activities that take place within the walls of the plant or are closely allied with it (e.g., material procurement, shipping, etc.) are sometimes described as *manufacturing* with a small *m*. In this vernacular the focus now turns from Manufacturing with a capital *M* to a discussion of manufacturing with a small *m*. The concern will be with the many actions and activities that are centered within the walls of the manufacturing plant. Attention will focus on the organization of operations and on various ways to control these systems. While concentrating on the small *m* of manufacturing, care must be taken to retain an awareness of the impact on the big *M*.

12

Creating and Controlling the Manufacturing Operation

The characteristics that a manufacturing facility must have are determined by factors such as the size and stability of the market that it will serve, the types of products that will be manufactured, the technology that will be used, and the people who will operate the enterprise. From the data and information that are collected on these and other parameters, the size and form of the plant must be determined. Success in achieving a design that has long-term feasiblity depends on the making of the proper choices from an array of alternatives that exist.

While the concurrent-simultaneous engineering process enhances the chances of successfully achieving a product that meets customer needs, the process is seldom able to proceed in an unencumbered manner. A principal constraint often occurs with the manufacturing facility. Because of the large investments associated with facilities, one of the following conditions may exist: an existing facility may need to be used; a limited amount of capital may be available for modification of an existing plant; if a new plant is be built, it must be completed within a specified budget; or limitations may exist on the types of technologies that can be employed. Customers may also impose restrictions (e.g., in terms of record keeping, frequency of inspections, delivery schedules, etc.). The planning of a manufacturing system becomes one of choosing among alternatives that are created by forces several of which may be outside the direct control of the manufacturing activity and many of which can influence the product design process. Limitations such as these emphasize the importance of establishing the broad outlines of the manufacturing system at the time that the product is first considered and of ensuring that the manufacturing representative on the concurrent engineering team is prepared to help the team create a design that makes best use of facilities while imposing the fewest restrictions on designers. As Marsing (1991) notes: "Clearly, effective long-term planning requires regular feedback of real changes in the environment, the capability of the factory and even new ideas that come from the organization. The long-term goals seldom change, but details in executing the plan to reach the goal often change."

The creation of a successful manufacturing system requires the development of detailed plans at various levels. Although the order in which issues are addressed will vary with circumstances and depend on the type of product that is to be produced, many of the following topics must usually be considered:

- The forecast of market demand for the product, both short term and long term, as well as expected fluctuations in this demand.
- The translation of market demand into a general production schedule.
- The geographical location of the plant.
- The process plans for the facility, including the unit processes to be employed in the plant, the material transport systems that will be required, and the material storage and retrieval systems that will be used.
- The physical layout of the plant.
- The selection of unit process controls.
- The information system that will collect data from the plant floor and provide access for its use in controlling the processes.
- Selection of the materials or subcomponents that are to be purchased, as contrasted to being made in-house.
- The delivery strategy of materials and components from suppliers.
- The identification of suppliers.
- Labor requirements for the facility.
- The procedure to be employed in launching the plant to full production.

A vast array of tools have been developed for accomplishing these tasks. Some of these are general and can be used with limited information; others require detailed data.

- Forecasting models, including regression analysis, exponential smoothing, cyclic forecasting, and Box-Jenkins methods
- Aggregate planning using such techniques as the graphical/tableau approach, an empirical approach, linear-programming optimization approaches, and parametric production planning
- Materials requirements planning (MRP)
- Manufacturing resource planning (MRP-II)
- Optimized production technology (OPT)
- Enterprise requirements planning (ERP)
- Computer integrated manufacturing (CIM)
- Enterprise integration framework (EIF)
- Computer-aided layouts and computer-animation of material flows
- Modeling and simulation

The economic costs and values of the various alternatives are critical in many of these analyses. In Chapter 13, modeling, simulation, and regression analysis are discussed as important tools for analyzing the manufacturing system. As will be found, these tools are particularly valuable in assessing the relative trade-offs among alternatives when the manufacturing activity is being developed, when modifications are being considered, or when the impact of certain changes in the marketplace is being examined.

Since the student will already be familiar with many of these tools or will find adequate discussions of these topics in various textbooks (Starr, 1989; Laufer, 1984; Bedworth and Bailey, 1982; Riggs, 1987; Tompkins and White, 1984), many of them will not be discussed here. Emphasis in this chapter is on the aspects of planning that deal with some of the more general issues that management must address during the creation and control of a manufacturing system. The discussion focuses on accomplishing process improvements, considerations as to whether to follow a market-driven philosophy versus a master schedule-driven philosophy, issues concerning make-buy decisions, and the use of statical process control (SPC) to control unit processes.

12.1   LEVELS OF INTEGRATION OF THE BUSINESS

As discussed in some detail in Chapter 9, the extent to which the enterprise is vertically integrated can have a large impact on the approach that is taken in controlling the system. An abbreviated list of key issues that must be confronted in deciding the proper level of integration include the following:

1. Financial
2. Technical expertise
3. Quality of the product
4. Responsiveness of the supplier
5. Control of the technology

The overarching philosophic question that must be answered by management is whether the benefits of having someone else focus on a particular manufacturing technology, provide the investment, and operate the manufacturing facility are sufficient to offset the reduced control that will exist if these tasks are not done by the enterprise. Some companies have focused on controlling everything. They have attempted to make the maximum number of parts that go into the final product. Others have chosen to buy a large fraction of their components from another manufacturer. As also mentioned in Chapter 9, practice varies from industry to industry. The companies that depend largely on outside suppliers frequently focus on the *system* as opposed to the design and manufacture of the individual parts of the product. They are often willing to give up the advantages of control for the lower capital investment that is needed.

It is perhaps worth emphasizing, as was done in Chapter 9, that the desire to control a particular aspect of the technology is often an important reason for choosing to manufacture a component. A careful analysis of the importance of these various issues is necessary during the early stages of planning a manufacturing operation. The means by which the operation will be controlled is strongly influenced by these decisions.

12.2 INTRODUCING NEW AND IMPROVED PROCESSES

The successful introduction of new or improved processes into the manufacturing environment requires that everyone clearly understand the potential risks and benefits that they afford. Management will generally reach the correct decision and undertake the proper actions regarding new and improved processes if the correct information is available to them on a timely basis. Timeliness is a crucial element of any successful project. A project that is started before all of the processes have been adequately proven can end in failure and incur large expenses. If a project is not completed on time, a market opportunity may be missed. Furthermore, a late project almost always costs more than a timely project. The discussion that follows is centered on knowledge and timeliness-all of which are critical for the effective planning and implementation of new or improved operations. Emphasis is placed on enhancing the overall level of understanding by the organization during the implementation phase. The presumption is that improved understanding leads to improved decision making. This will result, of course, only if many of the issues discussed in earlier chapters are recognized and resolved.

12.2.1 New Processes

In the concurrent-simultaneous engineering environment, a decision concerning a new manufacturing facility must be made in tandem with the decision to introduce a new product. The timing of a project is usually determined by working backward from the desired date for the introduction of the product to the market to the date at which the project must be started. Answers to the following questions are then determined:

- How much time will be needed to design the first prototype of the product?
- How much time will be needed to create the prototype?
- How much time will be needed to test the prototype product and to revise the design to ensure that the final product design will meet objectives?
- How much time will be needed to design the manufacturing facility?
- How much time will be needed to procure the equipment for the plant?
- How much time will be needed to install and prove-out the facility?
- How much time will be needed to launch the facility?

While the stages in the development of the product and the process are reasonably common to all products, the amount of time that is needed to accomplish each of the above will vary from one product to another. Care must be taken to provide the opportunity to repeat one or more of these steps, should that prove necessary. The answers to these questions create much of the information needed to prepare a critical path method time chart.

This discussion has taken place within a context that would imply that the processes that will be used in the manufacturing operation are well understood and therefore the machines that will be needed and the interactions that will exist among the various parts in the factory are well known. What happens when an all-new process is to be introduced into the manufacturing facility, either because of the special requirements of the product or because an opportunity exists to improve the system substantially through the use of new technology? Foster (1982) offers an interesting example of the problems that can arise in this situation.

> One example is Oxirane Corp., a company with a short but enviable history of developing novel processes for the chemical industry. In the 1970s, Oxirane announced plans to introduce a new, cheaper method to produce ethylene glycol (used in antifreeze and polyester) and vinyl acetate (used in paints and adhesives) simultaneously. Intending to tap the market potential quickly, the company skipped the pilot-plant phase of development, counting on correcting significant corrosive problems inherent in the process while the plant was starting up.
>
> But the problems could not be solved. The plant was never opened, and Oxirane went out of business. Preliminary research might have provided the answer, but even if it did not, at least it would have defined the magnitude of the problem. Alert managers could have redirected the development process or terminated the project.

It is imperative that experience be gained with a new processes before it is introduced. This may be in a laboratory environment or in a pilot facility. If the new process affects only a single station, it is often possible to create an arrangement that will provide a realistic test and some limited operating experience. The degree to which this test is realistic depends on how appropriate it is to isolate the single station from its environment. If it must interact with other stations or equipment, making the isolation unrealistic, it will be necessary to find some way to determine the effect on the total system. For example, how important is the station's material handling system, and how will this be assessed? Will parts that are supplied to this station have a quality that accurately represents the quality level of those in production? Will engineers be operating the equipment in the pilot facility while nonskilled trades personnel will operate it in the plant? Will operation of the pilot facility allow a reasonable estimate of the quality of the *overall* operation? Can a reasonable estimate be determined of the fraction of the time that the operation will be operating error free? It is often difficult to get answers to all these questions without building a complete system, a solution that is often economically infeasible. The effort, then, is to build the least costly testbed that can provide realistic experience with the proposed new processes.

If the new process was developed in the research activity and if it is the enterprise's desire to control the technology by limiting outside access to it, it will be necessary for the in-house manufacturing activities to provide satisfactory answers to all of the foregoing questions. It must be recognized that a responsible group will not be willing to accept an unproven process whose failure could jeopardize production. The test facility must be developed such that it is possible to guarantee that the total system will operate properly. This demands that the operating people be involved in the development program; otherwise, they may be hesitant to accept responsibility for operating it. If the new process is to be included in a system that will be supplied by a vendor, this supplier must be involved in the development effort. It is no more feasible to "throw a new process over the wall" than it is to "throw a new design over the wall." Teamwork is essential in the early stages of developing a new process. If the operating people are early made aware that a new process looks encouraging and that its feasibility needs to be explored, they can offer valuable insight into the means by which many of the key questions are answered. If the process is to be successful, it is essential that the operating people resist adopting the attitude that the risks of using *any* new process are so great that it cannot be considered *and* that the R&D group resist adopting the attitude that only a great new innovation is worth pursuing. Team participation is the key-in the identification of projects that are genuinely attractive for exploration and in recognizing the difficulty in bringing all new processes into use.

It is clear that much of the activity surrounding the development of a new process must take place before the concurrent engineering team begins the design of the product and the associated production process. The team cannot be expected to include untried processes in their design plans. This places an extra burden on the system to undertake the timely exploration and development of new processes in anticipation of a future need. As Bowen noted (Chapter 11), this requires exploration of alternatives prior to the formal initiation of a project. If the system waits to start the development of a new process until the concurrent engineering team needs it, there will not be sufficient time. If only the old, tried, and proven processes are permitted to be implemented, the ultimate result will be a noncompetitive manufacturing activity.

12.2.2 Improving an Existing Operation

In the improvement of an existing operation-or the upgrading of an operation to accommodate a new product-many of the issues mentioned above will arise. A critical concern is usually with the compatibility of the proposed improvements to existing operations and whether incorporation of a new process will negatively affect a portion of an existing operation. The answers to these questions are derived through many of the same steps that are followed in establishing an all-new process. The proposed change must be tested and evaluated in an environment that is sufficiently representative of production that the results can be expected to apply to the final operation.

The pilot operation that will be pursued prior to introduction of a new operation into an existing facility can sometimes be accommodated within the manufacturing activity. It might be possible, for example, to develop a branch of a production line that utilizes the elements of the full manufacturing facility both ahead and behind the experimental pilot line. If the production of current products is not disrupted by this, it might be possible to explore the new technology as a parallel operation. If the physical disruption to the existing line is small, it may be possible to explore the improvements during selected times such as off-shifts or on the weekends when the operation is not in full production.

For an improvement that can be characterized as a simple upgrade of an existing machine, the vendor providing the new facility can be asked to verify its performance. Thus the responsibility for ensuring that the new facility will not jeopardize or negatively affect the performance of other activities in the facility can be left to the vendor who will be supplying the new machines or instrumentation.

12.3 PRODUCTION CONTROLS

In a complex manufacturing operation, controls are essential or chaos will result. Foo and Kinney (1990:415-416) illustrate this with a description of the situation that existed at the AT&T Denver Works:

Prior to the implementation of a MRP system, the Denver Works was planning and releasing material and shop orders based on forecasts. Customer orders were assembled-to-order from a pool of work-in-process and finished sub-assemblies. However, since the forecasts generally did not match actual orders, informal systems were developed to expedite material to meet production schedules. This created considerable churn in the process and greatly contributed to the "hidden costs" of the factory. The deployment of an MRP system provided discipline and a common language to the material management function, provided a tool to more accurately plan and control material, and allowed the factory to accept and plan for additional orders with a minimum of disruption.

One of the critical decisions that the management of an operation makes is the *philosophy* that will be followed in the overall control of the operation. Much of the control of operations in U.S.-based plants has been accomplished with a master scheduling system. In contrast, many of the large-scale discrete parts manufacturers in Japan have operated with a just-in-time philosophy. In planning a new operation, an upgrade of an existing operation, or in attempting to improve the efficiency of an existing operation, it is important that a clear understanding of the limitations and advantages of these radically different operating philosophies be understood and that the appropriate control strategy be adopted.

The two systems are often distinguished as being pull driven (JIT) as contrasted to push driven (master schedule). In a pull system, production is initiated in response to an immediate demand. In a push system, production is initiated in response to a perceived future demand.

12.3.1   Just-in-Time Production

In its most idealized form, JIT is an operating philosophy that recognizes the importance of responding to the customer, whether that customer is the ultimate consumer of the product or the next station in the production system. The intent in a JIT system is to produce the number of units required, at the required rate, when needed by the customer. It is an order-driven system. At every stage throughout the manufacturing process, demand triggers actions. A part moves from station C to station D only when station D demands it. Production of a unit is not initiated until an order arrives.

Toyota, who is given credit for successful development of the just-in-time production system, spent more than a couple of decades developing it. The system did not attract great attention outside Japan until the mid-1970s. With the first oil embargo and the rapid success of the Japanese car manufacturers in the newly developing world marketplace, companies around the world rushed to implement the techniques that Toyota and other were using-often with unfavorable results. What most of these companies failed to recognize was that successful implementation of JIT requires more than the adoption of a set of techniques. The techniques, as important as they are, must be preceded by an operating philosophy that creates an environment in which JIT can flourish.

Many people call the Toyota production system a Kanban system: this is incorrect. The Toyota production system is the way to make products, whereas the Kanban system is the way to manage the Just-in-Time productions method. In short, the Kanban system is an information system to harmoniously control the production quantities in every process. *Unless the various prerequisites of this system are implemented perfectly*

> *(i.e., design of processes, standardization of operations and smoothing of production, etc.), then Just-in-Time will be difficult to realize, even though the Kanban system is introduced* [emphasis added]. (Monden,1983:4-5)

Other information must be provided to the production facility to operate a JIT system.

> Before entering the phase of dispatching the jobs by Kanban, overall planning throughout the plant must be made in advance. For this purpose, Toyota will inform each process and each supplier each month of a predetermined monthly production quantity for the next month's production so each process and each supplier in turn can prepare in advance its cycle time, necessary workforce, necessary number of materials, and required improvement point, etc. (Monden, 1983:23-24)

Monden discusses a number of prerequisites that must be implemented if a JIT production system is to be achieved. These include:

1. High quality must be maintained at each step of the operation. Defective products must not be allowed to move forward through the system. This suggests an objective of zero defects.

2. Production schedules must be smoothed. For a Kanban system to operate, large fluctuations in demand must be eliminated or smoothed. This suggests an objective of zero surging.

3. Responsiveness to customer demand requires that tool changes be accomplished in a short time. This implies objectives of zero lead time, zero set-up time, and a capability to operate efficiently at a lot size of one.

4. Unnecessary tasks must be eliminated. This implies an objective of eliminating all actions that do not add value (e.g., zero material handling).

5. High reliability must be achieved for each process. This implies an objective of zero breakdowns in the facilities.

An analysis of these conditions for the successful operation of a JIT manufacturing system led Nick Edwards, a JIT expert with Coopers & Lybrand (Fallon, 1986), to identify the "seven zeros of JIT":

- Zero defects
- Zero lot size
- Zero surging
- Zero lead time
- Zero set-up time
- Zero handling
- Zero breakdowns

The desirability of implementing these as operating objectives was discussed in earlier chapters. The need for high quality, the importance of continuous improvement through employee involvement, the need to listen to the customer, and the need to utilize well-proven processes is important whether or not a JIT production system is being utilized. One can choose to utilize these good operating procedures without following a JIT system. One *CANNOT* implement a JIT system successfully without also implementing these procedures. Thus, they become a prerequisite to a successful JIT system. It is this fact that many manufacturers failed to grasp as they rushed to copy the Toyota production system. As the name states, it is a system. It involves all aspects of the manufacturing process. It is not simply a Kanban system for releasing materials at each stage of the operation.

JIT attempts to manage the system in such a way that lead times are minimized and that waste is eliminated. Monden (1983:vi) states that in 1980 Toyota Motor Company had achieved a turn ratio of inventory that was 87. This is a level that is unheard of in most Western companies.

The results of an extensive study of the factors that influence the success of JIT implementation has been carried out by Sakakibara et al. (1993). A total of 16 important criteria for success of a JIT system were identified. These are listed and discussed briefly in Appendix A12.1.

12.3.2 Master Schedule Production

With a production system that is controlled by a master schedule, the manufacturing system is scheduled from the top down. The manufacturing plan must be derived from what is perceived to be the future market demand. This creates a push system for the manufacturing activity. From the perceived future market demand, there is created a detailed schedule for each operation throughout the manufacturing activity. Material orders are placed, deliveries are scheduled, and the requirement for each operation is determined. MRP and MRPII are designed to provide the capabilities to achieve these objectives. (See Appendix A12.2 for a brief description of these.)

The master demand schedule is often developed for a month in advance, with the detailed production plans being issued to the factory floor on a weekly basis. The complexity of most manufacturing operations is such that the detailed instructions to a single process cannot be changed without changing the instruction to all elements of the system, even when there is a recognized change in market demand. Thus, detailed changes in operations on the floor of the factory tend only to be made on a weekly basis.

There is little wrong in concept with this operating philosophy as long as the final orders are an accurate reflection of what was assumed for the market demand and as long as any fluctuation in the demand, either because of changes in the marketplace or through unforeseen changes in the operations on the factory floor, occur so infrequently that the weekly production schedule can ignore them. Unfortunately, this assumed stability seldom exists. Foo and Kinney (1990:416) make the following observation about the production of PBXs at AT&T's Denver Works.

> The master production schedule (MPS) is the driver of the MRP system and the proper implementation of the MPS ensures successful material management. Therefore, the decision on what to master schedule is crucial. For assemble-to-order systems such as PBX's, there is almost a limitless number of end-item possibilities made from combinations of the basic components and subassemblies.

Master scheduling at the end-item level for PBXs in unacceptable since the almost limitless number of end-items are impossible to forecast accurately and master schedule with any hope of validity. Furthermore, the many interdependencies complicate the engineering change process since the linkages between design drawings need to be correctly traced and the appropriate changes recorded in a maze of documents. This results in an environment which is error prone, not cost effective and has long product realization intervals because of the need for re-work. In addition, the hierarchy and drawing dependencies ensure that the BOM will be multi-level which for a "weekly bucketed" MRP system ensures that the manufacturing interval will, at best, be the equivalent (in weeks) to the number of BOM levels. For example, a 5 level bill of materials (BOM) in a weekly "bucketed" MRP system means that the manufacturing interval will be 5 weeks even though the product may be manufactured in 2 weeks.

If the changes in demand for the product occur more rapidly than the number of multilevels in the BOM, the result will be that products will be built that are not needed by the market or the planned production will be insufficient to meet an expanding market need. The consequence will be an inventory of parts that are not needed or, alternatively, a lost market opportunity. If these rapid fluctuations persist over a long period of time, the final consequence will be either a loss of market share or a need to expand the manufacturing facility to enable it quickly to make the additional parts that could not be manufactured under the previous plan. In either case the cost can be very large-in the first instance in lost sales or in the second an investment in an expanded facility to accommodate an added capacity that is used only to compensate for the inability of the firm to respond quickly to fluctuations in demand.

Note that both of the undesirable consequences above are the result of variances in the system-a variance in the market demand from that which was assumed and a variance in the operation of the facility. Karmarkar (1989:125) states this in the following way:

> Of the many standard assumptions made by MRP, the fixed lead times are the most troublesome. Why is MRP so susceptible to getting lead times wrong? The best answer is that production lead times vary depending on the degree of congestion or loading within the shop. The fallacy in MRP is that its releases produce the very conditions that determine lead times, but these lead times have already been taken as known and fixed in making the release.... There is another way of looking at this. A single lead-time must suffice in MRP for all situations faced on the floor. Consequently, the number has to be set high enough to cover all variations up to the worst case. If an order is ever late, people have an incentive to increase the planned lead time in the system so that the delay does not occur again.... Similarly, orders will usually be released too early and will often complete early, thereby increasing inventories in the system.

12.3.3   Comparison of JIT and Master Schedule Production

These two operating philosophies may seem to be completely incompatible, with JIT being basically a pull system and the master schedule system being a push system. They are, however, much more compatible than is implied by this description. A JIT operating system must have a long-term plan that reflects expected market trends, thus requiring information similar to that used in the preparation of a master schedule. A master schedule system must have a means of initiating replenishment of inventory on the floor, a process that is straightforward for a JIT system.

Monden makes the following observation concerning the comparison of Kanban system with MRP (1983:64-65).

> The Kanban system requires that an overall production schedule be circulated throughout the plant before actual production schedule begins. Such an overall plan is called a *master schedule* in MRP. This master schedule is very important for MRP, because it is a target to be rigorously maintained. In the Kanban system, the overall plan does not strictly target production, but merely sets up a loose framework that prepares the plant-wide arrangement for materials and workers at each process.
>
> Consequently, in the MRP system there must be a review at the end of every planned production interval-or bucket-that compares planned production with actual performance. If the review discovers a discrepancy between planned and actual performance, then remedial action must be taken. Since these are bucket sizes of at least a week, the master schedule must be revised weekly.
>
> The Kanban system does not require any comparisons between planned and actual performance at the end of a production interval-i.e., one day-because such comparisons must necessarily evolve out of the daily actual production process and the daily dispatching of production by Kanban. If the daily production plan-the sequence schedule-requires revision, such revision will be based on the dealer's daily orders and will reflect daily market conditions....
>
> However, the Kanban system can be compatible with MRP. After MRP made the master schedule, the Kanban system could be applied as a dispatching tool of production within each bucket.

Foo and Kinney (1990:420) describe the final stages of the implementation of JIT and MRP at AT&T's Denver Works as follows:

> [W]ith the successful implementation of the MRP system, the Denver Works had reached a point where the material demands were balanced with material suppliers. The shop floor was also operating efficiently through the implementation of JIT/TQC. However, the various factory operations were still not fully integrated. Since a goal of the Denver Works is to minimize, at the lowest cost, the total interval for manufacturing the end-item systems, a missing link was an execution system that not only initiated circuit pack starts but more importantly synchronized all operation in the factory. That execution system was called Integrated Pull Manufacturing (IPM).
>
> IPM synchronizes and integrates MRP planning, the manual Kanban systems into the equipment feeder shops, the storeroom and circuit pack manufacturing operations with the final assembly schedule. .... Critical concepts considered during the design of IPM were:
>
> 1.   MRP is for planning while Pull is for execution.
> 2.   Smooth and efficient manufacturing requires a level load which is driven by final assembly.

Karmarkar (1989:127) has examined the circumstances that determine which control system is best and offers the following advice to those who are attempting to determine the focus for their systems.

The key to tailoring production control lies in understanding how the nature of the production process drives the choice of control method.... For a continuous-flow process, ongoing materials planning is not essential and JIT supply techniques work well. Order releases do not change from week to week, so a rate-based approach can be used. At the shop floor level, JIT materials-flow discipline combined with pull release-kanban, for example-is effective.

In a repetitive manufacturing environment with fairly stable but varying schedules, materials planning can be a combination of MRPII and JIT methods. Order release may require MRP calculations if changes are frequent or if it is necessary to coordinate with long lead times or complex materials supply and acquisition. Pull methods work well on the shop floor.

As we move to more dynamic, variable contexts-like job shop manufacturing-MRP becomes invaluable for planning and release. Pull techniques cannot cope with increasing demand and lead-time variability. Shop floor control requires higher levels of tracking and scheduling sophistication. Materials flow is too complex for JIT.

Finally, in very complex environments, even job release requires sophisticated push methods. Where these are too expensive, the only option is to live with poor time performance, large inventories, and plenty of tracking and expediting.

Karmarkar has developed a very nice representation of the circumstances under which various control options should be chosen. These are presented in Table 12.1. As was noted, MRP tends to be favored for planning the materials that are needed and for releasing those materials into the manufacturing process when the product mix becomes complex. The JIT system is favored for the control of the shop floor and for the control of the processes when the product system is more closely aligned with a continuous process.

A combined JIT-MRP system-a hybrid system-offering the advantages of both has many attractions. A natural question to ask is whether there are any disadvantages. Hartland-Swann (1987) suggests that there are some concerns that need to be addressed as one attempts to find the proper combination of these two. The limitations with the implementation of MRP are the following :

1. Inaccurate master plan
2. Inaccurate inventory records
3. Inaccurate lead time
4. Insufficiently frequent updating
5. Shortages

The limitations with the JIT system are the following:

1. Long procurement lead times
2. Demand variability

Table 12.1   Production Controls Tailored to the Production Needs of the System

	Materials planning	Order release	Floor control
Pull (continuous)	JIT	Rate based	JIT/pull
Hybrid (batch repetitive)	JIT/MRP	Pull or MRP	Pull
Hybrid (batch dynamic)	MRP	MRP	Pull or order scheduling
Push (job shop)	MRP	Order scheduling	Operation scheduling

Source: Adapted from Karmarker, 1989.

It is Hartland-Swann's conclusion that using MRP for planning and JIT for shop control can be combined effectively if the execution and control functions are decoupled and if the physical arrangement of facilities on the shop floor are reorganized to smooth the flow of materials and to establish some form of cellular manufacturing. The successful use of JIT in combination with MRP will only be successful, however, if the enterprise is also willing to embrace the principles and work hard to implement the prerequisites enumerated in Section 12.3.1. Unless these are actively pursued as a long-term objective for the enterprise, it is unlikely that JIT or a combination of JIT and MRP will be successful.

12.4 INTERNATIONAL STANDARDS FOR QUALITY

The complex mixture of people, materials, machines, and products makes the control of a manufacturing system difficult. In the view of the consumer, the control must *assure* that the marketplace receives only high-quality products. For the manufacturer, the controls must create a system that *produces* high-quality products. It is not sufficient to inspect quality into the product. The system must produce high quality.

As the world marketplace has evolved, a proliferation of requirements has developed for meeting the quality requirements imposed by companies. In recent years an attempt has been made to create a system of registration and certification that will assure that anyone complying with these requirements will satisfy conditions that are deemed essential for meeting agreed-upon levels of quality. The effort is meant to assure that international trade will not be hampered by national requirements for unique standards of quality.

The international leadership for this effort has come from the European Union. The EU has decreed that a company will not be allowed to do business in Europe, including the importation of products, unless it demonstrates that it can meet the standards contained in EN 29000. The International Organization of Standards (ISO) has adopted ISO 9000 as the international standard that will assure compliance with EN 29000. Companies around the world have moved rapidly to comply with ISO 9000. Appendix A12.3 presents a description of the contents of ISO 9000.

12.5 CONTROLLING THE OPERATIONS

As efforts continue toward developing more comprehensive analytical descriptions of manufacturing systems, it can be expected that these will enhance the capability of controlling, as well as predicting, system performance. It is not possible, however, to await the completion of all of these. Existing manufacturing systems must be improved and those systems that are just now being planned or are currently under construction must be controlled in such a way that they produce high-quality products.

Although there is no single unique way in which the various subsystems of the manufacturing operation can be structured or managed to give a single-best response, there are a set of tools that have been found to be particularly useful in achieving high quality. The purpose here is to introduce these tools, explore the types of questions that can properly be addressed with them, and to seek to understand how they can be used to improve the management of operations.

The objective of controls is to ensure that the system is performing as desired. Controls are needed to assure proper operation of the individual machines, to ensure the quality of the final product, to provide an adequate response of the system to changing conditions, to maintain costs within the levels required for meeting profitability goals, and to ensure that the environment in the plant is safe and that working conditions are satisfactory. Responsibility for providing these, and other, controls is spread throughout the organization. Recent experience with participative management programs demonstrate clearly that the workers are best able to control the shop floor. The shop workers, being constantly involved with the machines and processes, are in the best position to understand the source of problems and to correct them. At progressively higher levels in the organization, the controls tend to focus more on system objectives-meeting financial objectives, meeting productivity goals for the system, minimizing waste, and so on.

12.5.1 General Concepts

Control of the operations begins with the processes. During design the important attributes will be established and the dimensions of the parts determined, X_i, and the variation that is allowed for these dimensions, ΔX_i. In some designs it will be acceptable for X_i to take on any values between $X_i \pm \Delta X_i$. In others it may be necessary that X_i have values between X_i and $X_i + \Delta X_i$ or between X_i and $X_i - \Delta X_i$. Since the ratio of $\Delta X_i/X_i$ usually influences the ease of manufacture and the final cost of manufacture, the design team usually seeks to make this ratio as large as possible. The design team must also determine whether parts having dimensions falling outside the design tolerances will be acceptable and, if so, the frequency with which they will be allowed. With the tolerances established the operation must be controlled in such a way that it will provide parts meeting these conditions.

If the design specifies that no part can be allowed to have a dimension that falls outside the tolerance range, the operation can be controlled in either of two ways. In the first, each part can be inspected and any part with dimensions that fall outside the range will be discarded. If the fraction that is discarded is small, this may be acceptable, *assuming that the inspection process does not add substantial costs*. Alternatively, it may be possible to create a system that will guarantee that all parts have the desired tolerances (e.g., a closed-loop feedback control system).

Closed-loop feedback control depends on relating a physical parameter of the manufacturing system to a critical dimension of the part. For a shaft with a desired diameter X_i, an attribute of the physical system, say Pi, might be the location of the end of the cutting tool. The depth of draw for a stamping might be related directly to the depth of penetration of a punch into the die. In a CNC lathe, this might take a different approach. Since the position of the tool holder is controlled automatically, subsequent parts will have larger and larger diameters as the cutting tool wears. Control is this situation could be achieved by measuring the diameter of successive parts and adjusting the location of the tool holder to compensate for the tool wear. In this circumstance, the diameter of the resulting product would need to be measured only rarely, just to ensure that the control system is operating properly.

In a closed-loop control system, Pi is constantly monitored and a means is provided to maintain the value of Pi at a predetermined value. The closed-loop system requires a sensor, an actuator, and a communication link that informs the actuator of

the need to modify *Pi*. A model of the process and its dependence on certain parameters must be available to establish the relationships between *Xi* and *Pi*. This can
be established through experiment, through a theoretical investigation, or through a
combination of the two. Since the closed-loop system operates in real time, issues
such as the time constant of the control loop, sensitivity of the loop, reliability of the
sensor, and the sensitivity of the system to noise must be considered.

In contrast to closed-loop control, it may be possible to use a system that operates under open-loop control. In such a system, no information, from either the
process or the product, is used to control the process in real time. For an open-loop
system to be effective, it is necessary that the process be so stable and so free of per-
turbation that it will operate reliably and reproducibly without external intervention.

Most, but not all, discrete-parts manufacturing is designed under the presumption that the process can operate under open-loop control in the short term. The more
the system approaches a continuous flow, the more likely that this will be the control
strategy. Feedback control is only provided for maintaining long-term performance
objectives. The presumption is that the process is sufficiently stable in the short term
that it does not require a constant input of information for maintaining control. Over
the long term, adjustments will be made to compensate for drift or wear of the machine, wear of the tooling, or for example, changes in the average properties of the
input materials.

The output from a system that is operating under open-loop control will be affected by fluctuations that occur naturally in the various elements of the system. The diameter of a turned shaft can vary for reasons such as that the properties of the material
being machined may vary from piece to piece; the precise orientation of the part as it is
held in the fixturing clamps can be subject to some variation; the speed with which the
spindle turns can change depending on the temperature of the gears, the frequency of
the electrical power source that energizes the spindle motor, and so on, and the temperature of the machine or the material being machined can vary. These and many other
events continually introduce a randomness into the performance of the operation. The
statistical distribution of these many factors will determine their collective impact. Although it is not required, it is usually assumed that the occurrence of any variation in the
inputs follows a normal distribution, thus implying that the distribution of values of the
attribute of the product that is being produced will also follow a normal distribution. Although other distributions may occur, it has been found that inaccuracies obtained by
assuming a normal distribution are not large enough to justify the complexity of using
others. The task for the operator is to find a means to determine the status of the operation and to judge what actions should be taken in the presence of random events that
introduce noise into the values of the attributes of the products.

12.5.2   The Normal Distribution

For convenience, it will be assumed that the probability of the occurrence of an event
is determined by the normal distribution having a mean, μ, and a standard deviation,
σ. The density function for the normal distribution is given by

$$f(x) = \frac{1}{\sigma\sqrt{2\pi}} \exp\left[-\frac{(x - \mu)^2}{2\sigma^2} \right]$$ (12.1)

The probability that a value lying between x and $x+\Delta x$ is $f(x)(\Delta x)$, where $f(x)$ is given by Equation (12.1). Figure 12.1 displays a normal distribution with a mean, μ. The locations of $\pm 3\sigma$ are shown on the graph. The fraction of the distribution contained within the various standard deviations can be found by integration of Equation (12.1) between the chosen limits. The results are summarized in Table 12.2 for values of σ between 1 and 6. Also provided in the table are the probabilities that a value will be found that is outside the $\pm (\lambda)\sigma$ limits. For the $\pm 3\sigma$ limits, a distribution that is normal will have 2.7 observations out of each 1000 that falls outside these limits. For the $\pm 6\sigma$ levels, 1.97 observations out of each 10^9 will fall outside these limits.

TABLE 12.2 Fraction of the Area under the Normal Distribution Curve between $\pm (\lambda)\, \sigma$

λ	$\int f(x)\, dx$	$1 - \int f(x)\, dx$
1	0.6826	3.17 in 10
2	0.9544	4.56 in 10^2
3	0.99730	2.70 in 10^3
4	0.9999366	6.34 in 10^5
5	0.999999426	5.74 in 10^7
6	0.99999999803	1.97 in 10^9

12.5.3   Statistical Process Control

As mentioned above, it is feasible to consider a manufacturing strategy that involves 100% inspection of all parts at every station. Such a strategy would be used, however, only when the cost of sending a defective part forward to the next station is so large that it must be prevented at all cost. It is not uncommon, of course, to utilize 100% inspection of the finished product before it is shipped to the customer, since the cost of shipping a defective product to a customer is considered to be sufficiently high that the inspection costs are warranted.

If a process is under open-loop control, it will produce parts with values of its attributes that follow a distribution that is assumed to be normal. A plot of the measured values of the desired attribute of a large number of parts will give a normal distribution of the form shown in Figure 12.2. If, however, the mean value of the attribute drifts during the time that the parts are made, a condition that would characterize a process that is not in control, the plot of these values would not follow a normal distribution. In all but extreme cases, the confirmation of either an in-control or an out-of-control process would require the making of a large number of observations.

An operator who is attempting to determine whether a process is under control by direct sampling is confronted with the problem of determining how an individual result should be interpreted. Since a single observation can fall anywhere on the distribution curve, a single measurement that is far removed from the mean can signal a problem with the process or may simply be the result of a statistical fluctuation, although this might have a low probability of occurrence. Determining which of these is correct again requires the taking of a large number of observations.

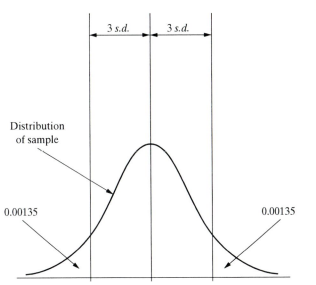

Figure 12.1 Normal probability distribution.

The intent of statistical process control (SPC) is to allow a determination of the status of the operation with a limited number of measurements under the circumstance that the process is subject to random fluctuations. SPC is an important tool for helping the operators on the floor analyze and control the unit processes. For SPC to be useful, it is necessary that:

- The process must be highly reproducible.
- Fluctuations in the process parameters must be random.

The periodic sampling of a key parameter, through the measurement of an attribute of the product such as a critical dimension, and the use of the SPC methodology allows the detection of drift even when the magnitude of drift is less than or comparable to the random fluctuations that naturally occur in the process. Since SPC has several uses, it is important to distinguish among these.

1. SPC can be used to determine the capability of a new process and to provide guidance in its modification in order to bring it under control.
2. SPC can be used to determine that the process being measured is under control, that is, that the fluctuations that are observed in the values being measured are the result of random events and do not result from some systematic shift in parameters that control the process.
3. SPC can be used to ensure that the process is capable of providing parts that are within the specifications established by the design team for the part.

Although these three uses are related, it is important to realize that *a process can be under control but could be providing parts that deviate from the specification by an amount greater than is permitted by the design.* Each of these uses is discussed in detail below using the concept of a control chart.

12.5.4   The Control Chart

The control chart provides a way of detecting whether a process is in control by the making of a limited number of measurements and by calculating the mean and the range of these measurements. Consider a manufacturing operation that is making parts that have a critical dimension denoted by X. SPC requires the collection of n samples (parts), at *regular* intervals throughout the manufacturing cycle. The value of X is determined for each part. The regularity with which the group of n samples is taken is determined by the extent of experience with the process, the stability of the process, and the degree of control that can be exercised over all the parameters that directly influence the process. Early in the introduction of a process, the sampling may need to be frequent. Later, as experience is gained, the frequency of sampling can be reduced. From these values a variety of control charts can be constructed.

The mean of the X values for the n parts contained in each group of samples is denoted by

$$\overline{X} = \frac{1}{n}\sum_{i=1}^{n} i\, X_i \tag{12.2}$$

The range R for the n parts in *each* group of samples, is given by

$$R = x_{max} - x_{min} \tag{12.3}$$

The discussion here will be limited to the $\overline{X}$ control chart and the R control chart. The values of $\overline{X}$ and R are plotted for each sample for a reasonable sequence of samples. The use of these charts in analyzing the process characteristics is given in the following example.

A lathe operation is machining a cylinder of steel. A sample is taken during the second and sixth hours of each shift. Table 12.3 contains the results of the measurement of the diameter of 10 samples with four cylinders being measured in each sample (n = 4). The mean, $\overline{X}$ and the range, R, of each of the samples is determined, using the definitions given in Equation (12.2) and (12.3).

For the example shown in Table 12.3, the data for the five days of operation are used to calculate $\overline{X}$, the mean of all 10 values of $\overline{X}$, and μ_R, the mean value of the 10 R values. $\overline{X}, \overline{\overline{X}}, R$, and μ_R are plotted in Figs. 12.3 and 12.4.

Although simple plots of the values of $\overline{X}$ and R are useful, their value is enhanced when they can be compared directly with the limiting values that would be expected for and $\overline{X}$ and R for the given physical situation. These limiting values are called the control limits for and $\overline{X}$ and R. The mathematical formalism needed to determine the limits for $\overline{X}$ and R can be found in a variety of books on statistics, including the book by Burr (1976). Only the results of these derivations are given here.

The significance of the control limits for $\overline{X}$ and R is straight forward. A value of $\overline{X}$ or R outside their respective limits indicates that the process may not be in

TABLE **12.3** Data for Calculation of Control Charts

Sample Number		Diameter (inches)	X		$\overline{X}$ Mean (inches)	Range
1	1.400	1.406	1.406	1.404	1.404	0.006
2	1.412	1.422	1.402	1.493	1.432	0.091
3	1.479	1.473	1.472	1.450	1.468	0.029
4	1.461	1.492	1.399	1.403	1.439	0.093
5	1.494	1.425	1.483	1.421	1.456	0.073
6	1.442	1.481	1.466	1.490	1.470	0.048
7	1.484	1.430	1.487	1.401	1.451	0.086
8	1.458	1.498	1.471	1.466	1.473	0.040
9	1.483	1.469	1.402	1.492	1.462	0.090
10	1.477	1.417	1.407	1.411	1.428	0.070

$$\overline{\overline{X}}=1.448 \qquad \mu_R=0.063$$

control and a modification in the process may be needed. Thus a knowledge of the value of the control limits becomes very important in ascertaining the status of the process. These limits are known as:

- Upper control limit for the $\overline{X}$ chart ($\mathrm{UCL}_{\overline{x}}$)
- Lower control limit for the $\overline{X}$ chart ($\mathrm{LCL}_{\overline{x}}$)
- Upper control limit for the R chart (UCL_R)
- Lower control limit for the R chart (LCL_R)

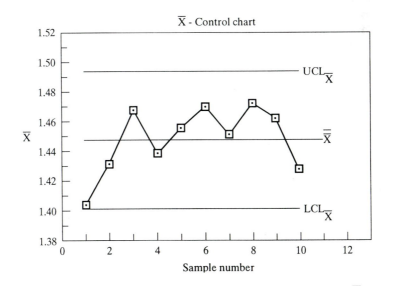

Figure 12.2 $\overline{X}$ control chart (data from table 12.3) $\mathrm{UCL}_{\overline{x}}$= 1.494, $\mathrm{LCL}_{\overline{x}}$=1.402, $\overline{\overline{X}}$ =1.448.

The equations for the determination of these limits are determined from the statistics of the distribution. The equations for the limits for the $\bar{x}$ chart are:

$$\text{UCL}\bar{x} = \bar{\bar{X}} + \frac{\lambda}{d_2} \sqrt{n} \; \mu_R \tag{12.4}$$

$$\text{LCL}_{\bar{x}} = \bar{\bar{X}} - \frac{\lambda}{d_2} \sqrt{n} \; \mu_R \tag{12.5}$$

where λ is the number of standard deviations that are being desired (3 for a $\pm 3\,\sigma$ spread, 4 for a $\pm 4\,\sigma$ spread, etc.), n is the number of measurements in each sample, and μ_R is the average of the range of X values. For the R chart

$$\text{UCL}_R = \left(1 + \frac{\lambda d_3}{d_2}\right)\mu_R \tag{12.6}$$

$$\text{LCL}_R = \left(1 - \frac{\lambda d_3}{d_2}\right)\mu_R \qquad \text{for } \text{LCL}_R \geqslant 0 \tag{12.7}$$

The values of d_2 and d_3 are determined by the number of observations in each sample, given in Table 12.4. With the values of UCL_X, LCR_X, UCL_R, and LCL_R calculated from Eqs. (12.4) to (12.7) and the data from Table 12.3, the control charts of Figure 12.2 and 12.3 are created. A blank control chart is shown in Appendix A12.4.

For the case in which the control limits are set at $\pm 3\sigma$, the constants in the equations above are known as A_2, D_3, and D_4. For n = 4 (Table 12.4), their values are:

$$\frac{\lambda}{d_2\sqrt{n}} = 0.729 = A_2 : 1 - \frac{\lambda d_3}{d_2} = 0.000 = D_3 : 1 + \frac{\lambda d_3}{d_2} = 2.282 = D_4$$

Values of A_2, D_3 and D_4 for any n can easily be calculated. Although values of A_2, D_3, and D_4 are frequently listed in tables (Burr, 1976), it must be remembered that the constants A_2, D_3, and D_4 are defined *only for the case* of $\pm 3\sigma$. For values of λ other than 3, Eqs. (12.4) to (12.7) must be used to calculate the control limits.

TABLE 12.4

Sample Size, n	d_2	d_3
2	1.128	0.853
3	1.693	0.888
4	2.059	0.880
5	2.326	0.864
6	2.534	0.848

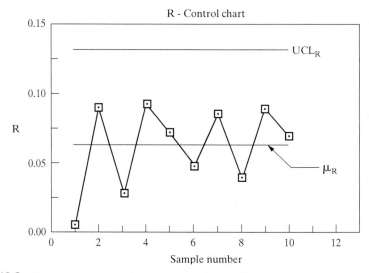

Figure 12.3   R control chart (data from table 12.3). $UCL_R=0.144$, $LCL_R=0.0000$, $\mu_R=0.063$.

12.5.5   Using SPC During Introduction of a New Process

During the introduction of a new process-the launch period-a process can be expected to undergo major changes and improvement. SPC can be an effective tool in determining whether the process is reaching a stage that can be characterized as being under control. Data are collected at frequent intervals and $\overline{X}$ and R charts are prepared. To bring the process under control, it is important to examine trends in $\overline{X}$ and R data. Identification and correction of procedures that result in reducing the range, R, of the metric that is being sampled is a critical step in improving the process. If this can be accomplished, the capability of the process will improve. A variety of techniques have been developed for treating the data during this early and usually somewhat unstable period to determine whether the process is approaching an equilibrium state.

12.5.6   Using SPC to Assure that an Operation Is Under Control

Data should be collected at regular intervals and a calculation made of $\overline{X}$, R, μR, UCL, and LCL for both $\overline{X}$ and R and the control charts for $\overline{X}$ and R prepared. The frequency of sample collection depends on the stability of the physical system. A very stable system requires infrequent sampling, whereas a system that is subject to frequent shifts must be sampled more often. Whatever the frequency, the sampling must take place at regular intervals.

For a process to be under control, $\overline{X}$ should lie at the midpoint of the desired range of values, and essentially all the values for $\overline{X}$ and R must fall within the limits of the UCL and LCL of the control charts. No clear trends should exist that would suggest a departure from the region that is defined by the control limits. In addition, and LCL. If any trends exist in the values of these quantities, either increasing or decreasing, this is cause for concern and actions should be taken to remove the tendencies, thereby improving the control of the processes. The existence of a value outside the control limits should be cause for concern. While there is a statistical probability of observing a single value outside the limits, the chance that close by values will all fall outside the limits is very small. More than one value outside the limits is a good indication that the characteristics of the process are changing. When there is evidence of a trend or when several values are close to exceeding the limits, it is important to determine the causes and to take appropriate action.

12.5.7 Using SPC to Assure That the Operation Is Capable of Providing Parts That Meet Specifications

Having demonstrated that an operation is under statistical control, it is now appropriate to inquire whether the demonstrated performance is adequate to meet the specifications that the designer has indicated is appropriate or required. Although some specifications may be asymmetric, as noted above, the symmetric case will be used for illustration here, namely a specification of the form of $X_0 \pm \Delta X$. A manufacturing system will be capable of providing parts that meet the design specifications if the following conditions are met:

1. The process is under control, as demonstrated by the $\overline{X}$– and R control charts.
2. The mean value of the attributes coincides with the nominally desired value of the attribute.
3. The distribution of the measured values of the attributes is such that the desired fraction of all parts will fall within the specified limits.

It is worth noting that of the three points above, the first can exist while either or both of the latter can be violated, as was discussed in Section 12.5.4

The second condition means that the process should be adjusted to make $\overline{X}$ the same as X_0. The third condition is somewhat less obvious. Because of random events there is always a *statistical probability* that some parts will be produced that have a value outside the specifications. The third condition states that the design-or the management-states that this number should not exceed a stated fraction of the manufacturing. For example, if it is desired that the number of parts *failing* to meet the specification *not exceed* 270 out of 100,000, the distribution of parts being manufactured must have a $\pm 3\sigma$ *capability that coincides with the specifications*. If it is desired that the number of parts *failing* to meet the specification not exceed 5.7 out of 10^7 parts, the distribution of parts being manufactured must have a $\pm 5\sigma$ *capability that coincides with the specifications*. Note that the distribution of parts that are *produced* by a given process is independent of the specification. The manufacturing activity is not at liberty to establish these limits unilaterally. They must provide a system that meets the specifications, or they must insist that the design be adjusted to be compatible with what can be accomplished.

To determine the capability of a process to meet a specified level of performance, it is necessary to start with the specifications that arise from the design. It is presumed, of course, that the specifications are not more restrictive than is required by the product. The assumptions that are normally made are the following:

1. The specification will state that a dimension must be maintained to some nominal value X_0, plus or minus some range ΔX.

2. The process will generate parts such that the dimension X will follow a normal distribution, characterized by the mean μ and a standard deviation σ.

3. The limits for acceptability of satisfactory parts is set at some multiple, λ, of σ.

To be acceptable, the parts must have a dimension falling within the limits $\mu \pm \lambda\sigma_{\bar{x}}$. Thus, $\lambda\sigma_{\bar{x}} = \Delta X$. Since the control charts are derived from averages on n samples, the value appropriate to the charts is

$$\lambda\sigma_{\bar{x}_n} = \frac{\lambda\sigma_X}{\sqrt{n}} = \frac{\Delta X}{\sqrt{n}} \tag{12.8}$$

To meet the specification, the process must yield values for the UCL and LCL of $\bar{X}$ given by

$$\bar{\bar{X}} \pm \lambda\sigma_{\bar{x}_n} = \bar{\bar{X}} \pm \frac{\Delta X}{\sqrt{n}} \tag{12.9}$$

The process must produce the value of μR for the R chart that will satisfy the specifications that the standard deviation of the process will be $\lambda\sigma x$. μ_R is given by

$$\mu R = \sigma X d_2 = \frac{\Delta X}{\lambda}d_2 \tag{12.10}$$

The values given in Eqs. (12.9) and (12.10) establish the limits on the control charts that will allow a determination of the capability of the process to meet the specifications established for the product. For the example shown in Table 12.3 this process would meet a $\pm 3\,\sigma$ level of 1.448 ± 0.091. If the specifications are for a variance less than this, the process will not achieve the desired level of performance. If the specifications allow a greater variance, the process will meet the design objectives. What should be obvious here is that the *design must establish both* ΔX and λ and that these two parameters are independent.

Values of the specifications should not be placed on the control charts. The control charts represent the message that the operation is giving the operator. It is telling the operator that the operation is in control or it needs adjustment. It is an independent matter to determine whether the operation is capable of achieving the design specifications. This will occur if Eqs. (12.9) and (12.10) are satisfied.

It was stated above that the operation will be capable of providing parts that meet the design specification if the mean value of the attributes coincides with the nominally desired value of the attribute. Although this is generally true, it should be obvious that it is possible to have a process that is sufficiently under control that all parts fall within the control limits even though the value of $\bar{\bar{X}}$ does not coincide with X_0. In such a situation, the process would be technically under control. To determine whether the deviation in the mean of these parts would have an adverse affect on the final product, the designers would need to examine the various consequences relative to the possible "stack up" of tolerances. It is possible that small deviations of $\overline{\overline{X}}$ from X_0 could be tolerated without adverse effects.

12.6   CONCLUDING REMARKS

Achieving a desired level of control for an enterprise involves much more than setting an objective for a given operation or determining the level of inventory that is desired. The philosophy that the management assumes is all important. Moving from one form of control of the operation to an entirely different form (e.g., MRP II to JIT), requires much more than declaring that it is time to change. Recognizing the implications of such a change is critical to achieving success. With continuous improvement being a constant goal of the organization, the use of tools such as statistical process control is essential.

REFERENCES

BEDWORTH, D. D. and J. E. BAILEY. 1982. *Integrated Production Control Systems: Management, Analysis and Design*, John Wiley & Sons, New York.

BURR, I. W. 1976. *Statistical Quality Control Methods*, Marcel Dekker, New York.

FALLON, M. 1986. JIT? Because It Works, *Managing Automation,* May:76-79.

FOO, G. and L. KINNEY. 1990. Integrated Pull Manufacturing-The Integration of MRP and JIT Systems*, Proceedings of the First International Conference on Systems Integration* (ICSI '90), IEEE, New York:415-425..

FOSTER, R. N. 1982. A Call for Vision in Managing Technology, *Business Week,* May: 24-33.

HARTLAND-SWANN, J. R. 1987. *Proceedings of the Second International Conference on Just in Time Manufacturing,* MRP or JIT: Which Is Best? Springer-Verlag, Berlin:81-92.

KARMARKAR, U. 1989. Getting Control of Just-in-Time, *Harvard Business Review,* Sept.-Oct.:122.

LAUFER, A. C. 1984. *Production and Operations Management*, South-Western Publishing Co., Cincinnati, OH.

MARSING, D. B. 1991. Taking Risks in Manufacturing in *Manufacturing Systems: Foundations of World Class Practice*, J. A. Heim and W.D. Compton (eds.) National Academy Press, Washington, DC.

MONDEN, Y. 1983. *Toyota Production System,* Industrial Engineering and Management Press, Norcross, GA.

RIGGS, J. L. 1987. *Production Systems: Planning, Analysis, and Control*, John Wiley & Sons, New York.

SAKAKIBARA, S., B. B. FLYNN, and B. G. SCHROEDER. 1993. A Just-In-Time Manufacturing Framework and Measurement Instrument, *Production and Operations Management,* 1(4).

STARR, M. K. 1989. *Managing Production and Operations*, Prentice Hall, Upper Saddle River, NJ.

TOMPKINS, J. A. and J. A. WHITE. 1984. *Facilities Planning*, John Wiley & Sons, New York.

QUESTIONS

12.1   When a new operation (e.g., a material handling system or a new unit process) is considered for introduction into an existing manufacturing operation, what are the questions concerning the feasibility of the new operation that must be explored prior to concluding that the new operation is appropriate? What are some of the actions that are undertaken to answer these questions?

12.2 When an all-new manufacturing facility is being considered, the planners are likely to examine a range of all new processes for their possible introduction into the facility. What questions concerning the feasibility of these all-new processes must be explored prior to concluding that they can be included in a new facility? What are some of the data that must be acquired or known prior to making these decisions?

12.3 The tensile strength data in Table Q12.1 were collected over a 30-day period on the sand cores used to make gray iron castings for engine components. The design specifications for tensile strength for these cores is 205 ± 50 psi. Using the data located in the table:

    1. Construct an $\bar{X}$ and R chart using each day as a data point. Add control limits and mean values to complete the chart. Assume that the desired control level is to provide a $\pm 3\sigma$ level of acceptance. The control chart in Appendix A12.4 can be used for this purpose.

    2. As manager of the core operation, how would you describe the performance of your operation during the past 30 days? If changes are needed, identify the objectives that you would have for them.

TABLE Q12.1

Day	Piece number 1	2	3	4	5
1	206	197	177	203	176
2	226	203	161	166	170
3	201	241	165	186	208
4	182	217	199	184	200
5	215	208	203	215	191
6	185	197	154	221	211
7	231	188	210	192	188
8	212	200	173	241	207
9	150	234	217	239	240
10	223	196	175	165	159
11	245	208	208	200	185
12	232	225	200	210	225
13	204	187	154	196	163
14	164	204	182	176	233
15	223	197	220	231	228
16	215	203	189	217	211
17	192	198	185	204	202
18	205	197	177	167	221
19	185	206	198	192	188
20	216	197	178	154	161
21	199	202	212	196	205
22	194	195	213	228	237
23	191	218	178	212	226
24	205	197	177	167	221
25	182	217	199	184	200
26	232	211	195	208	163
27	215	222	207	211	219
28	204	159	114	188	163
29	140	157	183	163	200
30	150	234	240	239	217

12.4 You have just learned that your major competitor has declared that their objective is to enhance the capability of their core making operation such that their yields of good cores will reach a ± 4σ level of acceptance. In determining whether you wish to attempt to meet the same quality standards in your facility, you must re-assess your capabilities. If your current operation is under control at a ±3σ level and is meeting design specifications, what will you need to change to be capable of achieving a ± 4σ level of performance?

12.5 How would you modify and describe each element of Figure 12.1 for a service organization (e.g., a bank or an insurance company)?

12.6 Is it appropriate to apply SPC to a service activity? Is it applicable to a fast-food restaurant or to a library? Discuss the limitations that a service activity presents and outline how you would go about solving any issues.

READINGS

FALLON, M. 1986. JIT? Because It Works, *Managing Automation,* May:76-79.

FOO, G. AND L. KINNEY. 1990. Integrated Pull Manufacturing-The Integration of MRP and JIT Systems, *Proceedings of the First International Conference on Systems Integration-* (ICSI '90,) IEEE, New York.

HARTLAND-SWANN, J. R. 1987. MRP or JIT: Which Is Best? *Proceedings of the Second Conference on Just-in-Time Manufacturing,* Springer-Verlag, Berlin:81-92.

KARMARKAR, U. 1989. Getting Control of Just-in-Time, *Harvard Business Review,* Sept.-Oct.:122.

APPENDIX A12.1
A JUST-IN-TIME MANUFACTURING FRAMEWORK

The framework for measuring the factors that influence a just-in-time manufacturing system has been studied by Sakakibara et al..(1993). Following a thorough review of the literature, visits to 12 U.S.-based plants that are practicing or implementing JIT, and extensive interviews of plant employees, they concluded that 16 dimensions are effective in describing a successful JIT system. These dimensions are given in Table A12.1.

TABLE **A12.1**   **JIT Components and Contents of the Measurements**

JIT Component	Contents Of Measurement
Setup time	Extent to which plant is reducing setup times to facilitate JIT
Lot size	Extent to which plant is actively seeking to reduce lot size
JIT deliveries	Extent to which vendors are making frequent deliveries and achieving quality certification
Supplier quality level	Extent to which plant is maintaining long-term relationships with suppliers and involves suppliers in the planning process
Multifunction workers	Extent to which workers perform a multitude of tasks
Group problem solving	Extent to which small groups of workers are encouraged to bring problems to team problem-solving sessions
Worker training	Extent to which management is committed to train workers to do multiple tasks and to maintain high skills in the workforce
Adherence to daily schedule	Extent to which adequate time is allocated to meet each day's schedule
Repetitive master schedule	Extent to which the daily production schedule is level and repetitive
Preventive maintenance	Extent to which PM practices have been introduced into worker's daily routines, PM established in each shift, and PM incorporated into the manufacturing strategy of the plant
Equipment organization	Extent to which manufacturing cells, machine and process layout, and equipment used for flexible floor layout
Simplicity of product design	Extent to which design specifications have been simplified, part count minimized, and customer orientation included in product design
Kanban	Extent to which the plant and vendors have been integrated into production through use of Kanban cards and containers
Pull system	Extent to which plant has introduced aspects of pull systems such as stopping to fix quality problems, efficient floor layout, and worker-directed production
MRP adaptation to JIT	Extent to which plant has integrated JIT concepts such as pull systems with MRP
Adaptation of accounting system	Extent to which the accounting system reflects or supports a JIT orientation

The validity of these measures was tested by examining three performance criteria of production systems: inventory turns per year, cycle time, and lead time to fill orders. Although it is clear that factors other than the 16 in the table contribute to performance of the system, the conclusion from this study is that these 16 factors are valid measures in terms of criterion-related validity. The message that this study sends is clear. If a JIT manufacturing system is to be installed, it is critical that actions be taken to ensure that proper attention is given to each of these 16 criteria.

REFERENCES

SAKAKIBARA, S., B. B. FLYNN, and B. G. SCHROEDER. 1993. A Just-In-Time Manufacturing Framework and Measurement Instrument, *Production and Operations Management,* 1 (4).

APPENDIX A12.2   MRP AND MRPII

MRP and MRPII are computer-based systems for the control of resources in the manufacturing environment. MRP, materials requirement planning, is used to assist in the management of inventory and production schedules. MRPII, manufacturing resource planning, uses feedback regarding current performace to adjust a master schedule for the total production system. As is readily evident, there is a strong connection between the two approaches.

MRP

A simplified representation of an MRP system is contained in Figure A12.1. MRP is a computer-based system that is designed to serve several important functions. First, it is expected to assist in assuring the availability of the materials and components that are needed to meet the desired manufacturing. Second, it attempts to achieve the lowest level of inventory that is practicable. Third, it is used to plan the manufacturing schedule.

An MRP is initiated with the creation of a master production schedule from anticipated or actual customer orders along with a production schedule that is achievable with the manufacturing facilities. The anticipated schedule for customer orders and the capability of the manufacturing system (e.g., constraints on capacity) are used to establish a production plan. This recognizes the capability of the sytem and the demands that the orders place on the system. From the planned schedule for production, the quantity and detailed specification for the material needs of the plant, namely the procurement of the specific materials and the providing of them when they are needed, are established. This plan recognizes any irregularity in demands and in complexity of the products that are expected to be

Figure A12.1   Schematic representation of the materials requirements planning (MRP) system.

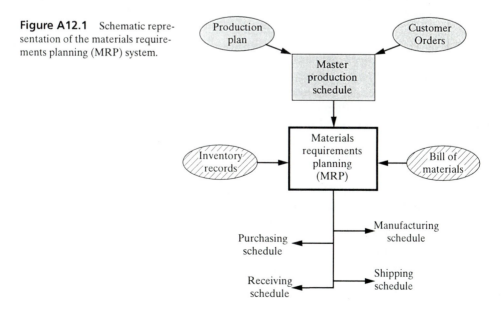

produced. Since this plan depends on the particular items that are to be produced and seeks to recognize any commonality of material or components that will be used by them, the bill of materials for the production plan is used along with a knowledge of the inventory of materials that is on hand. From this plan of material needs, schedules for the purchase of materials and for the receiving of those materials are issued along with the anticipated manufacturing schedule for the plant and the schedule to be followed for shipping finished parts. The manufacturing schedule gives the release of orders to the floor of the plant and includes any changes in due dates that occur for open orders and the plan for the release of future orders. An update of an MRP is required at regular intervals.

As is evident, the system is driven by the master production schedule. If it is inaccurate or changes rapidly, it will be difficult to achieve the desired result. Similarly, the effectiveness of the system depends on accurate information on inventory levels. This requires a regular updating of the information on the types and quantity of *all* materials in the plant. Although these systems have been found to be effective in reducing the quantity and investment in inventory as compared to earlier systems (e.g., fixed-order-size-systems), they do not necessarily result in the lowest possible costs. Since MRP concentrates on minimizing the inventory, it does not explicitly include the accomplishment of this at the lowest costs. The replenishment of inventory as required by the production schedule often results in frequent orders of materials in quantities that are below the optimal. This can increase transportation and handling costs. Improving on this situation led to the development of MRPII.

MRPII

MRPII extends the concept of MPR by including the financial components of the system and by providing the capability for the system planners to perform simulation of system performance. A principal element of MRPII is the existence of feedback mechanisms that permit changes to be made in the master production schedule based on conditions that exist in the production environment. Thus MRPII attempts to take into account changes that may be experienced in engineering, manufacturing, finance, puchasing, shipping, and marketing. It provides a means of incorporating into the master production schedule important variations in previously assumed parameters. Figure A12.2 offers a simplified version of MRPII.

The inputs to the production plan, strategic decisions of the management, sales demand, sales forecast, and inventory records of finished goods and materials serve as inputs to the production plan. This, in turn, leads to establishment of the resource requirements plan and the master production schedule. From this, the materials requirements plan is generated and from this, the schedules for manufacturing, purchasing, shipping, and receiving. The key innovation in MRPII from MRP is the creation of sources of information on the status of the system and use of this information for modification of the master production schedule and the material requirements plan. This provides a set of feedback loops that allow detailed modification of the plan to accommodate changes that may take place in customer demands, availability of production resources, availability of materials, and so on.

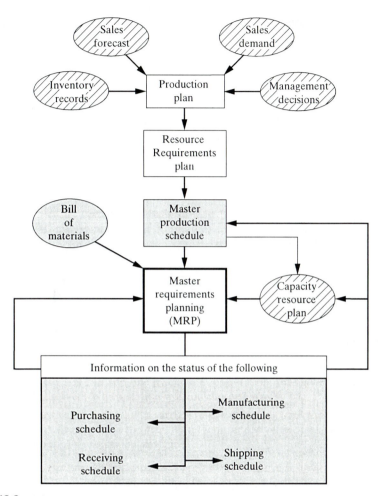

Figure A12.2   Schematic representation of the manufacturing resource planning system (MRPII).

REFERENCES

BEDWORTH, D. D. AND J. E. BAILEY. 1987. *Integrated Production Control Systems*, John Wiley & Sons, New York.

GUNN, T. G. 1982. The Mechanization of Design and Manufacturing, *Scientific American,* 247(3):87.

STOCK J. R. AND D. M. LAMBERT. 1987. *Strategic Logistics Management*, Richard D. Irwin, Homewood, IL.

VOLLMAN, T. E., W. L. BERRY, and D. C. WHYBARK. 1988. *Manufacturing Planning and Control Systems*, Dow Jones-Irwin, Homewood, IL.

APPENDIX A12.3
THE ISO 9000 SPECIFICATION FOR REGISTRATION
AND CERTIFICATION

With the creation of a common market that embraced many countries, the European Union recognized that trade among its members would be greatly enhanced if all countries within the community had a common set of standards for product quality. This goal was to be achieved in 1992 by the requirement that all companies selling products in the European Union meet a common set of standards. The details of the standards were established by the International Standards Organization (ISO) and were issued in 1987 as the ISO 9000 Series. These standards have been adopted in the United States by the American National Standards Institute (ANSI) and are designated as ANSI/ASQC Q90 Series.

Through a process of registration and certification, the ISO 9000 standard is supposed to ensure that a company that is in compliance has demonstrated that it possesses the capability to ensure that it has established and can maintain an adequate system of quality management to provide products having a predetermined level of quality and specifications. Compliance with the standard is typically determined by a third-party audit. The auditor must be qualified and accepted by the registering organization. A register is maintained of companies that have complied with the standard.

The ISO 9000 is a series of standards that specify the requirements for quality management systems. There are essentially five standards associated with the ISO 9000 series.

- ISO 9000 are quality management and assurance standards. It sets guidelines for the selection and use of the four remaining standards.
- ISO 9001 are standards for quality systems. These provide standards of quality assurance for companies that engage in design and development, production, installation, and servicing of products.
- ISO 9002 are standards for quality systems for companies that restrict their activities to the production and installation of products.
- lSO 9003 are standards for quality systems for companies that restrict their activities to final inspection and test.
- ISO 9004 presents guidelines for quality management and quality system elements.
- ISO 9000 and 9004 are primarily advisory in nature.

The table of contents of the ISO 9002 specification , which follows, indicates the details of these standards.

REFERENCES

PEACH, ROBERT W. 1991. State of Quality System Certification in the United States, 45th Annual Quality Congress Transactions, May 20:708.

SAWIN, S.D. AND S. HUTCHENS, JR. 1991. ISO-9000 in Operation, 45th Annual Quality Congress Transactions, May 20:914.

PURI, SUBHASH C. 1991. DEMING + ISO/9000 A Deadly Combination for Quality Revolution, *45th Annual Quality Congress Transactions,* May 20:938.

TABLE A12.2   ISO 9002 Table Of Contents

Introduction	4.8 Process Control
1.0 Scope and Field of Application	4.8.1 General
1.1 Scope	4.8.2 Special Processes
1.2 Field of Application	4.9 Inspection and Testing
2.0 References	4.9.1 Receiving Inspection and Testing
3.0 Definitions	4.9.2 In-Process Inspection and Testing
4.0 Quality System Requirements	4.9.3 Final Inspection and Testing
4.1 Management Responsibility	4.9.4 Inspection and Test Results
4.1.1 Quality Policy	4.10 Inspection, Measuring and Test Equipment
4.1.2 Organization	4.11 Inspection and Test Status
4.1.3 Quality System	4.12 Control of Nonconforming Product
4.2 Quality System	4.12.1 Nonconformity Review and Disposition
4.3 Contract Review	4.13 Corrective Action
4.4 Document Control	4.14 Handling, Storage, Packaging, Delivery
4.4.1 Document Approval and Issue	4.14.1 General
4.4.2 Document Changes/Modifications	4.14.2 Handling
4.5 Purchasing	4.14.3 Storage
4.5.1 General	4.14.4 Packaging
4.5.2 Assessment of Contractors	4.14.5 Delivery
4.5.3 Purchasing Data	4.15 Quality Records
4.5.4 Verification of Purchased Products	4.16 Internal Quality Audits
4.6 Purchases Supplied Product	4.17 Training
4.7 Product Identification and Traceability	4.18 Statistical Techniques

Control Chart

Plant	Dept.	Operation	Date control limits calculated	Engineering specification	Part no
Mach no.	Dates	Operation		Sample size/frequency	Part name

Averages (X bar chart)

Ranges (R chart)

Date																												
Time																												
1																												
2																												
3																												
4																												
5																												
Sum																												

Readings

344

13

Analyzing and Describing the Manufacturing Operation

In many ways Ptolemy (90-168) spoke as a prophet. For he enlarged the uses of mathematics in the service of science. While he drew on the best observations made before him, he emphasized the need for repeated, increasingly precise observations...He showed the crucial scientific talents-shaping theories to fit available facts and testing old theories by new facts. (Boorstin,1983)

An effective manager of a manufacturing system must understand relationships that affect the performance of their systems. For many people this understanding has been acquired through experience. They learned to operate systems by working with them. They observed and learned what was possible and what was impractical. Although experience will always be important, sole reliance on it presents serious limitations. Experience often offers little insight into the course of action that should be adopted in responding to situations that have never before been experienced; it is often unable to provide quantitative estimates of the impact of proposed changes in the system; and it has limitations in assessing the impact of new technologies on the system. This chapter is concerned with some of the analytic tools that can be used to augment practical experience and to provide useful insight into the operation of systems.

To create an analytical representation of the discrete manufacturing system, a model of the physical system must be created that is amenable to analysis. Allen (1987) offers the following comments in support of the advantages to be gained through the development of good models.

Horace Freeland Judson (1980), who has a chapter on modeling in his outstanding book *The Search for Solutions*, says: "A model is a rehearsal for reality, a way of making a trial that minimizes the penalties for error." He goes on to show that modeling is one of the most important activities in science. The authors of *Quantitative Systems Performance* (Lazowska et al. 1984) say, "A model is an abstraction of a system: an attempt to distill, from the mass of details that is the system itself, exactly those aspects that are essential to the system's behavior".

Good models of a discrete manufacturing system are often difficult and complex because of the discrete nature of the processes that are involved and the fact that the processes do not occur at regular time intervals. The absence of continuous processes makes it difficult, or impossible, to describe the systems by linear or nonlinear differential equations, integral equations, or difference equations that have their genesis in fundamental physical principles such as thermodynamics, fluid mechanics, statistical mechanics, and the conservation of energy and mass. While fundamental physical laws are not violated in discrete systems, it is the discreteness of events and the random or stochastic occurrence of the events that dominates the performance of these systems.

The randomness of the occurrence of events results from many factors. Machines fail and must be repaired; the human beings who are involved do not always behave in the same way throughout the work period; material properties are not always constant, schedules are subject to interruptions; the delivery of goods may be irregular, and, noise in sensors and controllers can introduce variability into the reaction of the machines. In combination, these variabilities introduce uncertainties into the system that bring into use the concepts of probability of occurrence of events. The formalism that is used to model these systems depends on probability theory.

It is important to draw a clear distinction between the analytical approaches that are used to describe the details of the phenomena that occur within a unit process and those used to describe the performance of a total plant. For many unit processes (e.g., a machining, welding, or stamping operation), the appropriate analytical approach is that characteristic of continuous systems. The physical processes are described by equations that are based in fundamental physical phenomena (e.g., the flow of materials under stress, or the rapid melting and solidification of material in a weld, or the deformation of material in a stamping operation, or the wear of a cutting tool). For the total plant, however, a realistic description can only be achieved by accounting for the *discrete* nature of events.

A number of techniques are useful in treating systems that have parameters affected by variability. Queueing models can be used to determine the equilibrium parameters of such systems. Simulation models can be used to examine their dynamic performance. Regression models can be used to capture the relationship among average parameters of an existing system. As Solberg (1992) observes:

There are many categories of modeling technique, including optimization, simulation, control theoretic, systems dynamics, queueing, and statistical methods. Each of these categories is defined by conventions, terminology, standard formulations, and methods. Specific models usually fall into one of these distinct categories, although one occasionally encounters hybrid models that cut across the boundaries. All of these modeling techniques have had their capabilities extended greatly over the past few decades. With the increase in power and availability of computers, we have been able to deal with many more parameters.

In using any of these techniques, the modeler is concerned with developing a representation that contains the most important physical characteristics of the real-world system. Thus the modeler must attempt to identify and find ways of representing the key relationships that exist among the various variables. Using these relationships, attempts are made to establish the performance of the system. Although it is natural to try to capture all the details that may persist in the real-world system within the model, Solberg (1992) notes that the power of a model or of a modeling technique is a function of validity, credibility, and generality. Usually, the simplest model that expresses a valid relation will be the most powerful. This is an admonition that should always be kept in mind as we discuss the need and desirability for modeling discrete manufacturing systems. The complexity of these systems is such that it is easy to become overwhelmed with the details.

In this chapter we discuss briefly some of the techniques used in model building and will use some of these techniques to illustrate the types of questions that the manager can reasonably expect to ask and the types of answers that can be derived. As has been mentioned in other circumstances, good data on the characterisitics of the systems must be available and used if the models are to be made representative of real-world situations.

13.1   NETWORK DESCRIPTION OF THE MANUFACTURING OPERATION

In creating a generic description of a discrete manufacturing operation, it is helpful to visualize the system as a network of operations (machines) and pathways (transfer devices) through which the material flows. A schematic of such a network is shown in Figure 13.1(a). The nodes in the network are represented by the circles and the different types of arrows identify the paths followed by three distinct parts. The operations contained at each node are responsible for carrying out a particular process on the parts that pass through the node. Each node therefore represents one or more machines that carries out a physical operation in a time that is determined by the requirements for the part.

The unit operations, designated as operations 10 through 90, might be thought of as milling, drilling, reaming, inspection, painting, and so on. While each

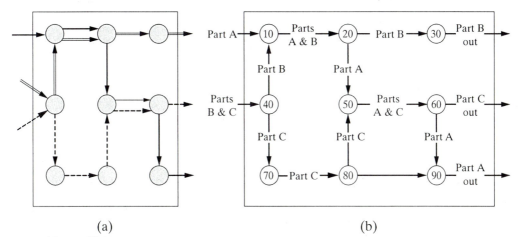

(a)   (b)

Figure 13.1   Network in a manufacturing operation; (a)general network representation; (b)network indicating identified operations and part names.

is indicated here as being unique, there might be some duplication of operations by various nodes. For the sake of simplicity, it is assumed that each operation shown in Figure 13.1(b) consists of a single machine (server) and that no provision needs to be made for material handling. These do not represent fundamental limitations and both will be relaxed in later discussions.

A unique path is followed by each part as it moves through the network. Part A moves along a path from operation 10 to 20, 50, 60, and finally 90. Part B follows a path that includes operations 40, 10, 20, and 30. Part C follows a path that includes operations 40, 70, 80, 50, and 60. This describes the operations that each part requires *and* the order in which they are accomplished. The other quantities that must be given to specify the properties of this network are the times that each part spends at each location (e.g., the processing times for each part at each location and the average arrival rates λ_A, λ_B, and λ_C of parts A, B, and C, respectively, to the plant).

13.1.1 Machine Utilization

The utilization level of a machine is the fraction of the total time that is being used to perform a useful function. For a machine that is servicing a single part with an arrival rate λ and a processing time T, the utilization of the machine is λT. Since the fraction of the time that a machine can operate can never exceed unity, $\lambda T \leq 1.00$. If the node services i part types, each with different values of λ and T, the limitation on the utilization is that

$$\Sigma_i \lambda_i T_i \leq 1.00$$

If the node has n identical servers, the limitation on the utilization is

$$\Sigma_i \lambda_i T_i \leq n$$

For practical reasons associated with time that must be spent in changing tools, maintenance, repair of failures, and so on, the utilization can never be assumed to be as high as 1.00. It is not uncommon to have maximum utilizations that range between 0.85 and 0.95.

13.1.2 Theoretical Limit for Production

The *theoretical limit* for the performance of a network (i.e., the production level for the network and the magnitude of the work in process), can be determined easily by analyzing the equilibrium flow through the network under the conditions that:

- There is no variability in the process times.
- There is no variability in the arrival of any parts to the various operations.
- There is no time required for machine setup.
- There are no breakdowns in any of the machines.
- The transit time for the parts between the stations is negligibly small.

For a system with these ideal characteristics, the maximum production will occur when the arrival rates are such that all machine utilizations are maximum, at which time one or more stations will be operating at 100% utilization. Increasing the arrival rate of parts beyond the value that yields 100% utilization will lead to a continual buildup of parts waiting to be served, thus leading to a nonstable situation. Thus the theoretical limit will be found for the system in which the arrival rate of parts to at least one station results in 100% utilization of the station.

For the example shown in Figure 13.1(b) and using the operating conditions listed in Table 13.1, the theoretical limit for network throughput can be determined by determining the values of λ_A, λ_B, and λ_C that lead to one or more machines with a utilization of 1.00. For the case where all arrival rates are equal, it is straightforward to show that the maximum throughput occurs for $\lambda_A = \lambda_B = \lambda_C = 0.154$/min., giving a total production of 27.7 parts/hr from the plant. For the assumption that $\lambda_A = \lambda_B \neq \lambda_C$, a somewhat higher theoretical output can be achieved. For this case the largest production can be shown to occur at $\lambda_A = \lambda_B = 0.154$/min and $\lambda_C = 0.167$/min. which yields a total production of 28.5 parts/hr. If no restriction is placed on the relative production of each type of part, the maximum theoretical production occurs with $\lambda_A = 0.094$/min, $\lambda_B = 0.250$/min, $\lambda_C = 0.167$/min., which yields a total production of 30.7 parts/hr. Since the value of the finished parts may differ, the values of the arrival rates that give the greatest theoretical *value* of production may be different from values that give the largest theoretical *number* of parts.

For the conditions that satisfy the theoretical limit, each part is produced when it arrives at a station; that is, there is no waiting for service. It follows, therefore, that the time that it takes for a part to traverse the network will be given by $\sum k Tk$, where Tk is the service time at station k. This simple calculation for the transit time is valid only for the theoretical limit, since a relaxation of any of the conditions listed in the first paragraph of this section will introduce randomness into the system, which requires that parts wait for service. In the more general case involving queues, the transit times must be calculated using techniques such as those described in Section 13.2.2.

13.1.3   Bottleneck Operations

The bottleneck node in the network is the node that limits production capacity. It is the node that is operating at the highest capacity, that is, it has the highest utilization. Thus an effort to increase capacity must focus on increasing the capacity of the bottleneck station, perhaps by adding another machine to the bottleneck station, by reducing the processing times through enhanced machine capability, by making a provision to operate this station extra hours, or by rerouting the work so that another less used station can share some of the load. While all of these alternatives may seem straightforward, the amount of flexibility in the plant to accomplish these changes is often limited by such practical considerations as space or the capacity of material handling equipment. Accomplishing significant changes may need to await a major upgrade in the plant. In the example above, operation 10 has the highest utilization and is the bottleneck node. Since only parts A and B pass through operation 10, this bottleneck does not serve as a limitation on the production of part C.

13.1.4   Line Balancing

Since the most efficient use of facilities occurs when all machines are operating near capacity, it is desirable to design the plant such that all machines are operating at

TABLE **13.1**   Processing Times for the
Operations of Figure 13.1(b)

Operation	Processing time (min.)		
	Part A	Part B	Part C
10	4.00	2.50	
20	3.00	2.50	
30		4.00	
40		1.50	3.00
50	3.50		2.00
60	4.50		1.50
70			6.00
80			4.50
90	2.00		

their highest practical utilization. A production facility in which this occurs is described as being balanced, and the process of developing the processing plan that accomplishes this is described as balancing the line. Although this may be the most desirable, it is not always feasible. Since it is the $\sum_i \lambda_i T_i$ that determines utilization for each station, it may be impossible to adjust processes and paths such that the same high value will be achieved for all stations.

13.1.5   Maintenance

Machines fail during use and must be repaired. Tools must be changed to maintain the quality of production. These and other unexpected events occupy time that could otherwise be used for production. The fraction of time available to accomplish these tasks is just the fraction of time that a single machine is not utilized for production, namely (1 - utilization). For a very reliable machine, or a system in which unexpected breakdown can be minimized by maintenance during nonproduction time, the fraction of time reserved for these tasks can be small, leaving most of the time for production. If machine failure is frequent, the amount of time available for production will be smaller and the utilization of the machine for production must be reduced. It is for these reasons that the practical limits for utilization is generally around 0.900.

13.2   QUEUEING MODELS

As a result of the discrete and random nature of the events that occur in a discrete-parts manufacturing activity, an *analytical description* of these systems requires the use of mathematical descriptions that are stochastic (i.e., models whose outcomes are determined by the probabilities of the occurrence of events at each stage of production). A proper accounting of the variability that occurs at each step is critical in understanding the performance of the system. One approach to accomplishing this is to use the formalism derived from queueing theory.

The arrival at a station, the waiting to be served, the servicing of the part, and the departure of the part are the steps that must considered, as depicted schematically in Figure 13.1(b). The buildup of queues and the need to await service arises from variabilities in previous events. Two arrivals that are closely spaced will force the second part to wait while the first is being serviced. Two arrivals that are far apart will leave the machine idle for a period of time. Even though the *average* of the arrival times is such that no waiting would occur between perfectly spaced events, the randomness will create queues of parts waiting their turn for service and will thus introduce delays in the performance of the processes.

Even though a system may contain processes that are subject to variances it is still meaningful to describe an equilibrium state for the system. The meaning of equilibrium can be visualized by imagining the comparison of a succession of snapshots that are taken of the plant. Because of the random nature of the processes, successive snapshots will show that the number of parts in waiting at each queue will vary somewhat from one picture to another. After accumulating a large number of snapshots and determining the number of parts in each queue in each snapshot, the average for the number of parts in each queue can be determined. If the system has been operating for a long time without changes in parameters, the *average* number of parts at each location will be found to remain constant. In such a situation, the system can be described as being in a *steady state* or as having reached a *condition of equilibrium* Under these circumstances, the expected values of a variety of quantities can be identified that are useful in describing the system. Examples of these are:

- The average interarrival times, and the average arrival rates, for the parts at the plant.
- The average number of parts awaiting service (i.e., the number in the queue at each station).
- The average time required to service each part at each operation.
- The average length of time that will elapse from the entry of a part into the system until its departure from the system, and so on.

Since the range for the values of these variables are determined in part by the randomness associated with the occurrence of the events, a proper description of the system requires that information be given about the statistical distribution of the events. This is accomplished by giving the variance of the distribution for each of the key events that affects the performance of the system.

For a system in equilibrium, the *output must equal the input*. Thus the *total output*, as measured in parts per unit time that are produced by the plant, will be $\Sigma i \lambda i$. The value of the total output of the plant will be $\Sigma i \, (\lambda i \, Vi)$, where Vi, is the value of the *ith* part. These two statements are true whether or not the processes have variances in their distributions. For the plant with variances, the values that are used for λi are the average values for each of the parts. The plant with randomness will have parts in queues which will increase the average amount of time that it will take for an entering part to complete its path through the plant (i.e., the average transit time of a part through the plant will be greater than the sum of the times spent in processing at each station). Methods to determine these averages are discussed in Section 13.2.3.

13.2.1 Little's Law: The Queueing Formula

This simple description of a queueing network can be used effectively to describe the movement of parts through a manufacturing facility. *Queueing theory provides one of the few laws that applies to a wide range of circumstances in discrete manufacturing.* The first general proof of this law was given by John Little (1961). A nice description of the law is provided in the following quotation from Little (1992).

> $L = \lambda W$ (Little's Law) is a mathematical tautology with useful mappings onto the real world. $L = \lambda W$ relates the average number of items present in a queueing system to the average waiting time per item. Specifically, suppose we have a queueing system in steady state and let
>
> L = the average number of items present in the system,
> λ = the average arrival rate, items per unit time, and
> W = the average time spent by an item in the system;
> then, under remarkably general conditions,
>
> $$L = \lambda W \qquad (13.1)$$
>
> This formula turns out to be particularly useful because many ways of analyzing queueing systems produce either L or W but not both. Then Eq. (13.1) permits an easy conversion from one to the other of these performance measures. Queues and waiting are ubiquitous in manufacturing: jobs to be done, inventory in process, orders, machines down for repair, and so forth. Therefore, Eq. (13.1) finds many uses.
> As a mathematical theorem, $L = \lambda W$ has no necessary relationship to the world. Given the appropriate set of mathematical assumptions, $L = \lambda W$ is true. There is no sense going out on the factory floor and collecting data to test it. If the real world application satisfies the assumptions, the result will hold.

This is an extraordinarily powerful law. It can be used in many circumstances. It offers an important "sanity check" on all sorts of data that may be collected in the system. In using this very powerful law, it is important always to remember that the law applies only to systems that are in *equilibrium* (i.e., to a system that has been operating for a sufficiently long time that it has achieved a state in which the probability of occurrence of events in not changing). As the system continues to operate, there will be no net increase or decrease in the average values of the quantities that describe the system.

Little notes in his original paper (1961) that the meaning of the term *system* is quite general, as is the specific meaning attached to the quantities L, λ, and W. The equation will hold as long as their is a consistency of meaning for all three terms. Many authors have explored the meaning of this relationship and have demonstrated that the equation can be used in a variety of other applications (e.g., relating the time-average cost structure of a system to the average cost per customer being served by the system) (Maxwell, 1970; Brumelle, 1971; Stidman, 1974; Heyman and Stidham, 1980; Rolski and Stidham, 1983; Glynn and Whitt, 1989).

Little's law can, and should, be used frequently to ensure that the quantities being used to describe a system are self-consistent. Since the equation relates the average number of customers (parts) in the system with the average time spent by each customer (part) in traversing the system, it is not necessary to measure both. Either L or W can be measured, allowing the other quantity to be determined through Eq. [13.1] In an actual situation, however, statistics on the system are frequently collected on both quantities, often independently, with various parts of the organization using the different statistics for their own purposes. When this occurs, it is critically important to ensure that the data are consistent. A violation of Little's law, Eq. (13.1), will immediately signal that something is wrong with one, or both, sets of data. Failure to make such simple tests effectively deprives the manager of an independent check on the consistency of the operations. Little's law is an example of a simple law that can offer insight into the operation of a system. Its use should be encouraged and expanded.

Little's law can be illustrated by using it to calculate the theoretical limits discussed in Section. 13.1.2 for the example given in Table 13.1. For the case in which $\lambda_A =$ 0.094/min, $\lambda_B =$ 0.250/min, and $\lambda_C =$ 0.167/min, applying Little's law to operation 90 for which part A spends 2.00 min being serviced, the average number of parts contained at station 90 is 0.188.

When more than one type of part is being serviced by a station, Little's law for that station becomes

$$L = \Sigma_i \lambda_i W_i$$

For operation 10 the average number of parts contained at station 10 is 1.00 for the condition $\lambda_A =$ 0.094/min and $\lambda_B =$ 0.250/min. For the path that part C follows in traversing the plant, the average number of units of part C being processed is 2.84.

While these calculations were made for the case that there was *no randomness in any of the processes*, Little's law is not restricted to this. Little's law can be applied to any equilibrium case, as long as the average value of correct value of λ is used and as long as a proper value of *L or W* is found in some way. For a network with variances present, accomplishing the latter requires detailed calculations.

13.2.2   Response of Single Station

Queueing theory can be used to develop a very simple formulation for the performance of a station for which the arrivals follow a Poisson process, with the service times being exponentially distributed, and having a single server. For such a situation it can be shown that the

$$\text{Response Time} = \frac{\text{Service Time}}{\text{1-Utilization}}$$

Response time in this expression is the same as *W* in Little's equation, service time is the average time spent in being serviced at the station, *T*, and the utilization $= \lambda T$. The performance for stations with other types of characteristics can be found in various books. (Taylor and Karlin, 1984)

13.2.3   Queueing Network Models

While the importance of Little's law can hardly be overemphasized, it alone does not provide answers to many of the questions that must be explored during the design and/or analysis of realistic manufacturing systems. Invariably, answers are needed to such questions as the length of the queues at particular stations, the impact on queues of changing the path to be followed by the parts, the impact on throughput of the addition of another machine, the distribution of the work-in-process (WIP) among the various queues, the transit times for parts through the plant, and so on.

Answers to these questions require numerical calculations that account for the probability of the occurrence of events throughout the network. Depending on the assumptions that are made in the models and the flexibility that is allowed the modeler in the specification of parameters that describe the system, the models can have substantial complexity and may require substantial computer capability for their solution. Some of the models [e.g., CAN-Q (Solberg, 1977)] assume that the system is closed. This means that the average number of parts within the system-the sum of the parts being serviced and those waiting in queues-is fixed. When a part leaves a closed system, another part is assumed to replace it at the entrance. The value of this parameter must be provided by the user of the model. Other models [e.g., queueing network analyzer (QNA) (Segal and Whitt 1989)] assume that the system is open, in which case the average arrival rate of parts to the system must be specified. The model determines the total number of parts in the system by calculating the average number of parts at each station, including those being serviced and those waiting in queues.

For either type of model it is necessary to provide:

- The routing of each part, including the stations that each part visits and the order in which the stations are visited.
- Average service times for each part type at each station visited.
- The number of servers at each station.

Since the occurrence of an event is assumed to be random, the probability of its occurrence will be determined by the distribution function that is characteristic of this class of events. In some models (e.g., CAN-Q) the system assumes that the interarrival and service times are always distributed according to the exponential probability distribution. In other models (e.g., QNA) the variances must be provided by the modeler and can be set to values that are representative of the particular process. If a transit time is assumed to exist between stations (i.e., if material handling processes are included) the mean transit time and the variability parameter for these must be given. For models that allow the flexibility, the following must be specified:

- The probability distribution of the interarrival times
- The probability distribution of the service times

The probability distribution functions are typically characterized by specification of:

- The mean of the process
- A variability parameter, c^2, for each process-arrival and service.

The variability parameter, c^2, is defined as the ratio of the variance of the distribution to the square of the mean of the distribution. The variance is given by $m_2 - m_1^2$, where m_1 and m_2 are the first and second moments, respectively, of the distribution function. Values of c^2 for various distributions are given in Table 13.2. A variety of other distributions can exist, for which the appropriate variability parameter must be calculated. The higher the value of c^2, the greater the variability of the processes.

Some modeling formalisms have the capability to analyze the impact of machine failures on the system. In this case values of the mean time between failures (MTBF) and the mean time to repair (MTTR) must be provided by the modeler for

TABLE **13.2**

c^2	Distribution function
0.000	Deterministic—no variance
1.000	Exponential—square of the mean equal to the variance
0.333	Uniform distribution with values falling between 0 and N with a mean of $N/2$
0.166	Triangular distribution with values falling between 0 and N with a mean of $N/2$

each station. As was discussed above, the variance of these parameters may be needed, depending on the particular model being used.

A general mathematical analysis of a queueing network with many stations, with more than one product each having different service times and arrival rates, random failure of machines, and allowance for the specification of probability distributions is extremely complicated. In certain special, limiting cases the equations can be solved analytically. In general, solutions to these network equations must be determined numerically, and certain assumptions are often made to ensure that the solutions converge rapidly, thus minimizing the computational requirements. A summary of the principal features of several queueing network packages can be found in a publication by Snowdon and Ammons (1988).

The models provide a calculation of the *equilibrium* values of many of the quantities that are important in evaluating a manufacturing system. It must be emphasized that these models have important limitations. First, they are applicable only to systems that are in equilibrium and cannot, therefore, provide insight into the *dynamic* behavior of a system (e.g., the short-term impact of a change in conditions, or the length of time required for the system to reach equilibrium, or the time necessary to recover from a shutdown, etc.). Second, these models assume that all buffers are of infinite size; that is, they do not allow for blockage of the line. All machines will continue to produce parts even when one machine in the network is down, thus allowing the buildup of large queues when the downtime is long. For a system with finite-sized buffers, blockage will occur when the buffers have been filled and the machines will be forced to stop work.

It should be noted that there are circumstances that cannot be treated adequately by these simple techniques. In a network in which the path taken by some parts revisit stations that were previously visited (i.e., there is feedback of parts to stations that have been previously visited), and when the service times for one type of product in the system is drastically different from others, instabilities can arise in the network that effectively prevent the system from ever achieving the definition of equilibrium used above. This is a circumstance that can exist in some semiconductor manufacturing where some parts make repeated visits to the same station (Whitt, 1993).

Despite these limitations, queueing network models can provide important insights into the potential levels of operation of a system and should be used to the maximum extent possible in understanding system performance, in designing new systems, and in assessing the impact that changes will have on an existing system. These models are extremely useful in exploring the implications of various actions that may be considered.

13.2.4 Questions of Interest

In assessing the capability of a facility or the value of a planned modification to an existing facility, it is often useful to explore a variety of "what-if" questions. These questions can usually be separated into two categories: those that can be explored with simple analysis and those that require numerical calculations. The former relate to those that are determined by the level of machine utilization. The latter involve effects that depend specifically on the stochastic nature of the events that occur in the network (e.g., the levels of WIP, transit times, varying maintenance strategies, the effects of different forms of material handling, etc.). These different kinds of questions are treated in the following paragraphs.

Questions that can be explored using machine utilization. Questions such as the following can be analyzed from an understanding of the level of machine utilization in the network.

1. How much improvement in the level of production can be achieved by adding an additional machine at the bottleneck station?
2. What is the impact on the level of production of adding another unique product to the plant?

For question (1) the elimination of one bottleneck will transfer the bottleneck to the station with the next-largest utilization. If this new bottleneck has a significantly lower utilization than the original bottleneck, it may be possible to enhance the level of production of the plant by increasing the arrival rates for the incoming materials. If, however, the line was initially well balanced so that the utilizations of *all* stations was initially high, the movement of the bottleneck from one station to another may offer little opportunity to increase the production level of the plant.

The impact on the production level of adding another unique product to the plant can be determined by calculating the arrival rates that the plant can accommodate following the addition of the additional part. In a realistic assessment, the maximum level of machine utilization should be taken to be less that 1.00. Although the choice depends on experience with the reliability of the machines in the plant, the maximum is often chosen to be near 0.900. The analysis to be followed will be similar to that used in Section 13.1.2.

Questions that involve numerical calculations. Answers to the following important questions require numerical calculations.

1. What is the impact of variability on the work in process (WIP) and response time of the plant?
2. If the service time of the machine at the bottleneck station can be reduced by $x\%$, what effect will this have on the WIP and the response time of the plant ?
3. What impact will a reduction in mean time to repair(MTTR) of $y\%$ have on the capacity of the plant?
4. Is it more important to reduce mean time between failures (MTBF) or MTTR of the machines?
5. What is the impact on WIP and the response time of the plant by replacing an unreliable machine with one that is more reliable?
6. What is the impact of using different forms of material handling equipment?

While specific answers to the questions above depend on the precise configuration of the manufacturing plant, the inferences that can be drawn from an analysis of the

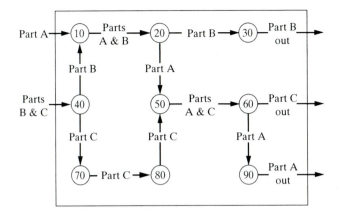

Figure 13.2 Network for Example.

example presented in Figure 13.2 and Table 13.3 are representative of what can be learned about more complex arrangements. The results presented in this section were obtained using the numerical approach developed by the Bell Telephone Laboratories known as Queueing Network Analyzer 3.3. Only a few seconds were required for each calculation, using a Sparc server 1000 computer.

Influence of variability. It was alleged above that the principal villain in creating queues and waiting for processing to take place is the variability that occurs in the many steps that a part must follow in being processed. The fluctuations that

TABLE **13.3 Characteristics for Examples**

Part number	λ_a arrivals/min	Order of nodes visited	Service times (min)
A	0.133	10	4.00
		20	3.00
		50	2.50
		60	3.00
		90	6.00
B	0.222	40	2.00
		10	1.50
		20	1.75
		30	4.00
C	0.200	40	2.00
		70	4.00
		80	4.20
		50	3.00
		60	2.00

occur in processing times, parts arrival times, and machine failures create a situation in which parts will be arriving at a station only to find that it is occupied. Queues of parts will result in parts that must wait to be serviced. It is this variability that creates the WIP in the plant. The importance of variability can be demonstrated by comparing the number of parts waiting to be serviced and the time spent at each node for a plant in which the variance parameter is varied between 0 and 1.0. The results of these calculations are given in Table 13.4. For these calculations the influence of machine failures was made negligible by using long mean times between failure and short mean times to repair.

The WIP at each station is the sum of the parts waiting to be serviced and the parts undergoing service. The parts waiting to be serviced are held in buffers at each station. Note that the average number of parts in service-the parts undergoing servicing in the machines-is not influenced by the variability, since the average number in a machine is given by the average utilization of each machine. Increasing the variability of the arrival rates and the service times greatly increases the number of parts *waiting* to be serviced at each station and therefore the total WIP on each path, but does not alter the average number undergoing servicing at each machine.

Although it is generally true that plant management tries to reduce the level of WIP in the plant, since it costs money to maintain WIP, there are circumstances where management will consciously increase the WIP in a plant. In circumstances in which machine reliability is a problem, it may be necessary to create buffers of parts that can be used by later stations during those times in which machines are being repaired. Reducing the effect of the bottleneck operation may also demand the creation of additional WIP. If reducing the effect of the bottleneck can be achieved only by operating the bottleneck station for longer hours than the rest of the line, it will be necessary

TABLE 13.4   The Influence of Variability on Plant Performance

Operation	$c_a^2 = 0.0$ and $c_s^2 = 0.0$			$c_a^2 = 0.3$ and $c_s^2 = 0.3$			$c_a^2 = 1.0$ and $c_s^2 = 1.0$		
	No. parts waiting	No. parts in service	Time at each node	No. parts waiting	No. parts in service	Time at each node	No. parts in waiting	No. parts in service	Time at each node
10	0.696	0.865	4.40	2.46	0.865	9.37	6.55	0.865	20.9
20	0.246	0.788	2.91	1.32	0.788	5.93	3.63	0.788	12.4
30	0.193	0.888	4.87	2.39	0.888	14.8	7.32	0.888	37.0
40	0.063	0.844	2.15	1.40	0.844	5.33	4.58	0.844	12.8
50	0.176	0.932	3.33	3.78	0.932	14.1	12.3	0.932	39.8
60	0.003	0.799	2.41	0.910	0.799	5.13	3.31	0.799	12.3
70	0.006	0.800	4.00	0.716	0.800	7.58	2.74	0.800	17.7
80	0.000	0.840	4.20	1.11	0.840	9.76	4.19	0.840	25.2
90	0.000	0.798	6.00	0.639	0.798	10.8	2.59	0.798	25.5

PART	Total WIP	Total time in transit	Total WIP	Total time in transit	Total WIP	Total time in transit
A	2.89	21.7	6.4	48.0	15.1	114
B	2.87	12.9	7.5	34.0	18.2	81.8
C	3.18	15.9	8.3	41.7	21.5	108

to supply parts for this station when the rest of the line is down, thus implying that they have been produced earlier and stored for use. Similarly, the output of the bottleneck must be stored until the complete line is reactivated. This will increase the total WIP. Because of the cost of maintaining WIP, a financial decision of whether it is cost-effective to increase the WIP can be made only by comparing the value of the increased production against the incremental cost, including the cost of the extra WIP plus the cost of operating the bottleneck station extra hours. Estimating the value of the total WIP is not straightforward. As a part moves through the plant, its value increases at each stage. Determining the value of the WIP therefore requires a complicated allocation of costs.

As required by Little's law, an increase in WIP must be accompanied by an increase in the average time spent for a part to move through the plant and the average time spent by parts at each station. Increases in variability of the processes increases these times.

The conclusion from this example:

- To improve plant performance, it is important to reduce the variability of all aspects of the operations.

Impact of a change in processing time for the bottleneck operation. The impact of a change in processing time for the bottleneck operation can occasionally be achieved by changing the type of tooling that is used. In the case of a metal removal process, the tooling that is available for use on lathe and milling operations is quite varied and allows a choice of machine parameters that will maximize the performance of the system. The following calculations of the impact of changes in processing time were made using the example from Table 13.4 for $ca^2=0.3$ and $cs^2=0.3$. It was assumed that the processing time for the bottleneck-operation 50-could be reduced by 10% by a choice of a new tool material. The result of this change is seen by comparing the two circumstances as shown by the appropriate columns in Table 13.5. As expected, the utilization of operation 50 dropped from 0.932 to 0.839, a decrease of 10%. Accompanying this reduction in utilization, the WIP at operation 50 decreased from 4.7 to 2.1, resulting in a decrease in the WIP for Parts A and C by more than 15%. The transit times decreased accordingly. The nonlinear response of the system to a change in the bottleneck operation is expected. It is seen that the bottleneck operation has changed from 50 to 30.

It might be argued that the tooling changes that have been hypothesized in this example should always be made. The answer, of course, is that such changes may not always be cost-effective. It is possible that the increased cost of the tooling needed to achieve this reduction in WIP would be sufficiently large that the value of the reduced WIP would not justify taking this action. In the present example, the situation may be even more complicated. In noting that the utilization of the new bottleneck is now 0.888, it is of interest to inquire whether a change of tooling would also be possible for operation 30 thus allowing a several percent increase to be made in the level of production of the plant. These questions can be explored only when detailed information is available concerning the value of the WIP, value of completed parts, potential for selling additional production, and cost of the tooling. The decisions of how to proceed must be based on the economic value of such changes.

The conclusion from this example:

- Although the effects of small changes in process times may be significant, their economic consequences must be examined in detail to arrive at the best overall solutions.

Impact of machine reliability. The influence of machine failure and repair is determined by examining the importance of the change in performance of the network, as measured by WIP and transit times, produced by varying choices of the mean time between failures (MTBF) and the meant time to repair (MTTR). The arrival rates and service times used in these calculations are given in Table 13.3. The variability parameters for the arrival and service functions were both chosen to be 0.3, since the variances in arrival and service times are somewhat controlled in a manufacturing operation. In contrast, the unexpected failure of a machine is much less under the control of the operators, and it is not unrealistic to expect that failure can occur over widely different time intervals. The exponential distribution of failures is more reasonable for this situation, in which case the variability parameter is chosen to be 1.0. As can be seen, as the MTTR increases relative to MTBF, the performance of the plant deteriorates. For the ratio of MTBF/MTTR greater than 20, the impact of machine failure becomes quite large. As would be expected, the effect is largest for operation 50, the bottleneck operation. Since both parts A and C flow through operation 50, the impact of machine failure is larger for them than it is for part B.

Since the influence of machine reliability depends on the MTBF/MTTR ratio, it is meaningful to inquire whether it is the value of the ratio that is critical or whether the absolute magnitude of each the variables is important. The impact of this can be seen by examining the effects for constant values of MTBF/MTTR but with different absolute values for these two individual parameters. The results of this are shown in Figure 13.3. Comparisons are made of the performance of the plant for a constant value of MTBF/MTTR = 14.3, the value corresponding to MTBF = 400 and MTTR = 28 in Table 13.6. While the absolute values of WIP and Transit Times are strongly affected by the absolute magnitude of MTBF and MTTR, it is seen that the relative impact on the three parts remains unaffected. The rationale for this is, of course, related to the phenomena that generates the influence of variability. A machine failure that requires long times to repair (i.e., long MTTR), even though it occurs less frequently, will have a greater impact on plant performance than will a failure that requires a shorter time to repair. The longer repair times allow longer queues to form, which cannot on average dissipate, thus increasing the average amount of WIP and the average transit times.

The conclusions from this example:

- Machine failure will affect drastically the performance of the plant when MTTR reaches a value that is a few percent of MTBF.
- The mean time to repair should be made as short as possible.

Impact of adding an additional server. As the bottleneck operation for the example described in Table 13.3, operation 50 has a high utilization level, at 0.932. The performance of the plant for the condition of MTBF = 400 and MTTR = 20 is displayed in the first three columns in Table 13.7 (this is repeated from Table 13.6). The performance of the plant is then compared with what it would be if the single ma-chine at operation 50 were replaced by two identical but somewhat inferior machines. The middle three columns give the results assuming that the service time using the two inferior machines was 50% greater than it was for the single machine. As can be seen, even with the longer service times, the presence of the two servers resulted in a utilization factor of 0.70 for the station, substantially lower than with one machine. As a result of the lower utilization, the performance of the plant is seen to be substantially improved. In the final three columns of Table 13.7, a comparison of the performance of the plant is given for the circumstances that the reliability of operation 50 for the two servers is assumed to be substantially worse than is the case for the balance of the plant. It is assumed that failures occur twice as often for the node (MTBF is half the size) and the MTTR is twice as long as for the rest of the plant. Despite this reduced reliability, the performance of the plant with two servers at operation 50 is better than when only one server is present. This exercise is intended to demonstrate that two ma-chines operating as equivalent servers may produce an effect that is better than a sin-gle server even though each of the two machines may require a longer service time and their reliability may be substantially less. The availability of the machines to share the workload produces these benefits.

The conclusion from this example:

- Adding a second server may prove to be an effective way of breaking a bottleneck, offsetting lower machine reliability or compensating for the

TABLE **13.5** **Impact of a Change in Processing Time**[a]

	Processing time (A) = 2.50 min Processing time (C) = 3.00 min			Processing time (A) 2.25 min Processing time (C) = 2.70 in		
Operation	No. parts waiting	No. parts in service	Time at each node	No. parts waiting	No. parts in service	Time at each node
10	2.46	0.865	9.37	2.46	0.865	9.37
20	1.32	0.788	5.93	1.32	0.788	5.93
30	2.39	0.888	14.8	2.39	0.888	14.80
40	1.40	0.844	5.33	1.40	0.844	5.33
50	3.78	0.932	14.1	1.33	0.839	6.51
60	0.910	0.799	5.13	0.939	0.799	5.22
70	0.716	0.800	7.58	0.716	0.800	7.58
80	1.11	0.840	9.76	1.11	0.840	9.76
90	0.639	0.798	10.8	0.650	0.798	10.90

PART	Total WIP	Total time in transit	Total WIP	Total time in transit
A	6.38	48.0	5.40	40.6
B	7.55	34.0	7.55	34.0
C	8.35	41.7	6.84	34.2

[a]Operation *50 contained in Table 13.3.

lower performance capability of a machine. The appropriateness of this depends on such factors as space availability, material handling capability,
and the economic value of such an approach.

Material Handling The introduction of a material handling system into the model
depends on the type of system that is to be used. If a forklift truck or a vehicle that
is to respond to requests for service is to be used for material handling, the unit can

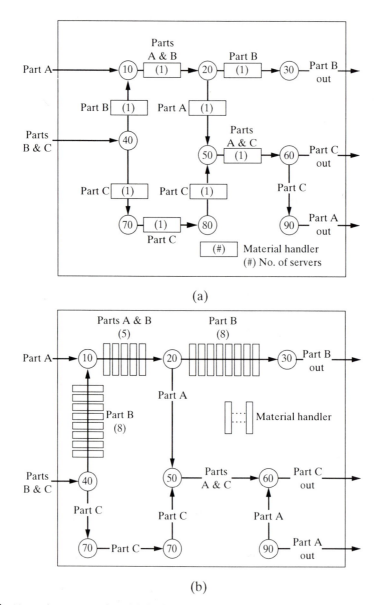

(a)

(b)

Figure 13.4 Network representation with the inclusion of material handling: (a) representative of the use of
forklift trucks for material handling: (b) representative of the use of a conveyor for the path that part B follows.

TABLE 13.6   Influence of Machine Reliability

	$c_a^2 = 0.3$ and $c_s^2 = 0.3$									
	MTBF = 400 MTTR = 0.1		MTBF = 400 MTTR = 10		MTBF = 400 MTTR = 20		MTBF = 400 MTTR = 24		MTBF = 400 MTTR = 28	
Operation	WIP at each node	Time at each node	WIP at each node	Time at each node	WIP at each node	Time at each node	WIP at each node	Time at each node	WIP at each node	Time at each node
10	3.33	9.37	4.92	13.9	9.89	28.0	13.4	38.0	18.3	51.7
20	2.11	5.93	3.13	8.88	6.18	17.5	8.14	23.1	10.6	30.2
30	3.28	14.8	5.56	25.1	13.5	60.8	19.6	88.5	28.9	130
40	2.25	5.33	3.36	8.01	6.73	16.1	9.03	21.5	12.1	28.7
50	4.71	14.1	10.2	30.7	41.9	126	97.3	292	639	1918
60	1.71	5.13	2.71	8.20	5.77	17.5	7.72	23.4	10.2	30.9
70	1.52	7.58	2.14	10.8	4.06	20.5	5.30	26.8	6.87	34.7
80	1.95	9.76	2.95	14.9	6.05	30.4	8.11	40.8	10.8	54.5
90	1.44	10.8	2.02	15.3	3.75	28.6	4.85	36.9	6.22	47.3
Part	Total WIP	Total Time	Total WIP	Total Time	Total WIP	Total Time	Total WIP	Total Time	Total WIP	Total Time
A	6.39	48.0	10.6	79.7	29.3	220	55.4	416	277	2080
B	7.55	34.0	12.1	54.6	26.8	121	37.7	170	53.2	240
C	8.35	41.7	14.5	72.4	42.1	210	81.0	405	413	2066

be treated as an additional node through which parts on that segment of the path must pass. The characteristics of this node will be described by an average service time, a failure mode, and parameters that characterize the variability of the performance of the node. In this sense the material handling node has characteristics that are typical of other nodes in the system. If the material handling system is a conveyor, it is possible to represent it as a series of j nodes with service times that are chosen such that jTj is the total time that would be required for a single node (e.g., a fork lift) to move the part between the two processing nodes. The network models for

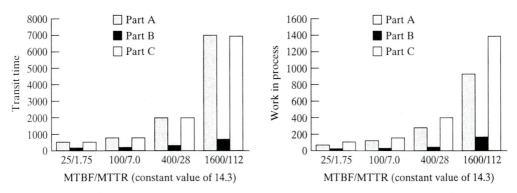

Figure 13.3 Impact of various values of MTBF and MTTR on WIP and transit times, with a contstant ration of the two being maintained at 14.3.

TABLE **13.7**

Impact of Multiple Servers

Oper ation	Operation 50 One Server Part A:T = 2.50min Part C:T = 3.00 min MTBF = 400 and MTTR = 20			Operation 50 Two Servers Part A: 3.75 min Part C: 4.50 min MTBF = 400 and MTTR = 20			Operation 50 Two Servers Part A: 3.75 min Part C: 4.50 min MTBF = 200 and MTTR = 40		
	No. Parts Waiting	No. Parts in Service	Time at each Node	No. Parts Waiting	No. Parts in Service	Time at each Node	No. Parts Waiting	No. Parts in Service	Time at each Node
10	9.02	0.865	28.0	9.02	0.865	28.0	9.02	0.865	28.0
20	5.39	0.788	17.5	5.39	0.788	17.5	5.39	0.788	17.5
30	12.6	0.888	60.8	12.6	0.888	60.8	12.6	0.888	60.8
40	5.89	0.844	16.1	5.89	0.844	16.1	5.89	0.844	16.1
50	41.0	0.932	12.6	1.54	1.40	9.13	10.1	1.40	35.6
60	4.97	0.799	17.5	4.77	0.799	16.9	7.63	0.799	25.5
70	3.26	0.800	20.5	3.26	0.800	20.5	3.26	0.800	20.5
80	5.21	0.840	30.4	5.21	0.840	30.4	5.21	0.840	30.4
90	2.96	0.798	28.6	2.88	0.798	28.0	3.92	0.798	35.8
Part	Total WIP		Total time	Total WIP		Total time	Total WIP		Total time
A	29.3		220	13.6		102	19.3		145
B	26.8		121	26.8		121	26.8		121
C	42.1		210	18.6		93.0	25.6		128

For all servers except operation 50:MTBF = 400 and MTTR = 20; for all servers: cs2 = 0.3.

these two cases are shown in Figure 13.4 (a) and (b).

The influence of the material handling system can be illustrated by a comparison of the following four cases.

- Case I: Network with no assumed material handling.
- Case II: Network with a single material handler (e.g., a forklift truck) provided to move material between the nodes. The network representation for this case appears in Figure 13.4(a).
- Case III: Network with a single material handler (e.g., a forklift truck) provided to move material on the path used by part B.
- Case IV: Network with a conveyor used to move material on the path that is used by part B. The network representation for this case appears in Figure 13.4(b).

The parameters for each of the four cases are given in Table 13.8. The service times for the material handling nodes were chosen to make their utilization modest. The consequences of the introduction of the material handling systems are summarized at the bottom of Table 13.8. In comparison to case I, the introduction of a truck-lift material handler, represented by case II, creates significantly greater WIP and a substantially longer transit time for all parts in the plant. When this type of material handling is introduced for the path that part B follows, a larger WIP and longer transit times result for part B and, to a lesser degree, for part A, since parts A and B share only one material handling operation. When the forklift type of material handler is replaced by a conveyor, as is case IV, the performance is improved for both parts A and B. Part C is affected very slightly because the queue lengths of operations 50 and 60 are affected slightly by the queues in the material handler between operations 10 and 20 and by the queue length at operation 20. A detailed comparison of the calculations for cases I and IV indicates that the amount of material at operations 10 through 90, the material being processed and the material awaiting processing, is nearly the same in both cases. The larger WIP for case IV arises from material that is contained on the conveyor. It is seen, therefore, that a conveyor system is capable of moving materials throughout a plant without significantly influencing the size of buffers that are needed at individual stations. While moving the material, the conveyor also acts as a buffer. While the conveyor offers the obvious advantage of improving performance, it is also less flexible in terms of changing location, changing capacity, and so on. The choice of which material handling system to use will depend on several factors, including the constancy of markets, relative costs of systems, and availability of space.

The conclusions from this example:

- Although it is possible to design a material handling system that will not seriously deteriorate the levels of performance for the plant, choosing the proper system depends on the needs for flexibility and the total costs to the system.

Response of the plant to external changes. External events create changing customer demands for products. Markets can change rapidly as new products are created and as customers react to the availability of new product attributes. The plant that can change quickly at the lowest cost will have an important advantage in the marketplace. A measure of the response time of a plant to external changes can be estimated by examining the average values of λ for the plant and the average WIP in the total plant. For example, if the average arrival rate for all parts is 0.200/min and the total WIP in the plant is 1000 parts, the time that it would take to exhaust all of the WIP could approach 5000 min. If parts that are partially processed are not to be destroyed when a change is required, a rough estimate of the earliest time that a new part can be expected from the plant would be about 80 hours. Obviously, a plant that is operating in a JIT mode, with its small amount of WIP, will have the capability to make a much more rapid response to an external change in market.

13.3   SIMULATION MODELS

To generate more realistic descriptions of the operation of large systems than can be obtained from the queueing models described above, it is usually necessary to employ models that explicitly treat the discrete, dynamic, and stochastic nature of the events. While continuous processes can be included in discrete-event simulation models, through the use of differential equations, it is more common to limit the models to the treatment of discrete events. Following the development of the model that is being used to approximate the physical system, a method of evaluating that model must be

TABLE 13.8   Impact of Material Handling on the Network

Part	Operations		Material handling							
			Case I		Case II		Case III		Case IV	
	Operation	Service time	No. stations	Service time	No. stations	Service time	No. stations	Service time	No. stations	Service time
A	10	4.00								
	Mtl.Handler		x	x	1	2.50	1	2.50	5	0.50
	20	3.00								
	Mtl. Handler		x	x	1	2.50	x	x	x	x
	50	5.00								
	Mtl. Handler		x	x	1	2.50	x	x	x	x
	60	3.00								
	Mtl. Handler		x	x	1	6.00	x	x	x	x
	90	6.00								
B	40	2.00								
	Mtl. Handler		x	x	1	4.00	1	4.00	8	0.50
	10	1.50								
	Mtl. Handler		x	x	1	2.50	1	2.50	5	0.50
	20	1.75								
	Mtl. Handler		x	x	1	4.00	1	4.00	8	0.50
	30	4.00								
C	40	2.00								
	Mtl. Handler		x	x	1	4.00	x	x	x	x
	70	4.00								
	Mtl. Handler		x	x	1	4.00	x	x	x	x
	80	4.20								
	Mtl. Handler		x	x	1	4.00	x	x	x	x
	50	3.00								
	Mtl. Handler		x	x	1	2.50	x	x	x	x
	60	2.00								

Part	WIP	Transit time	WIP	Transit time	WIP	Transit time	WIP	Transit time
A	6.91	52.0	21.0	158	7.90	59.4	7.51	56.5
B	8.15	36.7	28.5	128	24.2	109	11.6	52.2
C	9.11	45.5	25.8	129	9.18	45.9	9.18	45.9

[a]MTBF = 100; MTTR = 1

chosen. A number of computer packages have been developed that calculate numerically those parameters that can be used to estimate the performance of the system. This can be better understood by reference to a quotation from Pritsker et al. (1989) concerning the SLAM II simulation network language.

> A SLAM II network consists of nodes and branches. A branch represents an *activity* that involves a processing time or a delay. *Nodes* are placed before and after activities and are used to model milestones, decision points, and queues. Flowing through the network are *entities*. Entities can represent physical objects, information, or a combination of the two. A set of attributes is used to describe an entity and to distinguish one entity from another. For example,

a part entity may have attributes describing its type, weight, and value. Entities are directed through the network from one node to another according to routing conditions placed on activities. The SLAM II network is a model of the operations and procedures of a system.

In a SLAM II network model, an entity originates at a CREATE node and is routed over the activities emanating from the CREATE node. For each activity, the entity is delayed in reaching the end node of the activity by the activity's duration. When reaching the end node, the disposition of the entity is determined by the node type, the attributes of the entity, and the status of the system as modeled by SLAM II variables. Typically, the entity is routed over one or more activities emanating from the node to which it arrived. Alternatively, the entity is held at the node until a server is available or a resource is allocated to it. The entity continues through the network moving from node to activity to node until a TERMINATE node is encountered or no further routing can be performed. When this occurs, the entity is terminated, that is, deleted from the network.

A simulation procedure is used to analyze a SLAM II network by generating entities and processing their movements through the network. As entities flow through the network model, observations are made of travel times, node release times and the status of servers, resources, and queues. Data collection nodes are inserted directly in the network model to collect observations on model variables. Statistics are presented for these nodes and for SLAM II variables.

Discrete-event simulation is used to determine the values of network parameters. The way in which it is used is determined by the type of question that is being asked. If it is of interest to determine the equilibrium state of a system, it is presumed that a long simulation run will yield parameters that are appropriate to equilibrium. The parameters of the initial state that are chosen for the simulation are assumed to be relatively unimportant. The task is to assure that the final state of the simulation is, indeed, characteristic of equilibrium. If it is of interest to examine the dynamic behavior of a system, the choice of the parameters to use in describing the initial state become critical. Multiple runs are made using different values for the initial state, the values of which are determined by selecting at random from a collection that reflects their expected initial statistical distribution. From the collection of values for all runs, an average value for each parameter of the final state is determined. Using the strong law of large numbers, these averages are expected to approach the expected value of the corresponding variable for a system with an initial state having the prescribed statistical distribution.

The models require detailed data for a large array of events. This has two important consequences. First, it is necessary to know or to have a realistic basis for assigning values to all of these quantities. In the analysis of the impact that a projected change might have on the operation of a large system, the initial data would likely be obtained from observations on the existing operating system. Changes could then be made in the parameters that would be altered by the new facility and a comparison could be made between the two states of the system. It is more difficult to determine the proper parameters for a new facility whose processes may not have been fully characterized. Some additional approximations may be required in this case.

The second consequence is that the task of programming the model can become quite onerous. The development of general simulation models that allow the model of the system to be specified has greatly simplified this task. Improvements continue to be made by many of the developers of simulation languages. A useful comparison of simulation languages for the manufacturing environment can be found in Law and Haider (1989).

The accuracy of the results of the simulation is determined primarily by two factors. The first is the accuracy with which the model represents the system being modeled. This depends on the ability of the modeler to capture the important relationships among the system parameters. The second is the degree to which the averages of the parameters determined from the multiple runs approximates the expected value of the system parameters. This depends on the number of runs that are made-the larger the number runs the greater is the statistical validity of the average-and the assurance that the collection of values that was chosen to describe the initial states properly reflective the initial distribution of the system. A large number of runs for a large system can require several hours of calculation time.

Simulation is a powerful tool for analyzing the behavior of systems. It allows an examination of discrete events in a changing system and of the time progression of events. Its ability to provide information of the dynamic behavior of discrete systems is of great importance in understanding the behavior of large complex systems.

13.4   REGRESSION MODELS

Regression models of manufacturing systems are used most often when an analytical relationship among the system variables, Sk, and their influence on some performance parameter, P, of the system, are not known or when the exact form cannot be predicted analytically (Riggs, 1987). This is a technique that is frequently used in forecasting (e.g., in estimating future market demands based on historical trends, where several variables are believed to influence the future). These market estimates are often used to estimate the future need for new or expanded manufacturing facilities. Such is the case when it is desirable to identify the dependence of the market on such factors as the level of disposable personal income of consumers, the costs trends of similar products, the demographic characteristics of the customers, and so on. The dependent variable is taken to be the annual demand, D, for the product. The independent variables are the various parameters, Sk, that will influence the demand: namely, demographics, disposable income, cost trends, and so on. The task is to determine the form of the relationship

$$D = F(Sk) \qquad (13.2)$$

If it is expected that D is linearly related to the Sk values, the data are fitted to a linear equation whose coefficients are determined by a least-squares analysis. Other forms of the relationship can be assumed, including polynomials of various powers, semilog or logarithmic. Various techniques have been developed for testing the *goodness of fit*: how well the resulting equation represents the data, including the coefficient of correlation.

While regression models are very useful in deriving the dependence among variables, some caution must be exercised. First, great care must be taken in using the models to project performance for values of the variables that are well removed from the region in which experimental observations exist. Second, any hidden dependency among seemingly independent variables must be exposed and eliminated (e.g., if two variables are considered to be independent, when in fact they both depend on a third unspecified variable, the final form of the relationship will be misleading). Third, the data used in the analysis must be of high quality (e.g., they must be accurate and unbiased).

While regression analyses have not been used extensively in manufacturing systems, recent studies of fabrication lines for integrated circuits suggests that this can be an important tool in understanding the dependencies of the output on the choice of operating parameters. It appears particularly useful in those cases in which quantitative models are inadequate to capture the important relationships (e.g., the dependence of product quality and product yield on specific operating parameters).

13.5 BENCHMARKING

The tools discussed above are extremely useful in developing a quantitative evaluation of key operating characteristics of a new facility, of the benefits that may be derived from modifying an existing facility, or of searching for ways to improve the operational efficiency of an existing facility. They can also be very useful in benchmarking the performance of a facility against those of a competitors. Although it is unlikely that any competitor would allow access to the detailed data needed to model or simulate *its* facility, some important estimates of competitor's capabilities relative to the performance of your enterprise can still be made.

Imagine for a moment that it has been found that an equipment vendor is offering facilities that have the possibility of reducing substantially the service times of several operations used in your facility. If it is known that your competitor is building a new facility, it is prudent to assume that they will use the best available technology. It is possible to assess the competitors likely advantage by comparing the outputs of the model of your current facility with the outputs that would result from substitution of the new processing capability. Accomplishing this type of analysis depends critically on the availability of information from firms that may be suppliers to you and to your competitors. Of course, if your competitors are developing all of the processes used in their plants, this information will probably not be available or will be much more limited in scope.

13.6 NOTES ON SELECTION AND USE OF MODELS

In concluding this discussion, emphasis is again given to the importance of quantitatively understanding the system that you are managing. Queueing theory and simulation are important tools for accomplishing this. The capability to employ modeling and simulation effectively depends critically on the validity of the models that are developed to describe the physical situation. Good predictions can never be derived from a model that does not represent the physical system under study. The elements of the physical system and their interactions must be understood. The model must be tested to verify that it accurately represents the physical system under study. If it should fail this test, it must be modified and retested. Furthermore, good data are needed to arrive at reliable predictions.

Some models are easier to use than others. As Solberg (1992) noted, it is best to start with the simplest representation of the system, learn from it, and then add other features. While these additions will probably complicate the model, they can remove some of the limiting assumptions that are contained in the more simple models. This suggests that queueing network models should be favored in the early stages of studying a facility, while simulation, with its greater demands for data and computer capability, should be utilized when detailed comparisons with existing systems are being examined. When analysis of dynamic situations is needed, simulation will be required.

The ease with which models can be manipulated, the speed with which results can be achieved, and the insights that they provide to the modeler should convince any manager that these techniques deserve to be included in the arsenal of tools that are regularly utilized. The manager should regularly investigate what has been learned from modeling, whether the latest capability for modeling exists in the organization, whether adequate data are being collected and updated to enhance the capability to improve the models, and what limitations exist to enhancing the organizations capability to use models. These are the tools that will help optimize an organizations performance. To ignore them is to reduce your capability relative to the competition.

While the value and importance of using models for describing and understanding systems has been emphasized, it must be recognized that these approaches have limitations. As noted earlier, in building models for complex systems it is almost always necessary to introduce assumptions in order to make the model tractable to solution, even with high-speed computers. Thus the validity of the model in representing a real-world system depends very critically on the ability of the modeler to identify and include the critical elements in the model. A second limitation that must be recognized is that these models are not capable of analyzing and *predicting* some of the critical operational parameters of manufacturing systems, for example, the quality of the product that will be produced. Neither is it possible to predict the fraction of parts that will be generated at each step that will require rework. Although it is possible to include rework stations, inspection stations and scrap rates in the models, the probability that a part will require rework after a particular processing step must be provided by the modeler. It is not currently possible for the queuing or simulation models described above to predict this. Finally, it is not possible to predict the impact of much of human behavior on the performance of a manufacturing operation. Average times can be included for a human being to do a task, but the benefit of employee participation, team efforts, recognition programs, and so on, cannot be predicted. Although it has been nearly two-thirds of a century since the Hawthorne experiments were carried out, the models that have been discussed in this chapter do not include the factors that were discovered in those early experiments.

13.7 THE *DESCRIBING AND UNDERSTANDING* FOUNDATION OF WORLD-CLASS PRACTICE

> Models provide a rational basis for predicting the impact of decisions before their implementation by [quantitatively] describing the important elements, interactions, and dependencies. Empirical models comprise valuable knowledge that provides a basis for engineering and managerial practice....[They] are useful for considering operational questions that confront the manufacturing enterprise and to explore, in a timely way, strategic alternatives for the firm.... Models can be immensely powerful competitive weapons when used to capture particular competencies of the manufacturer and then leveraged throughout the enterprise that developed them. They offer an important means

of accomplishing organizational learning as they extend their use well beyond a particular control activity. Models should also be considered as a basis for evaluating continuous improvement efforts and changes made in the manufacturing system.

FOUNDATION: World-class manufacturers seek to describe and understand the interdependency of the many elements of the manufacturing system, to discover new relationships, to explore the consequences of alternative decisions, and to communicate unambiguously within the manufacturing organization and with its customers and suppliers. Models are an important tool to accomplish this goal. (Heim and Compton 1992)

REFERENCES

ALLEN, A. O. 1987. Back-of-the-Envelope Modeling, *EDP Performance Review*, July:1.

BOORSTIN, D. J. 1983. *The Discoverers,* Vintage Books, New York:20-21.

BRUMELLE, S. L. 1971. On the Relation Between Customer and Time Averages in Queues, *Journal of Applied Probablity*, 8:508-520.

GLYNN, P. W. AND W. WHITT. 1989. Extensions of the Queuing Relations L = λ W and H = λ G, *Operations Research*. 37:634-644.

HEIM, J. A. AND W. D. COMPTON (eds.) 1992. *Manufacturing Systems: Foundations of World -Class Practice*, National Academy Press, Washington, DC: 58-60 .

HEYMAN, D. P. AND S.. STIDHAM JR. 1980. The Relation between Customer and Time Averages in Queues, *Operations Research* 28:983-994.

JUDSON, H. F. 1980. *The Search for Solutions*, Holt, Rinehart and Winston, N.Y.:112.

LAW, A. M. AND S. W. HAIDER. 1989. Selecting Simulation Software for Manufacturing Applications: Practical Guidelines & Software Survey, *Industrial Engineering*, May:33-45.

LAZOWSKA, E. D., J. ZAHORJAN, G. S. GRAHAM, AND K. C. SEVCIK. 1984. *Quantitative System Performance: Computer System Analysis Using Queueing Network Models*, Prentice-Hall, Inc, Upper Saddle River, NJ.:4.

LITTLE, J. D. C. 1961. A Proof of the Queueing Formula L = λ W, *Operations Research*, 9:383-387.

LITTLE, J. D. C. 1992. Are There "Laws" of Manufacturing? *Manufacturing Systems: Foundations of World-Class Practice*, J. A. Heim and W.D. Compton (eds.), National Academy Press, Washington, DC .

MAXWELL, W. 1970. On the Generality of the Equation L = λ W *Operations Research* 18:172-74.

PRITSKER, A. A. B., C. E. SIGAL, AND R. D. J. HAMMESFAHR. 1989. *SLAM II Network Models for Decision Support*, Prentice-Hall, Inc., Upper Saddle River, NJ:33-34.

RIGGS, J. L. 1987. *Production Systems: Planning, Analysis, and Control*, John Wiley & Sons, New York, Chapter 3.

ROLSKI, T. AND STIDHAM, S. JR. 1983. Continuous Versions of the Queuing Formulas L = λ W and H = λ G, *Operations Research* Letters. 2:211-215.

SEGAL, M. AND W. WHITT. 1989. A Queueing Network Analyzer for Manufacturing, in *TELE-TRAFFIC SCIENCE for New Cost-Effective Systems, M. Bonatti (ed.)*, Elsevier Science Publishers B.V. (North-Holland), Amsterdam.

SNOWDON, J. L. AND J. C. AMMONS. 1988. A Survey of Queueing Network Packages for the Analysis of Manufacturing System, *Manufacturing Review* 1 (1):14-25.

SOLBERG, J. J. 1977. A Mathematical Model of Computerized Manufacturing Systems, *Proceedings of the 4th International Conference on Production Research*, Tokyo, Japan.

SOLBERG, J. J. 1992. *The Power of Simple Models in Manufacturing in Manufacturing Systems: Foundations of World-Class Practice*, J. A. Heim and W.D. Compton (eds), National Academy Press, Washington, DC .

STIDMAN, S. JR. 1974. A Last Word on L = λ W, *Operations Research* 22:417-421.

TAYLOR, H. M. AND S. KARLIN. 1984. *An Introduction to Stochastic Modeling*, Academic Press, Inc., New York.

WHITT, W. 1993. Large Fluctuations in a Deterministic Multiclass Network of Queues, *Management Science:*1020 39 (8).

QUESTIONS

13.1 (a) Verify the theoretical limits as given in Section 13.1.2 for the plant described in Figure 13.1 and Table 13.1. (b) Calculate the theoretical limit for the WIP for the plant.

13.2 What would be the theoretical limits for λ_A, λ_B, and λ_C for the circumstance in which they are not necessarily equal, for the plant described in Figure 13.1 and Table 13.1 when operations 10 and 60 are provided with two servers each? Would you advise adding both of the additional servers?

13.3 Your task is to recommend modifications to the plant shown in Figure 13.1 and Table 13.1 that will make it possible to increase the production of part A to 21 parts/hr. The production rate for part B must fall between 7 and 8 parts/hr and for C between 9 and 10 parts/hr. If you find that it will improve the level of production, you may assume that operations 30 and 90 use the same machines and can be combined. Similarly, operations 50 and 80 can be combined into a single operation, however, since a tool change will be necessary for part C if the two tasks are done on one machine, the service time must be increased by 2 minutes for part C over that required on two separate machines. No machine should have an utilization factor that exceeds 0.900. Give part paths, number of servers at each station, production levels, machine utilization levels, and WIP. The objective, of course, is to use the least number of machines to accomplish the production levels desired.

13.4 Two different independent failure modes for machines may be experienced. These may have different MTBFs, MTTRs and probabilities of occurrence. Suppose that short-term stoppages are frequent and that changes take place rather quickly, (e.g., tool changes, tool breakages, confirmation of machine settings, etc.). A second mode could be more catastrophic, with infrequent failures but with long times required to accomplish a repair. What would you expect the resultant behavior to be for a system that is subjected to these two types of failure mechanisms of its machines simultaneously?

13.5 You are the manager of a manufacturing facility that is making three distinct products on a production line that is operating at 80% of its maximum capacity as measured on a three-shift 5-day-a-week schedule. You have just received a telephone call from your company president telling you that the marketing department has identified a customer that needs a new product that is closely related to your current products but of a rather different design. Since your facility is the only one within the company that is a candidate to manufacture this new product, you are being asked to determine whether you can make this product at a volume that would essentially utilize the currently unused 20% capacity of your line. You are, of course, expected to manufacture this at the same profit margin that you are making on current products. What do you need to know to answer this question?

Figure Q13.6, on the following page, describes the flow of parts through a small manufacturing plant that is currently making four distinct parts. The value of each completed part is the following:

Part A $10.00
Part B $20.00
Part C $10.00
Part D $30.00

The service times for each part at each station is the following:
For the conditions that:

- No machine will have a utilization greater than 0.900
- No part can have an arrival rate less than 6/hr.

13.6   Determine the values of the arrival rates for the four parts that will give the highest value for the output of the plant. What will be the weekly value of production assuming that the plant operates 80 hr./wk.

13.7   Assume that you can add one machine at one station, thus giving that station two equivalent servers. Which station will you choose? Why?

13.8   Determine the new values of the arrival rates for the four parts that will give the highest value of the output for the plant. What is the added value of weekly production for the plant with this additional machine?

READINGS

LITTLE, J. D. C. 1992. Are There "Laws" of Manufacturing? *in Manufacturing Systems: Foundations of World-Class Practice,* J. A. Heim and W.D. Compton (eds.), National Academy Press, Washington, DC .

SOLBERG, J. J. 1992. The Power of Simple Models *in Manufacturing in Manufacturing Systems: Foundations of World-Class Practice,* J. A. Heim and W.D. Compton (eds), The National Academy Press, Washington, DC .

PRITSKER, A.A.B. 1992. Manufacturing Capacity Management Through Modeling and Simulation *in Manufacturing Systems: Foundations of World Class Practice,* J. A. Heim and W.D. Compton (eds), National Academy Press, Washington, DC

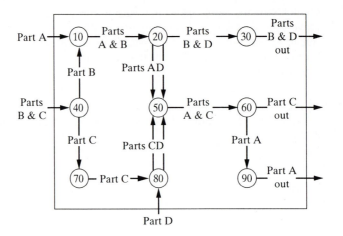

Station		Service Time (min)	Station		Service time (min)
Part A			Part C		
	10	4.00		40	2.00
	20	2.00		70	4.00
	50	1.80		80	3.10
	60	3.00		50	1.75
	90	6.00		60	2.00
Part B			Part D		
	40	2.00		80	2.00
	10	1.50		50	1.50
	20	1.60		20	1.00
	30	3.00		30	2.00

Figure Q13.6 Question 13.6-13.9

PART

7

Management of Technology

By endorsing and practicing the principles enunciated in the previous nine foundations of world-class practice, manufacturers will have taken the actions that will place them among those who can compete successfully in the world marketplace. As success with these approaches becomes more common and as more companies embrace them, this group of foundatiations will, in total, form the lowest common denominator of performance for successful enterprises. It becomes important, therefore, to search for the ways in which enterprises can achieve competitive advantage over those who are similarly following these foundations. The answer lies in technology.

Rather that using technology as a means of compensating for shortcomings, an enterprise that has embraced and is practicing the first nine foundations of world class practice will find that technology can become the defining difference between otherwise successful firms. The management of technology, including its acquisition, utilization, and control, will become a key ingredient of success for the enterprise that is commited to remaining competitive. As the tenth and final foundation of world-class practice states:

> FOUNDATION: World-class manufacturers view technology as a strategic tool for achieving world-class competitiveness by all elements of the manufacturing organization. High priority is placed on the discovery, development, and timely implementation of the most relevant technology and the identification and support of people who can communicate and implement the results of research. (Heim and Compton, 1992)

In part VII of this book, we presume that the enterprise has or is in the process of adopting the principles contained in the first nine foundations of world-class practice. In these circumstances, the management can turn its attention to the things that are necessary to identify and use effectively the technology that will make them most competitive. If they are to be successful in this, they must be prepared to confront the circumstances that led Richard Foster (1982:24), to make the following assertion: "The record is uncomfortably clear. Technology leaders tend to become technology losers."

The material dealing with the management of technology has been divided into three chapters. In Chapter 14, we explore the justification and current patterns that exist with regard to approaching technology as an investment. The assessment of technological opportunities is investigated in Chapter 15. Chapter 16 is concerned with the techniques that are needed to manage various types of technical programs successfully .

REFERENCES

HEIM, J. A. AND W. D. COMPTON (eds.) 1992. *Manufacturing Systems: Foundations of World Class Practice*, National Academy Press, Washington, DC:8 .

FOSTER, R. N. 1982. *A Call for Vision in Managing Technology*, Business Week, May 24:24-33.

14

Investments In Technology: Why? When? By Whom? And How Much?

> This historic and irreversible change in the way of doing things we call "innovation" and we define: innovations are changes in production functions which cannot be decomposed into infinitesimal steps. Add as many mail-coaches as you please, you will never get a railroad by so doing. (Schumpeter:1935)

Technology became a powerful force in society with the appearance of the Industrial Revolution. The limitation on growth that the strength and endurance of human beings and animals had imposed on society was thrown aside as energy sources and machines were developed and harnessed. The benefits that technology has allowed have been many.

The decision to pursue a particular technology is determined by many factors, including the enterprise's needs for technology in its forthcoming products and the opportunities that new technology offers for all new products and processes. Since the opportunities for pursuing interesting avenues are seldom limited by the technological opportunities, the capabilities of the organization and the relevance of the possible technologies to the enterprises strategy must restrict the directions that are pursued. The critical issue for the enterprise is making the best selection from among the wide range of opportunities. To accomplish this, it is necessary to determine who will be involved in making the decisions, the time frame that is acceptable for application of the technology, a cost/benefit analysis for the system of a possible successful technology, and the potential cost of *not* proceeding with the exploration. These are some of the issues that make the management of technology so difficult. Erickson et al. (1990:74) describe the issues facing management as follows:

> The management of technology is analogous to the management of investment. The development and use of technology must be guided explicitly by the business strategy of the firm; at the same time, technological developments should help define the opportunities and threats to which the strategy should then respond. Thus, the strategic management of technology involves a dialogue-a process through which both the strategic targets of the enterprise and the goals of its technology program are regularly reviewed and revised.
>
> In looking more thoroughly at that process, it is important to clarify what "management of technology" really entails. The management of technology encompasses the management of research, product and process development, and manufacturing engineering. Put simply, research expands the firm's grasp of science and engineering skills. Development makes this knowledge relevant to part of the firm's business. Engineering translates technology into products that are useful or desirable to customers. It is important, however, to think of these functions as forming a spectrum. In fact, one pitfall in managing technology is to see these as separate functions to be managed in a compartmental fashion. Many Japanese companies have shown the power of integrating all phases of the product creation process. By contrast, some U.S. manufacturers have suffered because they separate product development from manufacturing engineering; this results in poor manufacturability, with cost and quality problems and delays in product introduction. Effective management requires integrating these phases of the product creation process.

Quinn (1986) has identified three options by which a company can compete in the world marketplace successfully:

- Anticipating and servicing specific customers' needs better than anyone else in the world.
- Innovating faster and more continuously than all competitors.
- Leveraging one's own capabilities by accessing and utilizing the world's best external science, technology and information sources.

The first 13 chapters of this book emphasized the importance of the first two of these three options and offered insight into how these objectives can be achieved. The last three chapters are concerned with the third of these options. Quinn goes on to comment on the nature of technological innovation. He observes that:

> Companies seeking strategic advantage through significant technological innovation need to recognize the tumultuous, long-term realities of how this process operates....The central problem is that innovation-first reduction of an idea to practice-is a process that rarely can be planned and controlled with the kind of analytical certainty managers associate with other operations. Instead, technology tends to advance in a bubbling, intuitive, tumultuous process-more akin to a fermentation vat than a production line.

Since technological innovation has become one of many tools that the modern company uses in achieving its goals and objectives, it is not surprising that the goals and objective statements given in Chapter 3 would reveal the importance that companies place on the technological environment. Intel includes in its list of corporate values the delivery of innovative products and services and the maintenance of an innovative environment. AT&T declares that in achieving its business objectives of moving and managing information it will combine communications and computers and supply switching systems, transmission equipment, and operations to the telecommunications industry. The Timken Company states that it will attain its mission by maintaining excellence through investment in technology with a continuing commitment to timely implementation of the results.

Technology does not exist nor should it be explored for its own sake. Rather, it is a tool for accomplishing the goals and objectives that the company has established. A company that decides that it is going to be the technological leader in its field will have very different needs than the company that decides that it will be a follower in technology. The former will need to invest heavily in R&D, will seek to obsolete existing products with technological advancements in a time frame that places the follower companies at a disadvantage, and will choose to move into new technical areas that are more risky but offer the potential of large payoffs. As Kantrow (1980) observes:

> Technology bears an integral relation to a company's strategic thinking by helping to define the range of its possibilities. At the same time, it provides a good portion of the means by which strategy, once decided on, is to be carried into effect....
>
> In short, innovative success appears to be a function of good communications, purposeful allocation of resources, top-level support within the organization, and careful matching of technology with the market. If these factors have a salient common denominator, it is that they are all key elements in defining and implementing corporate strategy.

Rosenbloom (1978) comments on the different responses that an enterprise must make to technical development: "Technology....is neither autonomous nor wholly "outside" the control of managers. It is more useful to view it as interacting with the firm, part of its capability, shaped by as well as a shaper of corporate action. At times, to be sure, its principal significance is as part of the environment, when it emerges as a threat or as opportunity to which the firm must respond strategically."

The aggressive use of technology to achieve competitive advantage is clearly evident in the strategy currently being followed in the microelectronics industry. A news announcement in the January 28, 1994, *Wall Street Journal* stated: "Intel Corp., keeping up the pressure on rival chip makers, promised to ship its next-generation microprocessor in 1995 and said it will boost spending on new factory capacity ..." The remarks by Andrew Grove, Intel's chief executive officer, were said to be Intel's most specific pronouncement yet about its new computer chip "and illustrated its plans to use innovation to maintain the No. 1 position in microprocessors." The Intel announcement was a clear challenge to the makers of the PowerPC chip-IBM, Motorola, and Apple Computer-and to companies that had cloned the design of its current microprocessor, the Pentium, which was only introduced the previous spring but was expected to account for about 15 percent of the PC business in 1994. IBM and

Motorola responded in the March 31, 1994, *Wall Street Journal* with an announcement "that they plan to roll out an even more souped-up version of their new PowerPC computer chip.... Analysts said yesterday's announcement came earlier than they had expected, and seemed to be motivated by Intel's accelerated launch of 90-megahertz and 100-megahertz versions of its fastest Pentium chip in early March." With a clear view of its future products and services, the processes that it will use to provide them, and the goals that it has for achieving and maintaining its position in the marketplace, a company can use technology as an effective strategic tool.

Technology is not limited to the products that are offered or to the processes used in making those products. As companies strive to be more responsive to the marketplace and to shorten the time spent in developing new products, the *tools* that are used in the development process are becoming increasingly more sophisticated and powerful: for example computer simulation, tools for providing rapid prototyping, and communication systems that enhance interaction among remote design groups. Management must recognize the value of such tools and encourage their development, acquisition, and use if the enterprise is to remain competitive.

Technology, then, offers a triple-pronged path of support for the strategic plan of the corporation. On one path, technology must be available to support the plans of the corporation for its near-term products and processes. Along a second path are found the many opportunities by which technology can help make the firm more competitive through improved procedures and operating systems. The third path offers opportunities for a complete rejuvenation of the products and processes of the enterprise through the use of radically different approaches. As the tenth and final foundation of world-class practice states:

> FOUNDATION: World-class manufacturers view technology as a strategic tool for achieving world-class competitiveness by all elements of the manufacturing organization. High priority is placed on the discovery, development, and timely implementation of the most relevant technology and the identification and support of people who can communicate and implement the results of research. (Heim and Compton, 1992)

14.1   TECHNOLOGY AS THE ENGINE FOR COMPETITIVENESS

Economists have long been interested in understanding the relative importance of various factors that generate economic development. Early observations suggested that development does not occur in a regular fashion, but seemed to occur in "bursts," thus indicating that external factors probably have an important impact on development. This cyclic nature of the development process presented an interesting challenge that was actively explored by early prominent scholars, including Kondratieff, Kuznets, and Schumpeter.

Kondratieff (1935/1951) was a pioneer in the study of the cyclic behavior of capitalistic systems. Although he was concerned with both short term and long-term effects, his study of the factors that generate *long waves* in the economic activities of national economies was of particular importance. His analysis of historical data established the existence of long-term cyclic behavior in such quantities as the wholesale price indices of France, England, and the United States, the rate of interest in France and England, English wage levels, France's level of foreign trade, levels of production of coal, pig iron and lead in England, and in a host of indices dealing with the intensity of use of other physical resources. The statistical analysis of the data indicated that the long waves had a period of between 48 and 60 years. The similarity in the cyclic behavior of these various factors and their existence in various countries suggested that something fundamental was occurring. Of the various characteristics that Kondratieff identified as being important, the following is of particular interest: "During the recession of the long waves, an especially large number of important discoveries and inventions in the technique of production and communication are made, which, however, are usually applied on a large scale only at the beginning of the next long upswing.

The last maximim that Kondratieff observed occurred about 1920. The next maximim in economic activity followed in about 50 years, as he would have predicted. Since his studies had suggested that the growth and decline of capitalist economies would continue to occur naturally, in conflict with Marx's contention that the capitalistic economies would ultimately fail, he was imprisoned by Stalin and perished before the cycle of the 1970s appeared.

Simon Kuznets was a contemporary of Kondratieff. He summarized the results of his early studies in an article published in 1975 (Kuznets,1975:344,346). In this he offers the following conclusion:

> The comparisons of the combined growth rates of inputs with those of output [of national economic accounts] derived in this analysis leave a positive residual, which is due, largely if not entirely, to the contribution of technological change.
>
> A simple summary of the contribution of technological innovations to modern economic growth can be suggested. First, technological innovations produced a thoroughgoing transformation of conditions of work and of life, while increasing enormously the volume of final goods and changing substantially some of their characteristics.... Second, mass application of technological innovations had, inevitably, some negative consequences.... Third, although it is difficult to reduce the contribution to a single quantitative estimate of all the costs involved in conventional inputs, nonconventional requirements and negative by-products, one cannot avoid the conclusion that the net contribution of technological innovations must have been positive and large.

Kuznets (1975:339) makes the following observations about the "rough phasing of the life cycle of a major technological innovation."

> First, the time spans are long: about twenty years for the initial application phase, and much more than twenty years for the properly defined phase of diffusion.... It is in the interplay of technological advance and organizational, economic, and social adjustments that the crucial feature of the innovation, the *application* of a new technological element, lies....
>
> Second, the duration of the life cycle of an innovation is partly a function of its magnitude, for both the technological task and the process of social adjustment are the more formidable, the greater departure from established technological practices and the greater the potential diffusion of the innovation. That makes it all the more important to emphasize that even major innovations come in large, coherent groups; and that for both a firm and an economy, the analysis of the innovation process and of the contribution of innovation to economic growth must take account of this *clustering* pattern.

It is, however, the name of Joseph Schumpeter (1935/1951) that is most often associated with the identification of technology as the engine that drives the capitalist economy. In his book *Capitalism, Socialism and Democracy* (1943) he makes the following assertions:

> The opening up of new markets, foreign or domestic, and the organizational development from the craft shop and factory to such concerns as U.S. Steel illustrate the same process of industrial mutation-if I may use that biological term-that incessantly revolutionizes* the economic structure *from within*, incessantly destroying the old one, incessantly creating a new one. *This process of Creative Destruction is the essential fact about capitalism* [emphasis added]. It is what capitalism consists in and what every capitalist concern has got to live in.

Schumpeter goes on to describe the role of technology as providing a "perennial gale of creative destruction." Solow (1957) was searching for "an elementary way of segregating variations in output per head due to technical change from those due to changes in the availability of capital per head." In his definition, *technical change* means *any kind of shift* in the production function.

While none of the early economic studies were prepared to offer a specific formula by which the value of technology to the economy could be measured, or in turn to estimate the specific value that an investment in technology would offer, their works establish clearly that an important connection exists between the two. Later studies have attempted to expand the examples of these phenomena and to provide a more fundamental understanding of the complex interactions that produce them. The works of Berry et al. (1993) and Cheng and Dinopoulos (1992) are examples of this.

14.2 ECONOMIC VALUE OF R&D

Research and development plays a prominent role in the fabric of technology. The definitions of the various phases, as given in Appendix A14.1, enumerate the broad spectrum of ideas and concepts that are contained in R&D and they attempt to clarify the various steps that are taken in the process of converting an idea into a product or process. These processes are costly and the decision to support any of them competes with investment opportunities that often offer more immediate benefits. The common question that policymakers, both governmental and industrial, ask is: What is the proper level of investment in R&D for our enterprise? Answers to this question have always been difficult to develop and have often relied on subjective judgment. Efforts to develop a more objective answer to this question has stimulated a number of fundamental studies that have sought to quantify the economic value of R&D.

Edwin Mansfield (1972) has summarized what was known at that time concerning the "relationship between R&D and economic growth and productivity increase." Mansfield analyzes the methodologies used and the fundamental assumptions that were made in the studies that were made following Solow in 1957 and until 1972. He concludes that "although the results are subject to considerable error, they establish certain broad conclusions. In particular, existing econometric studies do provide reasonably persuasive evidence that R&D has a significant effect on the rate of productivity increase in the industries and time periods that have been studied."

---

*Those revolutions are not strictly incessant; they occur in discrete rushes which are separated from each other by spans of comparative quiet. The process as a whole works incessantly, however, in the sense that there always is either revolution or absorption of the results of revolution, both together forming what are known as business cycles.

The most careful analysis of economic data and the attempts to determine the quantitative values of R&D were done by Denison (1974) and Kendrick (1981). A summary of their studies is given in Table 14.1. Denison and Kendrick each conclude that 67% of the growth in industrial productivity has accrued through advances in knowledge (Denison) and technological advance (Kendrick). As Lewis (1982) observes: "By its nature, technological progress contributes to productivity growth.... A nation's productivity grows to the extent that the majority of firms advance their individual practices by using knowledge from all sources relevant to their needs-from fundamental scientific progress at universities to the experience of the firm's own employees."

Denison uses the term *advances in knowledge* to mean the physical properties of things and how to make, combine, or use them in a physical sense. It also includes *managerial knowledge*–knowledge of business organization and of management techniques, with the meaning of these terms to be construed in the broadest sense. Although it is natural to associate advancement in knowledge, as defined by Denison, and education, as noted by Kendrick, with the role of R&D, Mansfield (1972:478) points out that technological change often stems from sources other than organized R&D. Thus it becomes difficult to determine the impact of the formal organization and investment in R&D with the improvement in productivity that is believed to stem from it.

E. Mansfield (1991:24) has summarized our extent of understanding of the returns from R&D:

> [S]tudies indicate that the rate of return from R&D has been high. In manufacturing, the estimated rates of return have often been about 30 percent, and in agriculture, they generally have been about 40 or 50 percent. As would be expected, they vary from one time period to another. For example, Zvi Griliches and Frank Lichtenberg (1984) estimated that the social rate of return from privately-financed industrial R&D was about 10 percent in 1959-63, about 20 percent in 1964-68, and about 35 percent in 1969-73. Also, they seem to vary from one industry to another. For example, Jeffrey Bernstein and M.I. Nadiri (1988)

Total factor productivity for 1948-1969 after Denison (1974)	
Advances in knowledge	67
Improved resource allocation	17
Scale of economies	24
Other (including government regulations)	-8
total	100
Real gross product/labor hour from 1960-1973 after Kendrick (1981)	
Technological advance (including changes in labor, education, experience	67
Labor reallocation and capital-labor substitution	39
Volume change (scale economies, capacity utilization)	16
Government regulations	-22
total	100
Source: Adapted from Lewis (1982).	

Table 14.1 Relative Contributions to Productivity in the United States. (From Lewis, 1982.)

have estimated that the social rate of return in 1981 was about 20 percent in the transportation equipment industry, about 30 percent in the chemical and electrical products industries, and over 40 percent in the machinery and instruments industries.

We can believe that the fundamental studies have established the importance of R&D to economic growth and that this growth arises from improved productivity in processes and in turn from an increase in the GDP/hour of labor expended. The question of how much an individual country or firm should spend on R&D remains unanswered. These questions will recur often in this and the remaining chapters of the book: namely, when to invest in technology, on what should the investment be made, and what organization should do the work? While answers to these questions generally have important economic consequences to the enterprise, the development of the correct answer often entails factors that can only loosely be described as being economic. The proper combination of all of these factors are necessary to achieve a well-balanced world-class organization.

14.3   LIFE CYCLE OF TECHNOLOGY

The concept of the S-curve was discussed in Chapter 10 as a way of estimating the extent to which a market or product was approaching saturation. Technologies also possess a life cycle that is describable by the S-curve. They start slowly-the primitive stage, followed by a period of rapid growth-the transition stage, and begin to reach maturity in which further growth is slow-the mature stage, as is illustrated in Figure 14.1.

Irrespective of the particular technology that is being considered, a certain commonality exists among the ways by which the maturation of a technology occurs. With identification of a new technology, a large number of people will be interested intellectually and will initiate research efforts. If the technology is of a fundamental nature, much of this may be done in university laboratories. As some of the broad fundamental questions begin to be answered, some of the possible practical applications for the technology will begin to emerge. During the primitive stage, a number of firms will be intrigued by the appearance of these new opportunities. To explore the limitations and clarify the opportunities for the technologies, their investments in R&D will rapidly increase and a variety of new products will be investigated. The preliminary ideas that created the technology will be clarified rapidly as a result of this diversified exploration of possible applications. Although it is not always the case, it is often appropriate to classify these as *revolutionary technologies* in that they offer the opportunities for major changes in direction, either in products or processes.

As a result of the large investments in R&D, a greatly improved understanding of the technology will have developed and the directions that should be taken to exploit the technology will have become clearer. This improved understanding reduces the risk in making choices about particular products and processes and encourages the making of large investments in plants and facilities with increased confidence that marketable products will result. Rapid exploitation of the technology will be attempted as companies that have invested heavily in the R&D attempt to capitalize on the new opportunities. Although R&D does not cease, the rapid growth in the technology arises from investments that are made possible by the firms with large resources, usually far fewer in number than were initially involved in the primitive

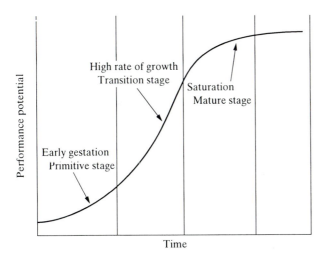

Figure 14.1 Life cycle of a technology with various stages of development identified as primitive, transition and mature.

stage. The industries that are using this technology are doing so because it offers them a competitive advantage. It is convenient to describe these as *competitive technologies*.

At later times, most of the principal areas that can profitably utilize the technology will have been identified and exploitation will be well under way. By this time the technology will have begun to mature. Because of early exploration of alternatives and heavy financial investments in plants and products using the technology, radical improvements in the technology will be less likely and the rate of improvement in the technology will begin to slow. The maturity stage is often characterized by a few large firms that dominate the use of the technology. In this stage the technology is basic to the business and is described as *base technology*. At this stage firms tend to compete through actively seeking to reduce manufacturing costs and to improve the quality of the products. R&D at this stage of the life cycle tends to focus on providing incremental improvements.

As the technology matures, there will come a time when an alternative will appear. The use of the older mature technology will begin to end, the older technology will no longer be exploited, and a new technology with new products will begin to appear. At this point the life cycle begins to repeat itself. As will be seen, the appropriateness of actions and responses to observed technological developments and opportunities depends on the life-cycle stage that the technology is in.

1. A technology that is revolutionary today can serve as a competitive advantage tomorrow before it ultimately becomes a base technology at a future time. The unwillingness of industry leaders to accept the possibility that a revolutionary technology can threaten their business, plus a disregard for the natural progression that begins with a discovery of the unusual but ultimately becomes commonplace, provides much of the explanation to Foster's lament that "technology leaders tend to become technology losers.

2. Companies tend to over-invest in the technologies that are basic to their industry, a circumstance that is common in a field that is utilizing mature technologies. Familiarity with the technology, a desire to protect an existing position and a conviction that there is always some likelihood that modest incremental improvement can be achieved in the

technology are all used to rationalize this practice. Many examples of this phenomenon can be found, but none is demonstrated more clearly than the lack of commitment for the technology of the transistor that the manufacturers of vacuum tubes demonstrated. As Schumpter might have observed, an existing mature technology encourages the emergence of a new technology that will *creatively destroy the existing technology.*

Paul Cook, former CEO of Raychem (Taylor:1990), gave the following response to the question: What are the biggest obstacles to innovation?

> For an organization to remain innovative, it has to be willing-even eager-to obsolete itself as fast as it can. So one of the biggest obstacles to successful innovation is success itself. All too often a company will develop an important new product and spend years asking itself the same questions-how can we make it a little better, a little cheaper, a little more sophisticated? Those are all important questions; there's always room for incremental improvement. But you can't let the entire innovative thrust revolve around making products faster, better, cheaper. A truly innovative company never stops asking more fundamental questions about its most successful products. Are there whole new ways to solve the problem-ways that might cut costs in half or double or triple performance?

It is important to recognize, however, that mature technologies can be the source of competitive advantage when applied to new or very different products, particularly for the company that has experience with the mature technology. Two examples of this are given by Erikson et al.(1990:77).

> The Sony Walkman...was a wildly successful new product based on comparatively mature technologies. The Walkman fortuitously combined Sony's work on the miniaturization of its tape recorder line and its work on lightweight headphones...
>
> Empire Pencil gained a major cost and quality advantage by using mature plastic extrusion technology as the basis of a new way to manufacture lead pencils. Conventional lead pencil manufacturing requires the use of fine-grained, high-quality wood, such as cedar, and a good deal of hand labor for assembly.... A development team ...realized that wood powder in a plastic binder could simulate the fine-grained wood. From there it was a straightforward step to produce pencil stock in a continuous extrusion process, with wood powder and a core of graphite powder in a plastic binder.

Although it would be incorrect to use the argument *that new uses for mature technologies will occur regularly* as a rationale for justifying large continuing expenditures on the base technologies, the possibility of it occurring emphasizes the need to continually search for new and exciting ways to use the mature technologies that each enterprise has mastered.

3.   The investment in base and revolutionary technologies must not be so large that the support for technologies that offer competitive advantage while not being remarkably new is starved. Since it is important to invest where the potential rate of growth for the technology is greatest, it is important to the health of a firm that competitively advantageous technologies receive adequate and sustained funding. For companies that maintain their competitiveness through the aggressive use of technology, it is common to find that the largest fraction of their R&D expenditures occurs in these intermediate technologies. As Erickson et al. note (1990:76): "The primary focus of industrial research and development is on extending and applying the technologies at the firm's disposal; they should be given the highest priority among the firm's investments in technology.

4. Investments made in the exploration of revolutionary technologies must be viewed as being of high risk but of potential high payoff. As Nelson (1984) observes, "Any investment in anticipation of a major breakthrough is a gamble." Although not every company will be able to afford making an investment in potentially revolutionary technologies, each must develop a mechanism that will allow identification of the fields that might generate the next generation of revolutionary technologies. Progress in these fields must be *monitored* continuously, even when the enterprise cannot afford to work on them directly. Awareness of scientific discoveries in other fields, understanding the environmental factors that will probably influence future technical directions, intuition about probable directions of future technology, and a willingness to invest in the exploration of unknown areas are all critical for a management that seeks to support the investigation of revolutionary technologies. The importance of developments in diverse fields on a technological innovation is given by Mort (1994).

> Digital xerography, enabling innovations such as printing, desktop publishing, quality fax and color, required the marriage of xerography with other disparate innovations: the computer, to provide digital electrical signals containing the image information; the laser, whose output intensity could be controlled by computers; and xerography, to convert optical information into marks on plain paper. These technologies were created with different time constants by diverse industries, so that their coalescence to form the newer innovations only became feasible beginning in the 1970s. Their integration continues to this day and is a prime example of what has been termed "technology fusion" and of its importance in the creation of new innovations.

5. A technology that is revolutionary and has not been used in any industry may require a long time to progress to a mature technology. If, however, a company recognizes the value of a technology that has matured in another industry and begins to use it, the time needed for the technology to reach maturity in the new environment can be very short. This can result in a surprisingly brief time advantage for the new user.

6. In the evolution from small entrepreneurial companies to large firms, loss of the corporate figurehead who experienced the creation of a revolutionary technology often results in the company adopting a more conservative approach to technology and, in turn, an unwillingness to support investment in the search for revolutionary approaches. The very different approaches currently being taken by Intel and IBM concerning the use of technology appear to be affected in a major way by the difference in the experiences of their current management.

7. The requirements for staffing programs in the three classes of technology can be quite different. Individuals with a deep understanding of a company's base technology often find it hard to admit that an alternative may offer important competitive advantages. People engaged in revolutionary technologies often are "free thinkers" who are viewed as undisciplined by the base technologists. Identifying people who are sufficiently unfettered in their commitments to a particular approach but sufficiently aware of the needs and constraints of the company and its industry can be very difficult. Since only a few such people may exist at any time in an industry, the effort to identify them and attract them to an organization often requires great effort on the part of the technical management.

8. The involvement of people throughout the organization in decisions concerning the content of the effort devoted to each class of technology is necessary but can produce very conflicting views that must be considered by the managers of technology. Since the operational people will have a plethora of problems that must be solved using the base technologies, they will often argue that investment in the competitively advantageous technologies would be nice but that the priorities must focus on the base technologies, thus implying that investment in revolutionary technologies is a complete waste of money. By contrast, members of a research laboratory may be enthralled with the possibilities of a revolutionary technology and resistant to spending time and effort to explore base technology questions. Creating an environment that encourages a useful exchange of views among all members of the organization about the relative benefits and risks of all technologies is an important part of the process of overcoming many of these deficiencies.

For the management of the company, the issue becomes when and how much to invest in a technology depending on its location in its life cycle. To use this effectively as a guideline, it is important to have a reliable means of determining the stage that the technology is in. The logistics analysis offers one such method.

14.4   LOGISTICS ANALYSIS OF THE TECHNOLOGY CYCLE

The life cycle of a technology can be treated in a manner completely analogous to that given in Appendix A10.1 for the growth of a product to fill a market niche. The logistics curve describes the maturity of a technology just as it describes the filling of a niche market by a product. For the case of products the logistics curve portrays the time growth of the number of units, $N(t)$, as the system approaches its saturation level of N_∞. The equation describing this is

$$\ln\frac{f}{1-f} = \alpha(t - t_0) \tag{14.1}$$

where

$$f = \frac{N(t)}{N_\infty} \tag{14.2}$$

For a technology, $N(t)$ can refer to the capability of products using that technology or could refer to the number of products that utilize that technology. The logistic plots for two distinct technologies, the capabilities of engines for aircraft, are given in Figure 14.2. The curve on the left is for piston engines and the curve on the right is for jet engines. According to these plots, the piston engine technology reached 90% of its capability in 1951 while the development of jet engine technology, which began in the early 1940s, reached 90% of its capability in 1981. Engine thrust is taken as a measure of the capability of the technology.

Modis and Debecker (1992) have discussed a complication that many logistics phenomena experience either very early in their life or near saturation. It is found, as shown in Figure 14.3 that the logistic pattern breaks into fluctuations that appear to be random and of a sizable amplitude. It is argued that chaos provides a mechanism for explaining these fluctuations. In the present context, the issue is to recognize that these fluctuations can occur and to attempt to ascertain the proper stage that characterizes the technology in the presence of these fluctuations.

14.5   THE TECHNOLOGY REPLACEMENT PHENOMENA

As technologies mature, other technologies replace them (e.g., transistors replaced vacuum tubes, jet engines replaced piston engines, workstations are replacing mainframe computers, plastics bottles are replacing glass bottles, fiber optics is replacing copper wire, etc.). Although it is clear that none of the older technologies disappear completely-small airplanes still use piston engines and some bottles continue to be made of glass-the major niche users of these products have converted to the new technology. As one technology approaches its level of maximum capability, a new technology appears. The S-curve of the former will cease as another technology replaces it and a new S-curve emerges. In looking at the total family of products that is affected by a class of technologies, the relative fraction of the family that is contributed by each technology can be analyzed using the logistics functions of each technology. This is shown in Figure 14.4 for the primary energy supplies for the world. The substitution of one form of energy for another beginning in 1860 follows this formalism (Marchetti:1987). The substitution of the VAX family of computers for the PDP family of products and the later substitution of the VAX technology by the RISC technology is also described by this formalism, as shown by Modis (1993). Similar analyses have been successful in describing the substitution of various technologies in the transport infrastructures in the United States (Nakicenovic, 1979).

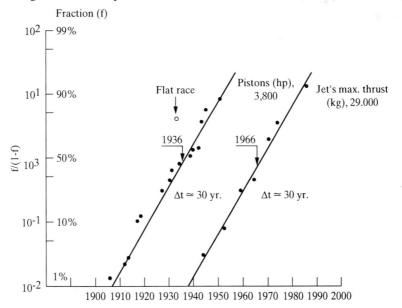

Figure 14.2   Logistic plot of the capacity of aircraft engines. (From Marchetti, 1987.)

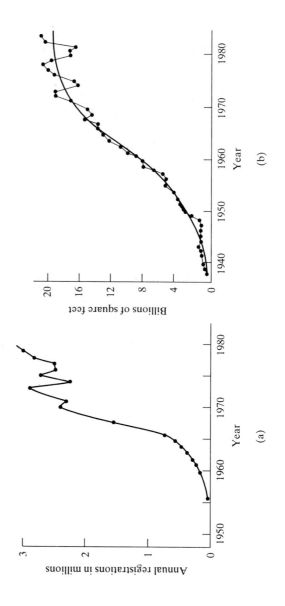

Figure 14.3   Logistic plots demonstrating chaotic behavior: (a) data for new car registrations in Japan (b) Annual plywood sales in the United States. (From: Marchetti, 1983, Montrey and Utterback, 1990.)

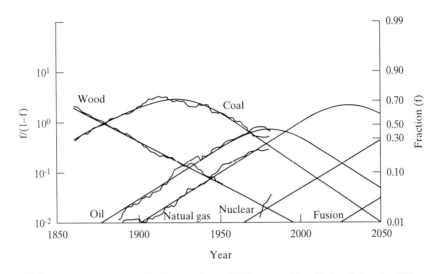

Figure 14.4   Primary energy substitution for the world. The curve identified as fusion should be interpreted as a future supply, possibly including solar energy thermonuclear fusion. (From Marchetti, 1987.)

14.6   ANATOMY OF THE TECHNOLOGICAL PROCESS

Technological progress frequently does not occur in neat, well-structured steps that follow a particular path or are subject to a simple set of constraints. It is the complexity of the process that makes the anatomy so rich and so important. The various stages that the process encompasses are often identified by the categories defined in Appendix A14.1. The statement given in Appendix A14.2 illustrates some of the difficulty arising from this categorization.

14.6.1   The Push versus Pull (or Evolutionary versus Revolutionary) Philosophy

In considering the anatomy of the technological process, it is useful first to inquire of the path that technological developments follow. Do they, for instance, generally start with an idea that was developed in the basic research laboratory, proceed to be explored in the applied research laboratory, and find their way neatly into a development organization? Or is the path more varied, with ideas arising in all stages of the process?

Developments that arise in basic research and proceed in a regular fashion through applied research to development and into a final product follow a path that is described by a *linear model*. In this model basic research is the fountainhead of ideas. New ideas are created as a result of the knowledge and understanding that are developed by basic research activities, whether located within the company or elsewhere (e.g., in universities, government laboratories, or other companies). Ideas are expected to *push* their way through the system and to result in new and interesting products and processes.

In contrast, in a situation in which ideas can be generated by any of the elements of the organization that are involved in technology, ideas are described by a pathway that follows the *parallel model*. Ideas for new products and processes can arise anywhere and can stimulate the organization to undertake their exploration and development. Basic research, and applied research and development are carried on in parallel with ideas, information, stimulation, and results flowing between and among the individuals and the groups in which they are located. The needs of the enterprise, whether identified by the customer or by elements of the company, are expected to pull the ideas through the organization and to generate products and processes in response to the needs.

In its most extreme form, an organization operating in the push mode could choose essentially to isolate its basic research activity from the rest of the company. The argument for doing this is that doing good basic research requires concentration by researchers. For them to concentrate, they must be free from distraction by the day-to-day problems that the company must constantly address. The topics that are being explored are generated primarily by researchers who are expected to be leaders in the technical fields in which they work. When the research activities develop a new idea or a new understanding of a concept, someone in the company is responsible for arranging a forum in which the researchers can present their ideas. Someone in the group is expected to carry out development of the idea. As is quickly recognized, this process possesses many of the elements described by the "throw it over the wall" paradigm.

By contrast, an organization operating in the *pull* mode would probably attempt to create problem-solving teams with all elements of the technical organization making a contribution. The needs of the company would identify questions that need to be solved, some requiring long-range research, others amenable to short-term development. Each group would be recognized as contributing important insight into finding the best solutions to the problems. As answers to the problems became available, they would be used by the team. Integration of all the technical activities will be an important goal of the organization.

Organizations that emphasize the push philosophy are likely to emphasize the importance of revolutionary developments for maintaining their competitiveness. The source of these revolutionary ideas is expected to be the research laboratories. By contrast, organizations emphasizing the pull philosophy are more likely to focus on evolutionary changes to existing developments as a basis for maintaining their competitiveness. The latter tend to emphasize continuous improvement as its objective.

What becomes obvious in this brief account of the extremes of philosophy is that both have merit and that neither can be excluded completely. There can be no argument about the importance of the *pull* of ideas through the company, with the continuous improvement that this implies. Similarly, there can be no disagreement with the contention that revolutionary developments take place with technology and that a lack of awareness and involvement in these can ensure that Fosters's lament that "technology leaders tend to become technology losers" will be proven correct. It is the need to balance these various forces that makes the management of technology so challenging.

With regard to the anatomy of the technological process, it is important to realize that the fundamental philosophy that is embraced by the organization will determine, to a very large degree, the way that the organization is structured and how it operates. In reviewing the philosophy followed by industrial research organizations, it is clear that the change from the *push* mode, which dominated between the 1950's and 1970's, to the *pull* mode being followed today has led to dramatic changes in the form of industrial research. Changing from a *push* to a *pull* philosophy for enhancing technological capability can be as dramatic for an organization as changing from a Master Schedule to JIT for managing a plant. The impact that these two very different philosphies concerning technology have on the organization, staffing and philosophy of work will be discussed in future sections.

14.6.2 Technological Bootstrapping

In an effort to determine the value of research, two studies were undertaken to trace the intellectual underpinnings of a number of specific developments. The first, named *Hindsight* and undertaken by the Department of Defense (Sherwin and Isenson, 1967), was an effort to determine the important elements of basic research that were instrumental in the creation of key weapons systems. The second, named *TRACES* and undertaken by Illinois Institute of Technology (1968) for the National Science Foundation, focused more on the intellectual origins of commercial systems. While each developed an impressive list of basic research topics that were considered to be essential to the success of the systems, the studies were not conclusive for reasons that centered on the following observation by Kuznets (1975:336)

> [The] sequence begins with an open-end, pre-conception phase-in which existing scientific and technological elements can, in retrospect, be seen as having supplied indispensable ingredients for the new technological conception underlying an innovation.... This phase is open-ended because the process of scientific and technological advance is continuous and it is difficult to establish the breakthrough basic to the innovation.

A combination of many intellectual steps, each made possible by previous steps, creates the body of knowledge that makes possible further progress in learning. This process of building continuously on previous achievements can aptly be described as technological bootstrapping. Because earlier knowledge has been created, advances can be made. Without this earlier foundation, the latter would have been impossible. Nelson (1984:6) notes this phenomena as follows:

> The precise path taken by technological advance is virtually impossible to predict, and there often are major surprises....[I]ndividual technological advances seldom stand alone. Almost always they connect intellectually and economically both to earlier advances along the same lines and to advances in other related technologies.
>
> Technological advances often are linked together because certain products form relatively tightly integrated systems....In a systems technology, an advance in one part of the system may not only permit but require changes in other parts. Thus a computer designed around integrated circuits is a very different machine from one designed around vacuum tubes.

An illustration of these interdependencies is given in Figure 14.5, which depicts a few of the technological developments that were made possible by the understanding of semiconductors that led to the development of the transistor. The *bootstrapping* of technology does not begin with the transistor, inasmuch as conceiving of a transistor required an understanding of the properties of semiconductors, which in turn depended on the description of matter made possible by quantum mechanics, and so on.

The strong interdependency of various pieces of knowledge and the impossibility of forseeing all the ramifications of a single advance in understanding and how it may ultimately affect other areas is a principal characteristic of the anatomy of the technological process. As Nelson noted, this places certain limits on progress. If the necessary understanding does not exist, either resources must be invested to create that understanding or the technological advance must await the creation of this understanding. The interdependency of many pieces of knowledge and the impredictability of the importance of a singe piece of technology is often offered as a justification for the support of basic research. Since the knowledge may be useful in unforeseen ways, it is argued, the search for knowledge is sufficient justification for the investment. Although this argument has some validity, it is generally not sufficiently persuasive to encourage the expenditure of large funds by a company when the pursuit of new knowledge in the principal incentives.

14.6.3   The Fraternity of Technologists

As a community, those who practice technology view themselves as members of technical fraternities. Much of this association begins in the colleges and universities, where specialization occurs in one of the disciplines of science or engineering. It is encouraged through employment-related professional organizations that represent groups interested in areas as specific as product design, electronic engine controls,

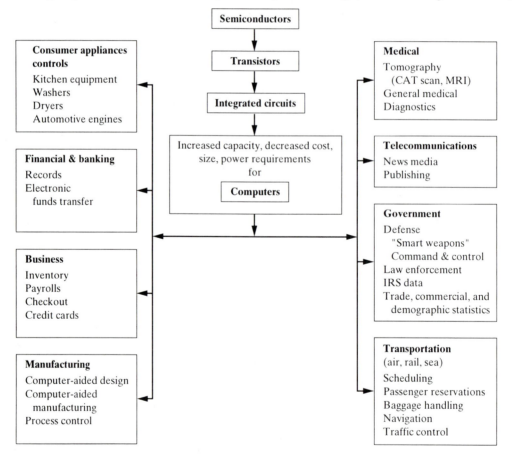

Figure 14.5   Representation of the bootstrapping that was made possible by the invention of the transistor. (Adapted from National Academy of Engineering, 1988.)

quality control, manufacturing, and testing. The professional organizations emphasize areas of principal interest to their members and provide intellectual stimulation through meetings and publications. Companies encourage their technical employees to attend conferences and seminars, to serve on committees, to be involved in the establishment of standards, and to publish the results of their research. These intellectual exchanges occur among people with different disciplinary training, from different companies, in different industries, and among residents of different countries. The sharing of common interests, the development of close personal relationships, and the improved understanding of the sources of success of certain directions of inquiry creates the fraternity of spirit that is important to the anatomy of technology.

The intellectual exchanges that occur through publications, meetings, and professional organizations lead to a leavening of the intellectual terrain and the creation of a common understanding of phenomena around the world. While individual companies will often have proprietary information and knowledge, any leadership resulting from this tends to disappear soon after the technology is used in production. The common understanding that is achieved by the technical community is sometimes ignored by the nontechnical community, with rather interesting consequences. This occurred, for instance, in the 1980s when Secretary of Commerce Brock Adams concluded that the U.S. government should "reinvent" the automobile since the Japanese manufacturers were producing a car that was more attractive to the U.S. public than were the cars build by U.S. manufacturers. It was viewed as self-serving and defensive that the technical communities in the United States, Japan, and Europe, collectively, stated that the technology that was in production was generally known and the fact that it was not being used by particular companies did not represent a technological deficiency. It was much later that it was acknowledged that deficiencies in many of the topics discussed in the first 13 chapters of this book contributed much more importantly to the lack of U.S. competitiveness than did the specific technologies.

14.7 FEDERAL GOVERNMENT'S ROLE IN TECHNOLOGY

Until rather recently the national technological agenda was dominated largely by the linear model of the process. The successes that members of the basic research establishment enjoyed in the wartime development of technological advances such as radar, jet engines, rockets, and the nuclear bomb led many to believe that further support of good science would be sufficient to ensure technological leadership for the country. If the country led in basic science, it was argued by such prominent people as Vanevaar Bush (1945), this would guarantee that industry would have technological superiority in products and processes (see also Appendix A14.2). Using this rationale, the United States developed a research university system that was the envy of the world. It developed a network of national laboratories. Many of its industries developed basic research laboratories. For the first three decades after World War II, the paradigm seemed correct. The United States, with the strongest defense establishment in the world, had committed to using technology as a source for superiority. The United States led the world in such things as the number of Nobel prize winners. It dominated the high-tech industrial world. Its markets were largely unchallenged.

The changes that occurred in the early 1980s were sudden and dramatic. Large deficits appeared in the foreign merchandise trade account, complete markets were lost to overseas manufacturers, and U.S. consumers developed an insatiable appetite for imports. The paradigm that the United States followed since the end of the war no longer seemed correct. Although it remained dominant in research, it was no longer competitive in much of the world marketplace. While still a leader in the creation of technology, it even lost its dominance of high-tech markets, as was illustrated in Figures. 2.5 and 2.7.

14.7.1   Encouragement to Industry

Just as the U.S. government was not largely responsible for the successes of the 1950-1980s, it could not be held totally responsible for the deficiencies that became apparent in the 1980-1990s. Although most of the changes that needed to be made were the responsibility of industry, the government played an important part in helping define the new paradigm under which technology is pursued. This will be particularly obvious in the discussion that follows regarding the sources of funding for R&D. Its role is also important in matters dealing with patents and copyrights, which will be discussed in Chapter 16, the encouragement of joint efforts between universities and industry, and the direct support of industry in precompetitive R&D. Although the actions of past and present federal administrations can be discussed with some accuracy, it is not possible to predict the directions that will be taken by future administrations. The philosophic orientation that each administration adopts is key to the types of actions that will be supported. Although there are many ways of illustrating this, it is vividly demonstrated by the sequence of actions that began with the signing of the agreement to create the joint industry-government Cooperative Automotive Research Program to further research on automotive-related technologies in the final days of the Carter administration, the cancellation of the agreement in the first weeks of the Reagan administration on the basis that it was improper for government to work closely with industry on commercial matters, and the creation of the joint industry-government "Clean Car" program to reduce automotive emission levels in the early days of the Clinton administration. Even for administrations that attempt to support industry through innovative programs, there are distinct limitations on their ability to influence independent actions. Nelson (1982;1984) has discussed the limitations that the free market imposes on the outcomes.

14.7.2   IR&D Funding of Research

In defense and space the federal government supports the research and development of products and is the largest, if not the sole, purchaser of the product. Since it is responsible for creating the technology and for using it in products in defense and space* the gov-

---

Space systems, as used here, refers to the launch facilities, propulsion systems, and the space station. The term is not intended to encompass such important uses of space systems as commercial telecommunications and the satellites that are used for their applications and similar commercial uses of space.

ernment is also responsible for developing a relationship with the industry that fosters the creation of the technology. Because of a complex procurement procedure, much of industry invests only modest amounts of its own funds in the R&D associated with the defense and space systems. Instead, many of the funds that are invested in R&D for matters that relate to defense and space are provided indirectly by the government through IR&D, the Independent Research and Development Program. The current situation regarding IR&D is described by the National Science Board (1993:114) as follows:

> DoD's Independent Research and Development (IR&D) Program enables industry to obtain federal funding for R&D conducted in anticipation of government defense and space needs. Because it is initiated by private contractors themselves, IR&D is distinct from R&D performed under contract to government agencies for specific purposes. IR&D allows contractors to recover a portion of their in-house R&D costs through overhead payments on federal contracts on the same basis as general and administrative expenses.
>
> Until very recently, all reimbursable IR&D projects were to have "potential military relevance." There has been some concern that the defense drawdown will serve to reduce the civilian R&D effort, not only in the form of commercial spillovers from weapons research but-more importantly-because of reductions in DOD procurement, out of which IR&D is funded. Given the importance of IR&D to industry's investment in critical technologies identified by DOD ...the rules for reimbursement have been eased and the eligibility criteria broadened. Reimbursement is now permissible for a variety of IR&D projects of interest to DOD including those intended to enhance industrial competitiveness, develop or promote dual-use technologies, or provide technologies for addressing environmental concerns.
>
> Between 1980 and 1992, IR&D for DOD and NASA varied between 5.0 and 7.0% of their total expenditures for R&D. In 1992, IR&D expenditures by industry was just over $4.8 billion, of which the government agreed that nearly $4.2 billion was legitimate. They reimbursed industry for $2.3 billion of IR&D programs. The balance was provided by industry.

14.7.3   Federally Funded Research and Development Centers (FFRDC's)

In addition to its role in providing defense for the country and the exploration of space, the federal government has responsibility for many areas, including the federal highway systems, the safety and control of the airspaces, safety of pipelines, medical research, and standards for many product areas. To fulfill its missions, it has chosen to establish a wide-ranging set of laboratories that are operated to explore, develop, and evaluate technologies that are viewed as essential or of potential interest. The structure of these is quite varied. There are in-house laboratories that are funded by and operated by various government activities. Examples of these federal intramural program laboratories are the U.S. Naval Research Laboratory, the Wright-Patterson Laboratories of the USAF, the National Instititutes of Health, the laboratories of the Bureau of Mines, and the Center for Disease Control. There are federally funded R&D centers that are administered by industry, some by universities, and some by nonprofit organizations. Examples of these include Lincoln Laboratory, which is funded by the USAF and administered by the Massachusetts Institute of Technology; Jet Propulsion Laboratory, which is funded by NASA and administered by the California Institute of Technology; and the Frederick Cancer Research Facility, which is funded by the Department of Health and Human Services and is administered by a number of industrial firms.

The Department of Energy, whose early forerunner was the Atomic Energy Commission (AEC), operates a group of laboratories that originated with the AEC and that were originally focused on the exploration of nuclear energy for military and peaceful uses. These so-called National Laboratories include Argonne Laboratory, Brookhaven Laboratory, Fermi Laboratory, Oak Ridge Laboratory, Lawrence-Livermore Laboratory, Los Alamos Laboratory, and Sandia Laboratory. Some of these are administered by individual universities, other by a consortium of universities, and others by industry. In a number of cases these laboratories have developed large research facilities that are available for use by scientists and engineers around the world (e.g., in support of the high-energy physics community). Since the earlier emphasis on nuclear weapons has diminished, the laboratories that focused heavily on military applications (e.g., Los Alamos and Sandia), are seeking to broaden their interests and to address problems that are of more immediate concern to the commercial sector.

14.7.4 Funding of University Research

The federal government funds basic research in essentially all fields of science, including the social and behavioral sciences, and engineering. The rationale for this use of federal monies rests on the premise that the technology base for this country is heavily influenced by a strong basic research effort, that it is frequently difficult for the doer of basic research to appropriate the benefits of the research, and that it is, therefore, an appropriate use of public funds to support these efforts for the national good. With the university's encouragement of scholarship and education, they are philosophically the least encumbered by the problems of markets, products, and profits. It was natural, therefore, that the universities would be the beneficiaries of the large federal funding of basic research.

In 1991 there was just over 200 institutions in the United States that were qualified to be listed as research universities, according to the classifications established by the Carnegie Foundation (1987). These are institutions that offered a broad range of baccalaureate programs, awarded at least 50 Ph.D.'s each, and received in excess of $12.5 million annually in federal research support. Half of this number received in excess of $33.5 million annually in federal research support. This collection of schools comprises the *research university establishment*.

The federal support of research at universities has been pivotal in the development of the outstanding research university establishment. By nearly every measure, the U.S. research university system ranks as the best in the world, whether in numbers of students enrolled, number of degrees awarded, breadth of programs that are available, number of publications, number of non-U.S. citizens who are attracted to them, number of Nobel prize winners who were trained in the United States, and so on. Whether the rational that led to the establishment and support of these institutions was totally valid or not, the success in creating a collection of high-quality institutions of higher learning cannot be denied. This, however, does not reduce the importance of the self-analysis that is currently under way in many of these institutions regarding their missions, their level of commitment to teaching, the nature of their involvement with industry, and their possible role in service to their alumni and older returning students. While the answers to each of these questions will have important implications for the future of the institutions, the overarching objective is to find the proper combination of answers that will allow the institutions to continue to prosper and support high-quality research programs while accepting a role in higher education that is broader than the traditional focus on research

14.8   PROFILE OF R&D FUNDING

14.8.1   Federal Funding of R&D

As shown in Figure 14.6, national funding for R&D activities in 1993 was $161 billion, representing 2.6% of the U.S. gross domestic product. In current dollars R&D spending increased by a factor of 6.1 between 1970 and 1993. In constant dollars the increase was only a factor of 1.7. While the expenditures by industry and the federal government have been comparable over the years, outlays by industry have exceeded those by the federal government in all years since 1979.

Total R&D expenditures are divided among segments of the technical community in a variety of ways, as illustrated in Figure 14.7. While the federal government provided 42% of R&D funding, it utilized only 10% of the funding in its intramural programs. The total amount of research performed by colleges and universities was only 13% of the total. Figure 14.7 (c) indicates the large fraction of R&D funding that is used for development programs, 59%, with basic research receiving only 16% of the total.

A comparison of the level of support of R&D among the countries that are principal commercial competitors of the United States is shown in Figure 14.8. Figure 14.8 (a) gives the total R&D as a fraction of GDP for the United States, Japan, Germany, United Kingdom, France, Canada, and Italy. Figure 14.8 (b) compares the fraction of spending for R&D for these seven nations when the expenditures related to R&D for defense are removed from the total. While the total relative expenditure for R&D is large for the United States, although falling behind Japan and Germany since 1986, the fraction of support going to nondefense has been substantially below that of Japan and Germany since 1970. This lower level of support by the United States for commercially related R&D has become a contentious issue for the government in recent years, stimulated by the realization that the leaders in commercially related R&D have become our most prominent competitors. The absolute spending for nondefense R&D by all of these countries, except for Japan, has remained a relatively constant fraction of

United States nondefense spending since 1975. Japan, on the other hand, spent 42 percent of the United States% investment in nondefense R&D in 1975 and had increased this to 68% by 1993. Not only is Japan's spending large compared to the United States, when normalized for the size of its GDP, its absolute investment is even larger. If technology is, indeed, the engine for economic growth, as argued by Schumpeter, Kuznets, and others, this relative decline of spending on R&D in the United States represents a serious matter of concern.

Figure 14.9(a) depicts the fraction of the national R&D expenditures that are provided by the various source of funds according to the character of the work that was done. Figure 14.9(b) depicts the fraction of the national R&D expenditures that went to various performing groups according to the character of the work that was done for the year 1993. In Figure 14.9(b) the funds for Federally funded R&D center performers are included in the sectors to which they are affiliated.

As is evident, industry is the principal source of funding and the principal performer of development and applied research. Basic research funding, however, is provided principally by the federal government and largely carried on by universities and colleges. Since it may be difficult to make the university research activity a part of the research team, this divergence between the funder and doer of basic research presents an interesting challenge for those organizations that are seeking to implement a pull philosophy by focusing the attention of all activities on areas that closely support their goals.

It should be noted that the statistics presented here do not include the amount of funds for basic research that is provided by state governments. Although there is increasing growth in this sector, stimulated by the desire to enhance state economic development through the creation of high-technology centers and businesses, this source currently represents a very small fraction of the national expenditure for R&D.

The conclusions that can drawn from this brief discussion of the national funding statistics of R&D are the following:

1. Industry and the federal government are the dominant sources for the funding of the national R&D effort-52% and 42%, respectively, in 1993.
2. Industry performs the major part of the national R&D effort-68% in 1993.
3. The federal government provides the major portion of the national funding for basic research-62% in 1993.
4. University and colleges perform the largest fraction of the national basic research effort-62% in 1993

14.8.2   Industry Funding of R&D

As Section 14.8.1 has demonstrated, the funding and performance of R&D by industry is largely concentrated on development and applied research, with less than 18% of the total of basic research funding being provided and performed by industry. Since industry uses both its own funds and a substantial fraction of the government's funds on development and applied research, it performs a higher fraction in these segments than it provides as a source.

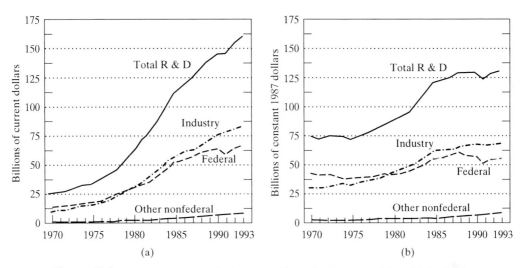

Figure 14.6   National R&D funding by source (a) funding in current dollars (b) funding in constant 1987 dollars. (From the National Science Board, 1993, p90.)

These statistics, representing all of industry, fail to offer insight into the variations that exist in the funding of R&D among different sectors of industry. Data on the spending of individual industrial sectors have been published annually by *Business Week* in a section entitled "R&D Scoreboard." The sectoral data are developed from

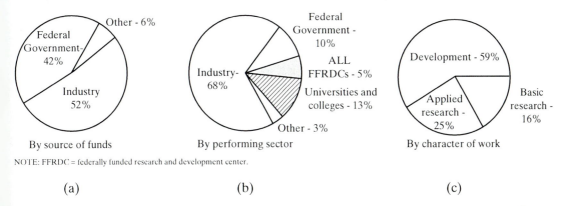

Figure 14.7   National R&D expenditures in 1993: (a) allocated by source of funds; (b) allocated by performing sector; (c) allocated according to the nature of the work. From National Science Board, (1993;92.)

the data that individual companies submit annually to the Security Exchange Commission through their required filing of 10-K forms. Although all publicly traded companies are required to make these submissions to the SEC, considerable variability is allowed in determining the contents of each category. Although care must be taken in making comparisons of absolute magnitudes among the entries, the trends that are displayed among various industrial sectors are meaningful.

Table 14.2 presents the totals for the various industrial sectors. The Scoreboard contains the data on nearly 900 companies, each with sales in 1994 of $91 million or more and R&D expenditures of at least $1 million or at least 1 percent of sales. R&D expenses include all dollars spent on company-sponsored R&D, excluding R&D under contract to others, such as U.S. government agencies. Profits refer to pretax income.

A number of interesting observations arise from an examination of the contents of Table 14.2.

1. Three sectors, cars and trucks, drugs and research, and computers, dominate the total R&D spending and together contributed 39% of the total industry spending in 1994.
2. R&D expenditure per employee varied between $969.1 (other consumer goods) and $33,198.1 (computer communications), with nine sectors spending in excess of $10,000 per employee.
3. No direct relationship is seen between R&D expenditures, either total or per employee, and sector sales.
4. No direct relationship is seen between R&D expenditures, either total or per employee, and sector profits.
5. R&D as a percent of sales varied between 14.1% (software and services) and 0.5% (steel), with the industry average being 3.5%.

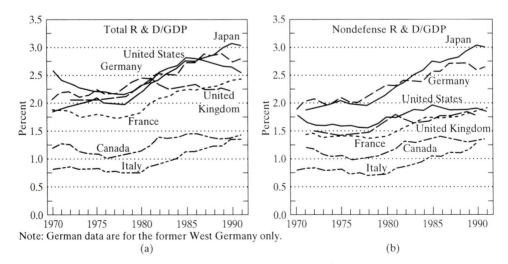

Note: German data are for the former West Germany only.

Figure 14.8   R&D as a percentage of gross domestic product for various countries: (a) total R&D as a fraction of gross domestic product; (b) nondefense R&D as a fraction of gross domestic product. (From National Science Board, 1993, p101)

6. R&D as a percent of profits varied between 5.3% (other consumer goods of consumer products) and 264% (system design of office equipment and services). The industry average was 39.0%.
7. Sectors spending more than 10 percent of sales on R&D were software and services, drugs and research, and computer communications.
8. Sectors spending 1 percent or less of sales on R&D were aluminum, other metals, containers and packaging, food, other consumer goods, oil, gas and coal, and steel.

Although no obvious relationships exists among any two of the headings in Table 14.2, it is interesting to note that five sectors ranked the highest in at least three of the four categories: Total R&D Expenditures, R&D Expenditures per employee, sales, and profits. The five sectors were:

1. Cars and trucks
2. Computers
3. Drugs and research
4. Chemicals
5. Telecommunications

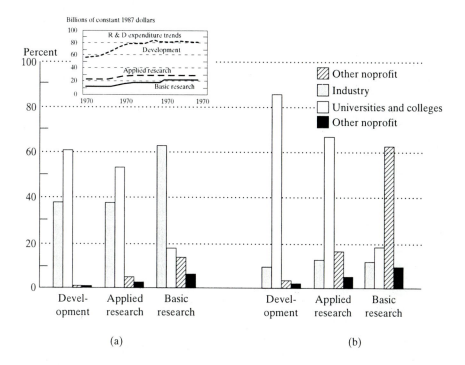

Figure 14.9   National R&D expenditures for 1993: (a) percent of funds according to the character of work as distributed by the source of funds (b) percent of funds according to the character of work as distributed by the sector performing the R&D. (From the National Science Board, 1993:95.)

TABLE **14.2** R&D Scoreboard 1994

Industry sector	R&D expenditures		Sales	Profits
	Total ($ X 10⁶)	($) Per employee	($ X 10⁶)	($ X 10⁶)
Aerospace	3,969.2	7,106.1	95,484.8	5,413.2
Automotive	15,255.5	9,257.7	406,747.0	28,470.3
Cars and trucks	13,691.0	11,568.0	342,023.0	23,488.9
Parts and equipment	1,196.5	3,289.4	50,382.1	3,734.0
Tire and rubber	368.1	3,658.1	14,342.0	1,247.5
Chemicals	4,973.5	10,289.1	133,690.4	14,087.1
Conglomerates	2,960.8	5,011.1	123,089.2	13,040.8
Consumer products	2,588.0	2,414.8	172,249.9	20,717.1
Appliance and home furnishings	362.2	2,465.3	21,010.8	1,290.1
Other consumer goods	680.8	969.6	95,852.0	12,865.4
Personal Care	1,545.1	6,936.9	55,617.1	6,561.6
Containers and packaging	188.8	1,299.5	25,187.9	970.2
Electrical and electronics	9,602.7	8,257.6	167,230.9	19,048.1
Electrical products	845.3	2,884.0	32,612.8	2,683.3
Electronics	3,707.1	8,426.2	68,336.1	6,623.9
Instruments	1,304.6	9,544.4	18,541.8	1,398.3
Semiconductors	3,745.6	12,729.1	47,740.1	8,342.6
Food	684.9	1,566.0	80,803.1	6,372.2
Fuel	2,405.8	4,267.4	344,424.9	23,481.9
Oil, gas and coal	1,699.9	4,116.4	325,188.0	22,067.4
Petroleum services	705.9	4,680.7	19,236.8	1,414.6
Health care	13,476.0	18,451.8	136,863.0	26,097.5
Drugs and research	9,564.4	24,059.6	81,803.1	18,074.0
Medical products and services	3,911.6	11,746.6	55,059.9	8,023.5
Housing	533.1	2,742.3	32,135.5	2,363.7
Leisure time products	1,528.9	6,478.2	35,930.8	2,781.6
Manufacturing	4,567.6	4,409.8	163,526.4	13,995.8
General manufacturing	2,001.4	4,583.7	63,621.3	6,314.7
Machine and hand tools	336.9	2,584.4	18,418.4	1,007.1
Special machinery	2,154.7	5,067.5	76,848.6	6,315.7
Textiles	74.6	1,696.2	4,638.1	358.3
Metals and mining	385.8	1,451.2	54,917.4	2,413.6
Aluminum	208.6	1,646.6	22,435.1	761.1
Steel	127.5	1,237.5	26,177.2	1,144.6
Other metals	49.6	1,375.3	6,305.1	507.9
Office equipment and services	16,746.3	15,898.8	234,720.6	17,946.2
Business machine and services	306.5	3,531.3	10,746.9	1,191.4
Computer communications	637.9	33,198.1	5,858.9	1,144.8
Computers	9,787.1	17,942.9	150,765.6	8,279.9
Data processing	294.2	7,209.5	4,917.7	717.7
Disc and tape drives	1,079.5	9,541.8	16,016.9	775.5
Peripherals and other	1,060.2	9,946.5	19,776.8	1,595.2
Software and services	3,218.1	27,646.9	22,887.2	4,109.8
System design	362.9	15,133.5	3,750.6	131.9
Paper and forest products	548.1	2,017.9	57,084.2	3,624.4
Service industries	213.9	1,431.6	16,455.5	771.5
Telecommunications	4,099.0	7,137.3	124,314.3	15,375.8
TOTALS	84,727.8	7,651.3	2,405,086.2	216,971.3

Source: Business week, 1995. July 3:1.

While these five sectors possess few common characteristics, they all have a strong interest in high technology, all depend on the use of high-technology tools for designing their products, and all are in fields that are subject to rapid change

It should be noted that economic conditions often lead to rather severe fluctuations in the summary statistics. For example, in 1992 other sectors would have been included in of the preceding paragraph, in particular aerospace, conglomerates[*] and medical products and services. In 1993 the list was the same as for 1994. It is also important to note that industrial performance was very strong in 1994. Although negative profits were experienced by aluminum, steel, computers, disc and tape drives, peripherals and others, and system design in 1993, these six sectors invested more than $1.4 billion in R&D.

The wide variations in spending observed among the 39 sectors listed in Table 14.2 suggest that the use of industry averages will be of little value in answering a query regarding the amount that a particular company should spend on R&D. A detailed examination of the practice of companies within a sector (which form the basis for Table 14.2 but are not given in the table) indicates that the variations within a sector is generally less than the variation among the sector averages. This suggests that members within a sector tend to develop a practice with regard to R&D expenditures that has some degree of consistency to the members of the grouping.

Chakrabarti (1990) has made a rather more detailed study of the relationship of the level of R&D funding to various measures of output, namely, the number of patents acquired by a firm, the number of publications in the technical literature, and the products and the processes that the firm has introduced. The study included 16 firms from the chemical industry, 32 firms from the machinery industry, 86 firms from the electrical sector, and 44 firms from the instrument sector. The criteria for inclusion of firms in the study was (1) annual sales in 1986 between $10 and $200 million, (2) annual R&D expenditure during the period 1977-1986 of at least $10,000, and (3) an annual growth of 10% in R&D expenses during the period. While these criteria effectively eliminate large firms and, in fact, a very large fraction of the firms included in Table 14.2, the results are interesting. For the sectors in this study other than chemical, R&D expenditure correlated with annual sales. R&D expenditures and patents are strongly correlated for the chemical and instrumentation sector but not for the electrical and machinery sectors. As far as publications are concerned, the larger firms published more. New products are correlated significantly with average R&D expenditures for only two sectors, machinery and electrical. It can be concluded from these studies that the technical behavior of firms in the four industries studied was quite different. The reasons for these differences, the factors that determine the levels of investment in a given sector and how these levels change with time are not understood, even within Chakrabarti's very restricted segmentation of industry. Inclusion of large companies only enhances the difficulty in developing a comprehensive understanding of these complex phenomena.

---

[*]The sector entitled conglomerates is composed of the following (1993): Actava Group, Allied Signal, Coltec Industries, Figgie International, General Electric, ITT, LTV, Pall, Pittway, Premark International, Rockwell International, Teledyne, Textron, Triac, and TRW.

REFERENCES

BERNSTEIN, J. AND M. I. NADIRI. 1988. Interindustry R and D Spillovers, Rates of Return, and Production in High-Tech. Industries, *American Economic Review*, Mar.:433

BERRY, B. J. L., H. KIM AND H. KIM. 1993. Are Long Waves Driven by Techno-Economic Transformations? Evidence for the U.S. and the U.K., *Technological Forecasting and Social Change*, 44:111.

BUSH, V. 1945. *Science and the Endless Frontier, A Report to the President*. U.S. Government Printing Office, Washington, DC.

Business Week. 1995. R&D Scoreboard: *1994,* July 3:1.

CARNEGIE FOUNDATION FOR THE ADVANCEMENT OF TEACHING. 1987. *A Classification of Institutions of Higher Education*, Princeton University Press, Princeton.

CHAKRABARTI, A. K. 1990. Scientific Output of Small and Medium Size Firms in High Tech Industries, *IEEE Transactions of Engineering Management*, 37, 1, Feb.:48

CHENG, L. AND E. DINOPOULOS. 1992. Schumpeterian Growth and International Business Cycles, *AEA Papers and Proceedings,* May:409

DENISON, E. F. 1974. *Accounting for United States Economic Growth 1929-1969*, Brookings Institution, Washington, DC.:79

ERICKSON, T. J., J. F. MAGEE, P. A. ROUSSEL, AND K. N. SAAD. 1990. Managing Technology as a Business Strategy, *Sloan Management Review*, Spring:73

GRILICHES, Z. AND F. LICHTENBERG. 1984. R&D and Productivity Growth at the Industry Level: Is There Still a Relationship?, *R&D, Patents and Productivity, Z. Griliches (ed.)*, University of Chicago Press, Chicago:465.

HEIM, J. A. AND W. D. COMPTON (eds.). 1992. *Manufacturing Systems: Foundations of World-Class Practice*, National Academy Press, Washington, DC:8.

ILLINOIS INSTITUTE OF TECHNOLOGY. 1968. *Technology in Retrospect and Critical Events in Science (TRACES)*, National Science Foundation, Washington, DC.

KANTROW, A.M. 1980. *Harvard Business Review*, July-Aug.:6.

KENDRICK, J. W. 1981. *Sources of Growth in Real Product and Productivity in Eight Countries*, 1960-1978, Office of Economic Research, New York Stock Exchange, New York: 31.

KONDRATIEFF, N. D. (1935/1951). The Long Waves in Economic Life, Review of Economic Statistics, XVII, 6:105 (1935) reprinted in Readings in *Business Cycle Theory, Richard D. Irwin, Homewood, IL.* (1951).

KUZNETS, S. 1975. Technological Innovations and Economic Growth, *Technological Innovation: A Critical Review of Current Knowledge*, Georgia Institute of Technology, Atlanta, GA. and San Francisco Press, San Francisco.

LEWIS, J. 1982. Technology, Enterprise, and American Economic Growth, *Science*, 215, 5, March: 1204.

MANSFIELD, E. 1972. Contribution of R&D to Economic Growth in the United States, *Science*, Feb:477.

MANSFIELD, E. 1991. Social Returns from R&D: Findings, Methods and Limitations,*Research*

Technology Management, Nov.-Dec.:24.

MARCHETTI, C. 1983. The Automobile in A System Context: The Past 80 Years and the Next 20 Years, *Technological Forecasting and Social Change* 23:3.

MARCHETTI, C. 1987. Infrastructures for Movement, *Technological Forecasting and Social Change* 32:373.

MONTREY, H. AND J. M. UTTERBACK. 1990. Current Status and Future of Structural Panels in the Wood Products Industry, *Technological Forecasting and Social Change* 38:17.

MORT, J. 1994. Xerography: A Study in Innovation and Economic Competitiveness, *Physics Today,* April:32.

MODIS, T. AND A. DEBECKER. 1992. Chaos-Like States Can be Expected Before and After Logistic Growth, *Technological Forecasting and Social Change* 41:111.

MODIS, T. 1993. Technological Substitutions in the Computer Industry, *Technological Forecasting and Social Change* 43:157.

NAKICENOVIC, N. 1979. Software Package for the Logistic Substitution Model, RR-79-12, *International Institute for Applied System Analysis,* Laxenburg, Austria.

NATIONAL SCIENCE BOARD. 1993. *Science and Engineering Indicators,* U.S. Government Printing Office, Washington.

NELSON, R. R. 1982. *Government and Technical Progress: A Cross-Industry Analysis* (ed.), Pergamon Press, Oxford.

NELSON, R. R. 1984. *High-Technology Policies: A Five Nation Comparison,* American Enterprise Institute for Public Policy Research, Washington, DC.

QUINN, J. B. 1986. Innovation and Corporate Strategy: Managed Chaos, *Technology in Society,* 7:263.

ROSENBLOOM, R. S. 1978. Technological Innovation in Firms and Industries: An Assessment of the State of the Art, *Technological Innovation: A Critical Review of Current Knowledge,* P. Kelly and M. Kranzberg (eds.), San Fransisco Press, San Fransisco, CA:228.

SCHUMPETER, J. A. (1942/1946) *Capitalism, Socialism and Democracy,* George Allen & Unwin, Ltd. London:81.

SCHUMPETER, J. A. The Analysis of Economic Change, The Review of Economic Statistics, XVII, 4 :2 (1935) and reprinted in Readings in *Business Cycle Theory,* Richard D. Irwin, Inc., Homewood, Ill. (1951).

SHERWIN, C. W. AND R. S. ISENSON. 1967. Project Hindsight, *Science,* June 23:1571.

SOLOW, R. (1957) Technical Change and the Aggregate Production Function, *Review of Economic Statistics,* 39, August:312.

TAYLOR, W. 1990. The Business of Innovation: An Interview with Paul Cook, *Harvard Business Review,* March-April:97.

QUESTIONS

14.1 Kondratieff's observations of the long-wave cycles in economic development were based on observations of events that took place between the late eighteenth and late nineteenth centuries. He predicted that a decline in economic development would probably begin in the United States in the early 1920s. Using a period of approximately 50 to 60 years, a decline in the late 1970s and a period of high activity beginning about now would be predicted. If this is true, list some of the technologies that were created in the 1970s that will contribute to the current activity. Make a guess at some technologies that might be expected to be candidates for development at the end of this cycle that might contribute to developments in the next cycle.

14.2 Kuznets describes the creation of the new as the "creative destruction" of the old. Identify five technological developments that effectively destroyed the technologies that were current at the time. Identify five current technologies that might be threatened by future technical developments.

14.3 The discussion of the push versus pull philosophy for technology, Section 14.6 suggests that an enterprise whose philosophy emphasizes the former will more likely focus on revolutionary technologies than will the latter, which will emphasize continuous improvement. Do you agree with this? Discuss your rationale and offer examples for either supporting or rejecting this categorization.

14.4 As a manager, would you support your employees participating in the larger fraternity of technologists? Why? If so, what actions would you take to encourage this?

14.5 As an executive of your company you have been invited by the U.S. Congress to testify on your company's position on the appropriatenbess of the federal funding for R&D. You have been asked to offer specific comments on the data given in Figure 14.5 (a) and (b). What points will you make in your testimony?

14.6 For the past seven years you have been an executive in a company that is a member of the general manufacturing sector as noted in Table 14.1. You have just accepted a position as an executive in a company that is included in the car and truck sector. Your first task is to offer a recommendation on the size of the R&D budget for the coming year for your company. How will you go about developing your recommendation?

READINGS

MANSFIELD, E. 1972. Contribution of R&D to Economic Growth in the United States, *Science*, Feb:477.

MANSFIELD, E. 1991. Social Returns from R&D: Findings, Methods and Limitations, *Research Technology Management,* Nov.-Dec.:24

SCHUMPETER, J. A. 1935, 1951The Analysis of Economic Change, *Review of Economic Statistics, XVII*, 4 :2(1935) and reprinted in Readings in *Business Cycle Theory*, Richard D. Irwin, Homewood, Ill. (1951).

APPENDIX A14.1 DEFINITIONS

The term *technology*, encompassing concepts that span the most practical to the most learned with a possible time frame for application that may be immediate to years away, is subject to misunderstanding and misinterpretation. As used in this book, it is used to encompass the body of engineering and scientific knowledge and methods concerned with the design, development, production, distribution, support, and disposal of products and services. It includes topics that normally come under the headings of basic research, applied research, and exploration. It includes hardware and software, individual machines and total systems, publications, patents, and know-how in the operation of processes and plants. It is not strictly restricted to physical elements related to these topics. Although these definitions span a domain that is greater than manufacturing, the generality of the concepts is such that they are applicable to manufacturing. Because of the importance of research and development to technology, the definition of these subheadings is particularly important.

Management of technology is the *system* by which technology is created, focused, and applied to satisfy a customer need. Since technology includes the concepts of invention and innovation, the distinction between the two is important. The following differentiates the two as used in this book:

- *Invention*. An act of creation-a new product, process, procedure, or system that offers a new way of doing some task or providing a new service.

- *Innovation*. The carrying forward from the idea or example to a commercially feasible product, process or service.

The definitions of the various phases in the R&D process are provided by definitions used by the National Science Foundation and the Department of Defense. The definitions used by the National Science Foundation for the various stages encountered during research and development are taken from the Publication NSF 87-314.

- *Research* is a systematic study directed toward fuller scientific knowledge or understanding of the subject studied. Research is classified as either basic or applied according to the objectives of the sponsoring agency. In *basic research* the objective of the sponsoring agency is to gain fuller knowledge or understanding of the fundamental aspects of phenomena and of observable facts without specific applications toward processes or products in mind. In *applied research* the objective of the sponsoring agency is to gain knowledge or understanding necessary for determining the means by which a recognized and specific need may be met.

- *Development* is systematic use of the knowledge or understanding gained from research, directed toward the production of useful materials, devices, systems, or methods, including design and development of prototypes and processes. It includes quality control, routine product testing and production.

The Department of Defense divides research and development into additional stages and assigns a numerical designation that is associated with specific elements in their budget documents. These are the following:

- *Research* (6.1). Includes all effort of scientific study and experimentation directed toward increasing knowledge and understanding in those fields of the physical, engineering, environmental and life sciences related to long-term national security needs.

- *Exploratory Development* (6.2). Efforts directed toward evaluating the feasibility of proposed solutions to specific military problems. Includes both applied research and the development of "bread-board hardware."

- *Advanced Technology Development* (6.3). Programs that explore "alternatives and concepts prior to development of specific weapons systems."

- *Advanced Development* (6.4). Programs that have begun development of hardware for test.

- *Engineering Development* (6.5). Programs that develop hardware for military use according to specifications established by the services.

APPENDIX A14.2 SCIENCE AND TECHNOLOGY IN INDUSTRY*

The present role of science and technology in industry has historical roots. To understand the contemporary atmosphere and problems, I want to spend a few minutes reciting or interpreting this history. This will be a narrow interpretation and I will confine myself to electronics in this country

Prior to the war-1940-there was a handful of firms that had what we would recognize as research laboratories, distinct units functioning as research organizations: Bell Telephone Laboratories, General Electric, Westinghouse. During the war many physicists left their laboratories and tackled engineering problems with great success and acclaim. This created a point of view which became widely accepted, the point of view that great engineering accomplishments come from research. Parenthetically, let me observe that the war effort did demonstrate that those who were well grounded in the basic sciences could make important and, at times, unique contributions to engineering.

The generation that was involved in the war is now the source of our elder statesmen; some were young enough then so that they are now in their prime while others are retired or about to retire. This group of researchers returned to their academic activity with an exposure to engineering, with experience in how to do engineering, in how to bend with the wishes and needs of the users, and with the beginning of an understanding of the interaction of society and technology, at times calling it society and science.

After the war, in the late 1940s and early 1950s, many other large industrial firms started new organizational entities dedicated to research. General Motors, Ford, and U.S. Steel are examples. Obviously, there were also many electronics firms. Underneath this large expansion in industrial research was the notion, very seldom articulated explicitly, that new products come about through an orderly procedure starting with science, moving through applied research, and finally appearing as some manufactured article, system, or new structure that is viable in the marketplace and has a beneficial impact on our society.

This point of view also prevailed in academic circles and took two forms that I want to identify. First, there was a profound shift in engineering school curricula, a shift to include more fundamentals, more science, and to eliminate courses such as drafting. Secondly, there was the evolution of a national science policy, which stated in its simplest form was: support scientific research and the ills of our society will be solved; also, the most efficient place to do research is in a university environment.

Now, after 20 years of continued expansion of research in universities, government, and industry, serious questions are being raised. Research has not produced all the anticipated benefits, at least not visible benefits for society as a whole. There has also been a restlessness among the engineers who wanted recognition. This led to the National Academy of Engineering. Engineers were restless also with the growth in the sciences and wanted systems approaches, case studies, etc. In addition to this

*Source:Piore, E. R. 1969. *Science and Technology in Industry,* The Interaction of Science and Technology, W. D. Compton (ed.), University of Illinois Press, Urbana:1-8.

reaction in the engineering community, there have been more disturbing reactions in industry and in government, where we've seen new statements of science policy. Even in the scientific community, voices have echoed this theme that science, or research, has failed society. The Department of Defense, an important source of funds for universities and industry, takes the Hindsight Study seriously and is slowly but surely eliminating support in many important areas of physics and chemistry. The White House issues statements that we must now apply the results of our research. There is a body of scientific opinion to the effect that our selection of research programs should be based on predictable benefits to society.

In going through this history, I have tried to be neutral and not expose my prejudices. I do have certain firm convictions that I will now articulate, but in order to get understanding and support for these convictions, it is necessary to understand the attitudes of one's peers and of those who supply the funds.

Large industrial companies have a spectrum of technical activities that extend from basic research to production. The sociologists of science classify these activities in a variety of ways by the use of phrases such as applied research, advanced development, development, production engineering, manufacturing engineering, and manufacturing. Points on this spectrum are best characterized by the freedom a technical person has in selecting his problem. The basic research person supposedly has the greatest freedom to select his problem. The technical person on the manufacturing floor supposedly has the least freedom. Very broadly, they are both in the same technical environment. They differ in their motivation, in their audience, in their rewards, recognition, and personal satisfaction. The motivation for the person's involvement is the rewards he receives from society or the audience to which he addresses himself.

I approach this spectrum of technical work on the basis that all the people involved work on real problems which have intellectual content, and that a high degree of training is required for the solution of those problems. I am not trying to create a hierarchy of snobbishness. A research-oriented person generally looks to the outside world for his audience and his recognition, not necessarily within the social structure in which he works. This is true in universities, government, and industry. A research person gets greater satisfaction from recognition from his peers with whom he does not live than from his peers who reside in his immediate vicinity. In spite of this, industry in our contemporary world does want research people in its laboratories and it makes certain demands on them—demands, most simply stated, for responsiveness and sensitivity to the technical environment of that industry; their choice of problems should be motivated by the man-made world of the industrial complex in which they work. One hopes that in observing the fabrication process they will find important problems, whether in development or in product engineering, and, simultaneous with their coupling to the outside world, they will do research that may have an impact on future products, processes, or systems within a given industrial complex.

Normally, the sociologists of science and technology concentrate on the last item and do not quite realize the importance of the total research operation and what are considered its less glamorous functions. Many manufacturing procedures are based on the state of the art. This is especially true in our contemporary world in the fabrication of transistors or more complex structures known in the trade as monolithics. There is a large element of alchemy, of trial and error, or 'black magic.' There is a lack of fundamental understanding of the processes involved. From time to time, these processes go bad and one is at a loss to find the reason. The normal procedure is to stop the line, clean everything thoroughly, paint everything thoroughly, and then start all over again. Very often everything goes well after the cleaning process. This obviously produces an element of uncertainty and discomfort, a certain nervousness. Having been stimulated by the existence of this feeling of panic, one looks to research to try to give some fundamental understand so that one can gain control, so that one can better understand the man-made world.

Quite apart from these catastrophic considerations, one can observe historically that, as one develops an understanding of a process and anchors it to first scientific principles, one acquires an opportunity to reduce costs. I will continue to stress this function of research in industry because it is the least understood by those who write about it. Normally we read about the orderly process of something coming out of basic research, going through applied research, and finally, through the other categories I have identified, into manufacturing. Our world is not that orderly.

We also design products and systems based on past experience, past products, intuition, and general knowledgeability. One can cite many historical examples. Coming back to the contemporary world to an area in which I have direct contact, one finds that computers–their structure, the architecture of the total system-evolve from experience, esthetic appeal, intuition, and some understanding of the problems that the computer will solve. This is true not only of the hardware configuration of the computer but also of the software. We are very far from starting with first principles of science, going through careful analytic procedures, and coming up with a design for the hardware and software of a computer. Progress is being made and some general theorems are being evolved, but we remain very far from the goal.

Thus, it is hoped that a person dedicated to research will be stimulated by these large gaps in fundamental understanding and will dedicate himself to trying to obtain better understanding, to see whether he can find first principles and determine whether a systematic intellectual base can be built. The mysteries of our world reside in some of the very structures we have built. Not all mysteries reside in the stars and in the natural systems.

There are some standard historical examples of this and I might as well throw them out. We had a great deal of communication going on before we began to put a structure under it with information theory. We had many steam engines serving and revolutionizing society in railroads and steamships before we had adequate structures known as thermodynamics and statistical mechanics.

What one wants from a research person is a *reaction to his environment, taste in the problems he chooses, and a great impatience with trivia*. Thus one provides the stimulus and gives the man the freedom to respond to it. Unfortunately, there are very few people who have the gifts to do research. And our universities have drilled into their graduate students that if they don't do research especially basic research, they somehow have failed in their purpose on this earth. And they do not learn the questions: is this a good problem? Is it a serious problem? Is it a solvable problem? So one of the first judgments industry must make as a young man joins it is: has he the taste and talent to do research? Can he be given freedom? Can he be productive-productive to society while getting satisfaction out of his work? This sort of judgment is very difficult and produces much irritation and aggravation.

The function of research in industry is to look at the totality of technical activities in a given firm and, from that totality, select areas that are amiable to solution and illumination. The validity of this characterization of research in industry can be observed in the mobility of people. Some time is spent in research, then there is a drift into development, and at times into the more commercial function. In contrast, a similar drift in academia presents a different pattern-department, dean, laboratory director, etc.

Research in the universities has a much more restricted set of stimuli as the scientist tries to look at the man-made world. This is why certain areas of research are dominated by industry.

It is very difficult for me to differentiate between basic research and applied research. I guess all the work currently going on in holography can be classified as applied research. Early laser work can possibly be classified as basic research, and much of the present laser work can be classified as applied research. I do not quite understand these classifications. I guess applied research is associated with making some device work or applying some principle without necessarily committing oneself to a specific system or specific product or black box. Possibly the difference is whether one publishes in the *Physical Review* or in the *Journal of Applied Physics*. The research person feels rewarded if he publishes on the outside and is acclaimed by his peers. The applied research person, I guess, hopes that some of his ideas actually will end up in some useful gadget, some useful application that will hit the marketplace, whether a commercial marketplace, a military marketplace, or a social marketplace. One finds that possibly one-third of the applied research finds immediate utility, for other considerations come into play before it can hit the marketplace. These considerations include the existence of a marketplace and the cost of the gadget, black box, or system. This interaction of technology and society is another topic.

I have commented on the following matters: the broad interests of those academic men who participated in the war, the questioning of science as the fountainhead of the solutions for problems facing society, the lack of agreement on what training to provide and engineer, and the role of research in industry.

These apparently diverse observations are related, and now I should like to draw some conclusions which I will state as problems.

How can one create the state of mind that a very productive, rewarding, and creative life can be led even though one is not a theoretical physicist? How does one erode the highway of intellectual snobbishness? How does one present the functions of research in the world? It is all-pervading and prevailing.

How can we make our young researcher aware of those problems our society faces to which he can make a contribution through his training and acquired skills while receiving satisfaction and the recognition of his peers? There are many volunteers to modify our foreign policy and social structure. A technically trained person by his training has no unique wisdom in these areas, but he can contribute in an area that requires his skills.

15
Creating the Technology Roadmap for the Enterprise

> The new always looks so small, so puny, so unpromis-
> ing next to the size and performance of maturity.
> (Drucker, 1985)

Confronted with rapidly changing environments, enterprises must have a very clear understanding of their technology needs/opportunities. The plans must be sufficiently detailed that they identify what technology is needed, when the technology is expected to be used, and who will be involved in the efforts to develop it. Resources must be committed to assure that the plans are met. The enterprise that expects to use technology as a potent competitive tool must assure that its technology plans are fully supportive of the overall business plan. The final result of this planning process should be a *roadmap* of the enterprise's technology needs that involves a consensus understanding of the risks that are associated with this plan in terms of the timing, costs, and capabilities of the organization.

The approach to accomplishing this roadmap will involve asking numerous questions concerning the needs for technology as well as exploring opportunities that technology may offer to the enterprise. To formalize answers to these questions, we use a number of matrices that encourage the exploration of the important questions. It should be recognized, however, that these are simply tools for assisting the enterprise and that their principal value is not in their completion but in the assistance they offer in assuring that questions are not omitted inadvertently.

It will be recalled that quality functional deployment (QFD) (Chapter 11) is a tool for use in concurrent engineering that seeks to assure that all relevant activities are encouraged to participate in the design process and to ensure that the final design is compatible with the needs and capabilities of all groups. Since the process of creating a technology roadmap for the enterprise must also involve all segments of the enterprise, the tool for accomplishing this is, by analogy, termed technology functional deployment (TFD).

The completion of the technology roadmap can be an arduous task. Products often involve a host of technologies. For a large enterprise that produces a variety of products, the list of relevant technologies can be daunting in length. A technology roadmap cannot address all of these. It is essential, therefore, that the enterprise make a reasoned judgment concerning those that are most critical to its continuing competitiveness and that these be included in the roadmap. The objective of this process must not be lost in the details: *the development of a consensus understanding of a technology strategy that fully supports the business strategy and the commitment of the necessary resources in the business plan to support development of the technologies.*

15.1 CLASSIFICATION OF TECHNOLOGIES

Although a particular technology can have dramatically different uses and values under different circumstances, it is helpful to think about technologies according to their possible impact on the competitiveness of the firm. The impact depends, of course, on the way the *firm will use* the technology. Three broad classes of technologies will be useful in later discussions.

Technologies that are basic to the Enterprise (B). Base technologies are those in which a firm must be competent in order to perform competitively in its *current business*. These might be called the "bread-and-butter" technologies of the company. For a maker of internal combustion engines, the base technologies would probably include the physics and chemistry of internal combustion engines, combustion, the system of introducing fuel to the combustion chamber, physical metallurgy, foundry processes, and metal machining. For the manufacturer of integrated circuits, the base technologies would include the making of masks, the processes of photolithography, ion implantation, diffusion and surface pacification, the computer simulation of device designs and the automation of chip testing.

Technologies that offer important new competitive advantages (CA). Technologies that offer new competitive advantages have the potential to provide the firm with all new products, existing products with greatly improved attributes, or greatly improved processes for manufacture. For the internal combustion engine manufacturer, the advantageous technologies might include the capabilities of providing much lighter engines through the use of new lightweight materials, more versatile engines that can operate on alternative fuels, or engines that use radically different combustion chambers that would minimize the presence of unburned hydrocarbons in the exhaust. For the maker of integrated circuits the advantageous technologies might include the use of synchrotron radiation as a source of photolithographic illumination to allow increased circuit density and the use of computer vision systems that provide 100 percent inspection for production defects.

Technologies that have the potential to be revolutionary (R). Revolutionary technologies are those that could revolutionize existing products, replace existing products with products that would appeal to very different markets, and perhaps require very different processes for production. New products may be created that would obsolete products being offered in current markets. Satellite communications revolutionized the transmittal of data between remote locations. The aircraft turbine engine certainly represented a revolutionary technology for the piston engine manufacturer. For the internal combustion engine manufacturer, a revolutionary technology might be the development of batteries that would make the electric vehicle a practical competitor for the internal combustion engine or a clean-burning two-stroke internal combustion engine with a substantially greater fuel efficiency than the existing four-cycle engine. For the manufacturer of integrated circuits, a revolutionary technology might be the development of biologically based computing as an alternative means of achieving greater capability and inherent learning over that provided by conventional silicon-based technology.

Three observations need to be made regarding these classifications.

1.   No reference is made in the three classifications as to whether they are more concerned with basic research, applied research, or development. Although some element of all of these can be present in any of the classifications, the potential uses of the technology in relation to the existing technologies determines the classification. The time frame in which the technology might be applied differs substantially for the different classifications, with (R) likely to be further in the future than for (CA), which, in turn, will be longer than for (B).

2.   The classification of technologies applies equally to product, manufacturing processes, services, disposal of used product,s and to all other activities of the enterprise. Although product and manufacturing processes are often closely allied and will often be using technologies with similar characteristics, other parts of the organization may be using more-or-less aggressive technologies. It is critical that the technology that is most applicable to the strategic needs of each activity be clearly identified when investment decisions are being made.

3.   At a given time, a technology that is considered basic in one industry may be classified as being competitively advantageous for another industry. As an example, in the late 1970s the automotive industry was faced with the need to improve the fuel economy of their vehicles while meeting stringent emission requirements. It appeared feasible to introduce digital controls, combined with information from numerous sensors, that would have allowed the system to adapt to many conditions and to perform in a more optimal manner. Since adaptive controllers had been in use in the aerospace and defense industries for a number of years, these technologies would have been classified as basic to these industries. The first use of programmable computer controllers for the engine was an exciting development and represented a competitively advantageous technology for the automotive companies in the early 1980s. Today that technology is widely used and would be considered to be a base technology for the automotive industry.

15.2   THE TECHNOLOGY FUNCTIONAL DEPLOYMENT PROCESS

The enterprise's needs for technology are driven by many issues, including the strategy that the enterprise is following for its products and processes, the approaches taken by its competitors, and the opportunities that technology presents. The number of decisions that must be made, the consideration of the relationships among possible choices, and the need to include many parts of the organization in arriving at decisions encompasses a complexity of alternatives that is as numerous as was the process of designing a product that will meet the needs of the customer. In analogy with quality functional deployment (QFD) for the design process, the process by which the decisions concerning the proper path to follow in the development of technology is described as the technology functional deployment (TFD) process. As a further analogy with QFD, the TFD process is intended to provide managers of technology with a procedure that will help them choose to support the proper technology at the proper time with the proper level of resources.

15.2.1   The Cycle Plan

Fig. 15.1 is a schematic diagram of a very simple version of a plan that would be characteristic of an enterprise that has three product lines, A, B, and C, each of which has a life cycle of three years. It is suggested in this figure that the work load for the enterprise is equalized by staggering the introduction dates of the three products. For illustration, it is also assumed that the time taken to effect an upgrade of each product is the same for all. While all of these assumptions result in a very simplified plan, the intent is to suggest that the enterprise has a long-term plan, in this example eight years, for the introduction of upgraded and improved products and that it expects to be working on an all-new product near the end of the time period. This plan for upgraded and new products is described as a *cycle plan* for an enterprise. It will be noted that the tasks that are to be completed during the period described as *upgrade of the product* are the same as those discussed in Section 10.3.1.

The products AI, BI, and CI include modifications, improvements, and additional attributes over those of A, B, and C, respectively. Products AII, BII, and CII are expected to offer additional improvements and attributes over their predecessors. The new attributes and improvements will each be met by the application of particular technologies. The amount of *reach* that each of these improvements represents over the previous version is generally determined by a combination of several factors, including the status of development of the technology, the time available to finalize and prove-out the product design, and the cost of implementation and the cost and time to create the manufacturing processes. Although it might be reasoned that the time needed to accomplish these tasks should not be determined by the timing shown on this cycle plan and should be left to the rate of progress of the project to determine, this is frequently not the case. The planners in the enterprise know that the marketplace expects or needs the new or modified product on a date that is close to those indicated on the chart. Marketing and sales will often advertise the introduction of a new product for some days or weeks or months before its actual arrival in the market. The workloads for the employees cannot be adjusted randomly since people cannot move to a new task until the last has been completed. All of these factors lead to a schedule that leaves limited flexibility for making major shifts. The result of this is that the cycle plan exerts great influence on the entire organization in terms of workloads, technology needs, and expenditures.

Since product D, shown on the cycle plan, is likely to be drastically different from those that are currently in production, it should be designed by the concurrent engineering processes described in Chapter 11. When the concurrent engineering team launches its efforts, it will demand evidence that any technology that is to be used in this

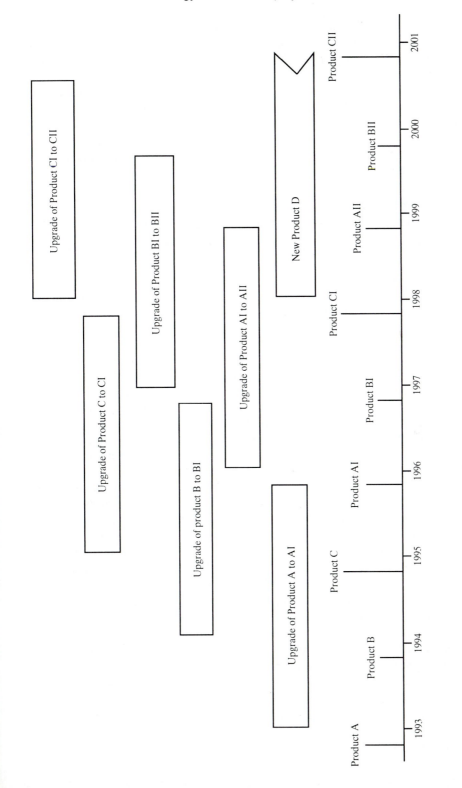

Figure 15.1 Schematic diagram of a product cycle plan for an enterprise having three different products, each with a lifetime of three years. The blocks indicate the time for the upgrade of the products and include all aspects of the product development cycle, including the redesign, prototyping, testing, and facilitization of the manufacturing plant, with the launch of the new product occuring just prior to year-end.

product be adequately developed and demonstrated. They cannot risk using an unproven technology in a final design. This presents a serious challenge to the managers of technology. Products that are needed near the end of the cycle plan will not have early detailed designs. Thus the needs of those products for specific technologies are likely to be uncertain. It is critical, however, that exploration of new technologies that may be useful to new products be started well before the technologies are needed. This creates a high priority on *anticipating* the probable future needs of products for technology. Only in this way will it be possible to have sufficient time to explore all problems and to develop sufficient confidence in the solutions to make it likely that concurrent engineering design teams will have the confidence to use the new technologies.

15.2.2   Technology Needs as Determined by Future Products, Processes, or Services

From the product plans, as displayed in the cycle plan of Fig. 15.1, it is necessary to identify the technologies that are needed at various times and to be certain that these needs are met. Completing matrix I (Fig. 15.2) for each new attribute for each new product that is shown on the cycle plan of Fig. 15.1 provides a means of accomplishing this in a systematic manner.

The information at the top of matrix I is needed to convey general information about the product and its focus in the market. The product description is a general statement of the features of the product. Its expected date of introduction must be consistent with that shown on the cycle plan. If the product is to be sold to a customer for inclusion in the customer's product, the identity of the customer is given. The name of the principal competitor for the customer's product and the identification of that product is given along with the identity of the best-in-class product against which the virtues of the proposed product can be measured. Also included are the estimated annual sales for the three years of the product's assumed life cycle. This is the information that someone who is knowledgable in the field will need to understand the general intent of the product.

Although it may be difficult to complete all the entries at the top of matrix I relating to the market features of a product that is expected well in the future, it must be emphasized that the completion of the columns of the matrix is critical for *all* planned products. As will be seen shortly, the information gleaned from a condensation of these charts provides the basis for planning both the short-term and long-range R&D programs of the enterprise. The completion of matrix I requires a careful and objective assessment of many factors that affect the possible success of development.

New attributes of the proposed new or modified product, process, or service: The description of the new attributes that are expected for the proposed product, process or service should be given in engineering terms for these attributes (e.g., improvement in fuel economy of a car in reduced gallons per mile, enhanced speed of welding in increased inches per second, detection of flaws of specific types in a time of X milliseconds, etc.) The more precise the statement of engineering attributes is made, the more accurate can be the discussion of technology issues.

Developments needed for new or modified product, process, or service: Since it may be possible to provide the attributes planned in column 1 in a number of ways, this column must provide a list of the developments that would provide, if

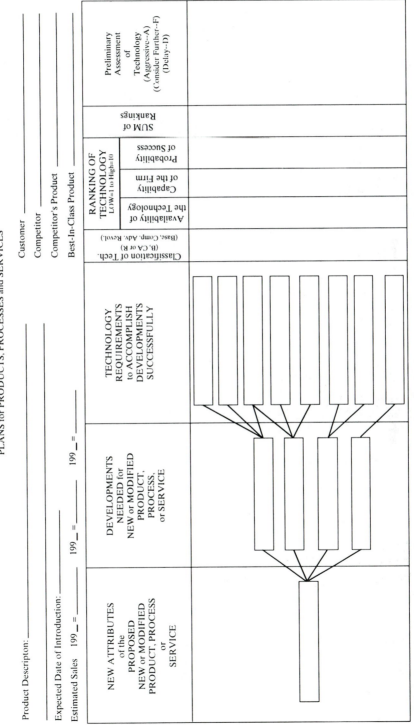

Figure 15.2 Matrix I: technological needs as determined by plans for products, processes and services.

completed successfully, the desired attributes. In the case of improved fuel economy this might be achievable by reducing the vehicle weight by 400 pounds, by using a turbocharged diesel engine, by a reduction of vehicle aerodynamic drag from 0.3 to 0.2, or by a combination of several or all of these.

Technology requirements to accomplish the developments successfully: A list of the specific technologies that are needed to accomplish the developments are given in this column. If a number of technologies might be possible for each development, these should be listed. For example, to reduce the weight of the vehicle it is possible to consider the use of aluminum or composite materials for the frame and body components. Alternatively, it might be possible to use an all-aluminum engine or mostly aluminum with titanium elements in parts of the engine. A third alternative would be to reduce the size of the vehicle. Achieving each of these will require the use of a different technology.

Classification of the technology: Each of the technologies are identified according to the earlier classification of base (B), providing a competitive advantage (CA), or revolutionary (R). The classification that is given relates to its use in the current products, processes, and services of the enterprise. It is quite possible for a technology to be given the classification of revolutionary even though it may have been in use in another industry for a number of years and would be classified as a base technology for that industry. In the example given above, a vehicle with a *body frame* made of composite materials would be classified as R since current experience exists only with composite body panels, while an aluminum engine could be classified as B, even though it may only have been produced by a competitor.

Ranking of the technology: In the next three columns a ranking on the basis of 1 to 10, with the highest being 10, is to be provided for three items that will strongly influence whether the enterprise will give further consideration to exploration of the technology.

Availability of the technology is to be interpreted as to whether the technology, or the basis for the technology, is available-perhaps at a competitor, in a university, or in another industry. The rank increases with the degree to which it is available. Obviously, if the technology is only available from a competitor, the ranking will be less than if it is available in another industry that is not a competitor.

Capability of the firm refers to the capacity of the enterprise to accomplish the development of the technology. Determining the ranking requires consideration of such factors as the existence of employees who understand the technology or its basis, the availability of knowledgeable personnel to undertake the tasks, and the proper facilities to complete the development.

Probability of success is an assessment of the likelihood of success. If the technology or its basis is not available anywhere, causing the current status rank to be low, or if the capability of the enterprise to accomplish the development is low, the probability of success may also be low. It is possible, however, for the current status to be ranked rather low, with both the capability and probability of success being ranked high. An example of such a situation would be the discovery of a new material or process that is revolutionary and for which in-house personnel will be capable of undertaking its further development.

The capability of the firm to complete the technology in the time allowed must be considered when ranking the probability of success. The issue is whether the development can be completed to the satisfaction of all concerned in a time consistent with the expected date of introduction of the proposed product, process, or service. Clearly, the further into the future and the more revolutionary the technology, the more imprecise this estimate will be. The natural tendency of technologists to underestimate the length of time to complete a development must be challenged constantly by the managers of technology, using the experiences of past developments and intuition derived from previous experiences.

The *probability of success* (1 - probability of failure) is used here rather than *risk* so that a highly desirable technology will get a high rating in this column, just as it would for the other rankings.

Sum of the rankings: The sum of the rankings given in the preceding three columns is to be entered here. The sum can vary between 3 and 30. A high ranking suggests that the technology deserves further consideration. A low ranking may suggest dropping the technology from consideration or considering a low effort that extends over a longer period of time.

Preliminary assessment of technology: Any proposed action that is to be taken regarding a particular technology can only be preliminary at this juncture. The immediate objective should be to try to reduce some of the alternatives. It will be helpful to eliminate any technology that seems highly problematical or that is duplicative of other technological options that are more attractive or that have unusually high risks. A discrimination among the various alternatives according to the following categories will be useful at this juncture: aggressive action is possible (A); further consideration is warranted (F); delay/eliminate further effort (D).

15.2.3.   Composite Technology Needs for Future Products, Processes or Services as Identified in the Cycle Plan

Each new or substantially changed product in the cycle plan will require the completion of matrix I. For a cycle plan even as simple as that shown in Fig. 15.1, this could entail seven unique matrices. To obtain a good overview of all the technologies that may be needed to support the cycle plan, it will be helpful to condense all matrices I into a form that is represented by matrix II (Fig. 15.3).

The first step is to translate the data from matrix I for all technologies *that have been given a preliminary assessment of (A) or (F)* onto matrix II. All the products, processes, or services that are contained in the cycle plan are listed along with their technologies and their classification and rankings. For the sake of illustration, matrix II suggests that to be successful some products will require the development of a new manufacturing process (process B, in the example) or a new service (service C, in the example). The example also suggests that the preliminary assessment of the technology needs for products D and F could not conclude that the development of a single technology was the most appropriate path. Instead, the preliminary assessment left two technologies for consideration for product D and three technologies for product

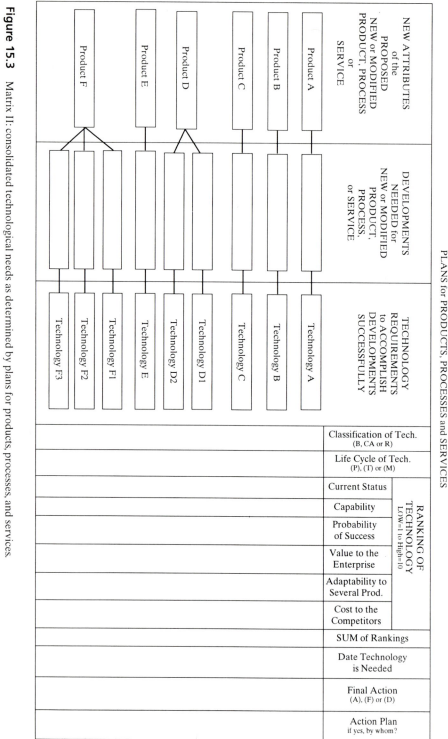

Figure 15.3   Matrix II: consolidated technological needs as determined by plans for products, processes, and services.

F. This is a circumstance that might easily arise if one of the technologies was classified as (CA) and the other as (R) and if the product was needed in the distant future. For near-term needs, it is unlikely that revolutionary technologies would be assessed as being (A) or (F).

In addition to the information that was translated from Matrix I, it is now possible to review all the proposed technologies and to make some additional choices. These include:

Life cycle of the technology: This is an assessment of the status of the proposed technology in terms of its life cycle and requires an assessment of whether it is in a primitive (P), transient (T) or mature (M) status.

Ranking of the technology: Columns 1 to 3 *under this heading* are copied from matrices I. In columns 4 to 6 *under this heading* a ranking is to be given on the probable impact of the technology on the organization. These were not included in matrix I, since a composite view of all technology is needed before this assessment can be properly given.

Value to the enterprise is a composite of (1) the estimated value of the technology to the customer, (2) the price that the customer is expected to pay for the product that contains the technology, and (3) the cost of including the technology in the product. A product with an expected high overall value to the customer that will allow an attractive price with a low production cost will deserve a high ranking. A product having a modest value to the customer with an accompanying high production cost must be given a low ranking. Although it is difficult to determine these quantities for developments far in the future or during early exploration of revolutionary technologies, it is absolutely critical that an estimate be developed of the competitive advantage that a successful completion of the technology would provide the enterprise. A small incremental improvement might rank high if this development is necessary to counter a competitive action already in the marketplace. It might rank low if it is judged to be simply a *nice thing to do*. The ranking given under value to the enterprise should answer the question: *Assuming successful development, what is its impact on the business?* Value to the enterprise should be ranked high only if the answer to this question is large.

Adaptability to several products, processes, or services is a recognition that a technology may have many or few uses. It may be applicable to a family of products or group of processes, or it may be strictly limited in its application. A technology that is broadly adaptable to several of the products listed in column 1 may be worth pursuing even though it will prove costly to a particular product. This is often the case for a new development whose costs will be reduced only through its use on a large number of products-as is experienced with developments whose costs can be expected to decline because of the *learning-curve phenomenon*.

Cost to the competitors is an estimate of the reaction of competitors to the development. Responses by the competitor may vary between investing heavily to reach your level of expertise, ignoring your actions, abandoning the market, bringing forward the date of introduction of a new product, modifying their technology to attack new markets, and so on. If the competitor is going to experience high costs in matching your developments, the likelihood of their undertaking this development will be reduced. Even if they decide to proceed, a high cost can put them at a disadvantage. A low cost of duplicating your development will probably not serve to inhibit your competition. *It is important that this entry be compatible with the entry under class of technology. If a technology is expected to offer a competitive advantage, the cost of duplication to the competitor will probably be high, otherwise any competitive advantage will be short-lived.* The challenge in making an assessment of the response that a competitor will give is to be realistic without underestimating the actions that a competitor may take. While an underestimation of your opponent can be disastrous, an overestimate can be very costly.

Sum of the rankings: The sum of the rankings given in the previous six columns is inserted here. It can vary between 6 and 60. A high ranking suggests that the technology is needed, can be made available, will be of high value, and offers the opportunity for a competitive advantage. A low ranking may encourage dropping the technology from further consideration or considering a low effort that extends over a longer period of time.

Date the technology is needed: The date to be entered here is determined from the data contained at the top of matrices I. Since the introduction date for the product that uses this technology is known, it is possible to determine when the details of the technology must be proven to allow the facilitization and manufacture of the product that will use the technology to occur.

Recommended final action, including priority order: The management is now confronted with the very difficult task of determining among the many technology opportunities and making a decision regarding which will be pursued and their relative priorities. To accomplish this, the management must fit the task of accomplishing what the product, process, and service needs are for its cycle plan with the resources that the enterprise possesses.

Action plan: Although it is likely that action plans for the development of products and processes that are in the cycle plan will be mostly handled in-house, participation by some vendors may be practical. In any case it will be necessary to make:

1. A determination of the availability of the key personnel to work on the various technologies, either in-house or at the vendors.
2. An assessment of the possible involvement of outside activities in the development of the technologies (e.g., vendors).
3. An assessment of the level of effort that will be needed, in terms of personnel and facilities, to complete the technology in the time allowed.
4. A preliminary allocation of resources available for the pursuit of technology needs as identified by (A) or (F).
5. A preliminary allocation of facilities, including test facilities, laboratory spaces, and so on that will be needed for the tasks identified by (A) or (F).

In the case of products D and F that are illustrated on matrix II, it may be necessary to eliminate one or more possible technologies from consideration or the determination that it is important to carry along the less critical technologies at a low level, either internally or through an external source. This might be the case for a revolutionary technology that has a somewhat uncertain probability of success but that would have great potential if it was successful.

15.2.4 Opportunities Presented by New Technologies

Not every attractive product or process development can be expected to be foreseen by the marketplace. The discovery of new phenomena, the appearance of new inventions, and the acquisition of new understanding of previously unanswered questions may suggest that a new technology can have exciting implications for new products or for products that have drastically different attributes from those that are currently in existence.

In an examination of more than 500 innovations in products and processes that occurred in the 121 companies whose interests encompassed five manufacturing companies, Marquis (1969) found that "one-fifth arose from someone saying, aha, maybe we can find a use for this *technical* idea." While the proportion of new innovations that are stimulated by market and production needs are substantially larger, a successful enterprise can never ignore the possibilities that new technologies may offer exciting alternatives. It is important for management to provide avenues by which its technologists can suggest opportunities for the utilization of technologies in new or different ways.

Matrix III (Fig. 15.4) provides a way of formalizing the discussion of these opportunities. While matrices I and II are primarily to be filled out by teams that include membership from sales and marketing, engineering, manufacturing, and servicing, it is likely that the initial compilation of matrix III will be the responsibility of the technologists with participation by the management of the operating divisions serving to refine the alternatives and generate a consensus of views on the opportunities.

Technological opportunities: The first column of matrix III requires a listing of opportunities that are viewed by the technology experts as being of likely interest to the enterprise. In most circumstances this list will *not* include the technologies that are listed in matrices I and II, since these will have been identified as being needed to meet the objectives of the cycle plan. If it happens, however, that a technology listed in matrices I and II offers the possibility of a product or process that is well outside those contained in the cycle plan, this would be listed in matrix III. Examples of the latter were given in Section 14.3-the Sony Walkman and the use of plastic extrusion by Empire Pencil.

New or modified products, processes, or services that the technology may allow: Column 2 requires an identification of the new or modified product, process, or service that the technology might allow if developed successfully. The entries in this column need to be as specific as possible and should include the same type of engineering descriptors as were used in column 1 of matrix I.

Level of knowledge of the technology: The third column requires a ranking of the level of current understanding of the technology. Since it is recognized that this ranking could be very imprecise for new and perhaps unproven technologies, new products, and so on, it is sufficient to give a ranking of either low, medium, or high. It should be noted that even though a technology may be unknown to your industry, it could be ranked high if it is currently in use in another industry.

Probability of successful development: In this column a ranking is needed of the probability of successful development of the technology. Since it is recognized that this ranking could be very imprecise for new and perhaps unproven technologies, it is

MATRIX III

ASSESSMENT OF OPPORTUNITIES PRESENTED BY TECHNOLOGY

PROPOSED TECHNOLOGICAL OPPORTUNITIES	NEW or MODIFIED PRODUCTS, PROCESSES, or SERVICES that TECHNOLOGY MAY ALLOW	Level of knowledge (High, Med. Low)	Prob. of Success (High, Med. Low)	Life Cycle of Competing Tech. (P), (T) or (M)	KEY PROBLEMS that must be SOLVED before TECHNOLOGY can be used SUCCESSFULLY	Assessment of Competitors Response	Desirability Further Consideration	Detailed Assessment? (If Yes, by Whom?)	Mode of Exploration Int. (I), Ext. (E) Joint (J)	Action Plan (if yes, by whom?)	Final Responsibility for Program	Review Authority

again sufficient to give a ranking of either low, medium, or high. As in the preceding column, even if a technology is unknown to your industry, it could be ranked high if it is currently in use in another industry.

Life cycle of the currently competing technology: Since the technology identified in column 1 will usually be new to the enterprise and since *current* products and processes do not utilize it, it is important to understand the level of maturity of the technology that will be challenged by this new technology. The location in its life cycle of the *currently used technology* should be indicated by (P), (T), or (M). A new technology frequently has a higher probability of competing with an existing technology if the current technology is mature. The new technology is less likely to be successful if it must compete with a current technology that is either primitive or in transition.

Key problems that must be solved before the technology can be used successfully: Any new technology will have problems that must be solved before it can be expected to compete successfully with an existing technology. The more revolutionary that the new technology is and the more drastic are the changes that it will create in existing products and processes, the less precise will be the statement of the problems that must be solved. It is here that the technologists must exercise great restraint and caution. It is very easy for the technologist to become enamored with a new technology and to fail to recognize the critical nature of some problems. In making this assessment, it is important to seek the advice of mature, realistic people who have experienced the successes and frustrations of generating new technology.

Assessment of competitors' response: The probable response of the competitor is important in making the final assessment. If the competitor has a large commitment to the existing technology, as is usually the case for a base technology, it may be that the response will be to attempt to protect the existing technology, almost at all costs. On the other hand, if the competitor is already exploring alternatives to the base technology, the urgency that would accompany the search for an alternative could be greatly increased.

Desirability of further consideration: The intent in this column is a determination of whether the opportunities offered by the technology are such that it is deemed worth further exploration. This offers an early opportunity to conclude that the risks of possible failure and the severity of the problems are so great that further consideration of the technology is unwarranted or premature. A conclusion that the possibilities of a technology are sufficiently exciting that further exploration is warranted does not mean that a full program will evolve immediately. A decision to consider a technology more fully provides the opportunity to explore more fully the possible benefits and to understand the probable pitfalls.

Is a detailed assessment needed? If a detailed assessment is needed to better understand the many features of the technology, this should be undertaken by those who are most knowledgeable about the technology and have the best long-term view of the technological needs and aspirations of the enterprise. A useful way of accomplishing this part of the evaluation is the development of a technology assessment, a description of which is given in Appendix A15.1. As discussed in the appendix, the intention of the technology assessment is to *assess* and not to predict.

This is not a document that attempts to predict what technology will be used when and for what products, but to assess the applicability of a technology for a particular product, process, or service in more detail than can be done with matrix III. In this way the possible risks and benefits that will be encountered in the development and use of the technology can be made explicit. This evaluation should relate the technology to the corporate business plan, assess the potential pitfalls in its implementation, estimate the time and cost in proving out the technology, and conclude with a recommendation for future action. An outcome of this assessment may be any of the following: abandon further efforts, monitor efforts being taken by external groups regarding the technology, or undertake a program to exploit the technology. If the conclusion of the technology assessment is other than abandonment, a plan of action will be needed. The content of this will vary depending on how the development will be undertaken. As an aid in developing a more detailed plan, it is helpful to complete the next column in matrix III.

Mode of exploration: Recognizing that various arrangements exist for the successful exploration of the technology, the intent here is to offer a preliminary judgment concerning the most appropriate approach. These include an in-house activity, either within an existing group or by a newly created team; an external group, such as at a vendor or at a not-for-profit research and development group; or through a joint program between an inside and external group (e.g., between a vendor of the component and you, the likely user of the new development).

The plan of action to explore new technological opportunities: The details of the plan will vary significantly depending on whether it is recommended that the study of the technology be undertaken by someone external to the enterprise or whether it is to be accomplished in-house. An outline of the content of each of these approaches is contained below. A joint program would contain elements of both of these. For a joint program, explicit statements of the responsibilities of each group must be developed as well as agreements reached concerning the rights of each group to the benefits of the development.

External Program: While the financial implications of simply *monitoring* progress that may be taking place in some laboratory or company will have financial implications different from those of the active creation of a program that is to be carried out by someone outside the corporation on your behalf, plans for these two will have many common elements, including:

1. Identification of the organizations that will be developing the technology.
2. Development of formal contracts, as necessary, for the undertaking of efforts.
3. Development of an understanding and procedure for monitoring progress.
4. Proposing, to the extent needed, appropriate means of providing partial or total support for the program.
5. Identifying the frequency with which reports on progress will be made to management, along with the frequency with which the technology assessment will be updated with new information.

In-house program: If the conclusion of the technology assessment is for the corporation to undertake a program of exploration and development of the technology, it will be necessary to:

1. Identify the organization that will undertake the exploration and development of the technology.

2. Create a preliminary personnel and facility schedule that will indicate the level of effort that is to be expended and the approximate dates at which sufficient information is expected that key questions can be answered.

3. Create a preliminary budget, including personnel, expenses, and facility costs with an estimate of the lifetime costs of the program.

4. If the proposed program competes with existing programs for personnel and facilities, an estimate of the impact on these programs must be given. If the proposed program has an impact on programs that are contained in the cycle plan, a means of offsetting this negative impact must be given or modifications to the cycle plan must be recommended and adopted before any work is begun on the proposed technology.

At this time the final two columns of matrix III should be completed:

Identification of final program responsibility: It is necessary in this column to identify the person or group that will have final responsibility for the program and the person who will be the program leader and the program spokesperson.

Identification of the program review authority: Review of program progress and program needs must be given to someone or to some corporate committee. If it is the latter, it is important that the dates on which the reviews will occur be approximately identified. Unless clear lines of responsibility and authority for the review of the program are established, a new program that must compete with established programs will often find that interest in it quickly wanes. It is too easy to destroy new programs that must compete with existing programs.

A word of caution: The people who are given program responsibility and who have final review authority must have an enthusiasm for the program. To assign a new program to people who see it as an unnecessary diversion of time and resources will probably doom it to failure.

15.2.5   Technology Needs in Support of the Development of New Engineering and System Tools

To an increasing extent, new products, processes, and services must meet a complex set of specifications. Performance requirements are being imposed by government regulations, quality and safety of the product are continuing to be priorities for the customer, and the incorporation of new technologies are being used to provide important competitive advantage. The challenge to the engineering community is to accomplish the necessary tasks in the shortest possible time, with the most

MATRIX IV

NEED for NEW ENGINEERING
and
SYSTEM TOOL

ENGINEERING and SYSTEM NEEDS	NEW TOOLS NEEDED to SATISFY NEEDS	TASKS REQUIRED to PROVIDE the NEW TOOLS	RANKING LOW = 1 TO HIGH = 10			Sum of Rankings	Desirability Further Consideration	Detailed Assessment? If Yes, by Whom?	Expected Date of Completion	Mode of Exploration Int. (I), Ext. (E) Joint (J)	Recommended Action Agressive (A), Consider Further (F) or Delay (D)	Action Plan (If Yes, By Whom?)	Final Responsibility for Program	Review Authority
			Adequacy of Technology	Availability of Technology	Expected Lifetime of Tool									

Figure 15.5 Matrix IV: need for new engineering and system tools.

error-free designs that will be competitive in the marketplace in terms of cost and quality. As has been emphasized repeatedly in past chapters, the next level of competitiveness is increasingly being determined by the corporation's capability to respond rapidly to the changing marketplace.

To accomplish these complex tasks in the shortest amount of time, the engineering community is constantly searching for new tools that will enhance their capabilities. In almost all cases these tools rely on the development or use of a new technology or the use of a current advanced technology in new ways. To remain competitive, the technology needs of the corporation in the support of these new engineering tools must be identified and proper effort expended in acquiring them. Matrix IV (Fig. 15.5) provides a convenient way of identifying the technologies that are needed for the development of these tools.

Engineering and system needs: Examples of the needs of the engineering and system include such items as a reduction in the number of iterations needed during a new design, improved models that allow better exploration of "what-if" questions regarding the implications of particular actions that affect several elements of the enterprise, improved capability to determine the optimal allocation of efforts among various segments of the corporation, reduction in the time required to produce the prototypes of early designs, reduction in the number of prototypes that must be created to create a final design, improved real-time analytic capability for the analysis of manufacturing data, and new means of communicating among diverse groups.

The tools needed to satisfy the needs of the engineers and system: The examples identified in the preceding paragraph suggest some of the following in terms of the tools that are needed: commercially available analytic programs (e.g., NAS-TRAN), discrete-element simulation programs (see Chapter 13 for extended discussion of applications to manufacturing systems), computer simulation programs that provide performance analyses for specific components and systems (e.g., aircraft engines), simulation programs that can predict the performance of a complete product (e.g., ground vehicles that are constrained to provide particular levels of performance at a desired fuel efficiency and levels of emission of pollutants), modeling programs that allow the simulation of product performance under extreme circumstances (e.g., energy absorption of structures during a crash), computer design programs of manufacturing processes that include non-linear properties of the material being processed (e.g., the design of metal stamping dies that will generate final products having well-controlled dimensions or the design of extrusion processes for fiber-reinforced materials), and systems that provide rapid prototyping of preliminary designs (e.g., stereo-lithographic systems). Entries in this column may include tools that are either dominantly software or hardware.

Tasks required to provide the new tools: While the tasks may be as simple as acquiring a commercially available piece of hardware or the building of a new facility, it is often the case that the tools are not available in the form that they are needed. Some of these may take years of dedicated effort of highly trained people to generate. For example, the computer design of stamping dies that explicitly treated the spring back of the resulting metal part took dozens of person-years to complete successfully. Even larger efforts were required to develop computer programs that would successfully predict the collapse of mechanical structures during a crash. Even when commercially available, there may be special needs in terms of the training of personnel, modification of programs, and so on.

Rankings: In the next three columns a ranking on the basis of 1 to 10, with the highest being 10, is to be provided of three items that will strongly influence whether the enterprise will choose to give further consideration to the exploration of the technology.

Adequacy of the technology applies to whether the current level of development of the technology is adequate for the new tool. Although commercial vendors frequently contend that their product is adequate for the task, care must be taken in accepting this claim.

Availability of the technology applies to whether the technology is controlled by someone who will refuse or will demand an unreasonable price for making it available or whether it can readily be made available for your use.

Expected lifetime of the tool refers to the probable period of time that the tool will be useful to the organization. A tool that can be expected to be used over a long period of time will deserve a high ranking.

Sum of the rankings: The sum can vary between 3 and 30. A high value suggests that the technology and the expected value of the tool over a long period of time is good.

Desirability of further consideration: Based on perceived needs, the fact that new tools are appropriate, that the tasks that are required to provide the tools are not excessive, and the sum of the rankings for the technology, a preliminary conclusion can be made as to the desirability of proceeding further. It is not uncommon, of course, to identify needs and possible tools that would be *useful* to the enterprise that are *well beyond the current technological capability* of the community. If this is the case, further consideration would probably not be warranted.

Is a detailed assessment needed? If the answer to this is yes, a technology assessment, described in Appendix A15.1, may be appropriate. As noted before, the individual or group who will undertake this assessment must be fully cognizant of the needs of the corporation and the capability of the technology for further development.

Expected date of completion of the tool: A realistic date must be established for completion of the tool. If acquisition is from a commercial source, this may be readily available. If the tool requires substantial development, the technical groups will need to make careful estimates of the time and effort needed. In the case of any tool, the expected completion date must include sufficient time to allow verification of the accuracy of its results, the training of people to use the tool, and the documentation of the tool, a task that is critically important for any software tool that is expected to be long-lived and widely used.

Mode of exploration: This column requests the identification, using the technology assessment, if one is available, of the possibilities of exploring the development of the tool with someone outside the corporation through a contract, jointly with an external activity, or internally. A decision regarding the most acceptable mode will depend on such factors as the availability of the proper internal technical capability, an understanding of your system by a group outside the company, the desirability and capability of controlling the dissemination of the results, and the probable costs of each mode.

Final action: Based on the previous information, a decision is needed of the desirability of pursuing the development of the tool or delaying further consideration.

Action plan: If it is the conclusion that further action is warranted, a plan will be needed that will provide preliminary details concerning timing and costs. The discussions at the end of Sections. 15.2.3 and 15.2.4 regarding the development of an action plan applies here.

Final responsibility for the program: The identification of who will be responsible for carrying out the final program must be identified. It should be noted that this may not be the same person or group who developed the action plan. It might, for example, be concluded that a new facility is needed, that it should have specific features and capabilities, and that the procurement and installation will be assigned to some person or group who can effectively manage this procurement.

Review authority: Recognizing that the development of some tools requires a long-term commitment, it is important that the person or committee who will have continual purview over the progress of the program be identified.

15.3 THE CORPORATE TECHNOLOGY ROADMAP

At this point preliminary technology plans will have been completed for the three major technological needs for the corporation: to support the cycle plan (Section 15.2.3), for the exploration of new technological opportunities (Section 15.2.4), and to support the development of new engineering and system tools (Section 15.2.5). This preliminary information is now consolidated into the technology plan for the corporation. This consolidation becomes the *technology roadmap* that the corporation will follow in the years ahead. The result of this consolidation is shown schematically in Fig. 15.6. This is a description of all the technologies that will be under study by the corporation in the time frame extending to at least the end of the cycle plan for the new products and perhaps longer for the exploration of new technological opportunities and for the development of new engineering tools technologies. It is at this point that final budgets and plans for each technology must be finalized.

It is from the preliminary allocations of resources among the tasks that it will be possible to judge whether sufficient resources will be available to fulfill the needs required in the cycle plan. *It is at this stage that consistency of recommendations with requirements must be assured.* If it concluded that all technologies cannot be pursued within the time frames that are needed, a review must immediately be undertaken of the cycle plan and a determination made of the changes that can be made in it that will allow the enterprise to accomplish what the revised matrix II demands. In other words, *it is absolutely essential that the technology plan be consistent with the product cycle plan and similarly, that the cycle plan be completely supportable by the technology plan.*

The landscape is strewn with the corpses of cycle plans that could not be met because the technologies were not available at the requisite time or because the plans were built without proper awareness of the difficulties in developing the technologies. What is all too familiar is a cycle plan that starts with very aggressive goals for its technology needs and thus ensures products with competitive advantages, a late recognition that the technology will not be available, a revised set of product objectives that

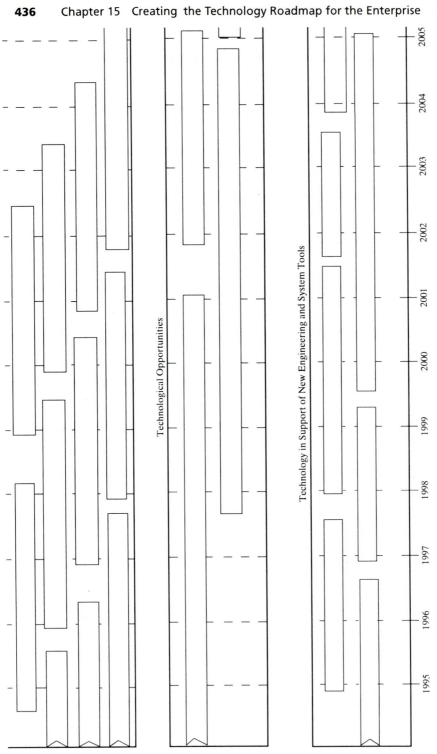

Figure 15.6   Technology roadmap for the corporation

can be met but are with products that are only marginally better than the current product, a revised product that is barely competitive, and a subsequent loss of market share. This can be avoided by recognizing that *the creation of this plan is the opportunity, indeed necessity, to ensure that the business plan and the technology plan are consistent. Without this consistency, neither will be successful.*

While seeking to support the cycle plan, it is also critical that the corporation properly assess and support the long-term opportunities that technology provides and that it provide the necessary technological tools for its engineers and its system. It is upper management's responsibility to ensure that a proper balance be maintained among the support of these three diverse needs for technology.

This picture of the technology plans for the next 10 or so years provides several important pieces of information for the management. First, it provides a view of the demands that the corporation will be placing on its technical community for support and innovation. Second, it permits a critical evaluation of the total resources that are needed and the capability of the corporation to provide these resources. Third, if circumstances in the marketplace change, if new opportunities arise, or if unexpected problems develop in completing any of the needed tasks, the impact of this on other tasks and later programs will be immediately visible.

If matrices II, III, and IV have been developed properly, the technologists and business operators will have agreed that the plan is realistic because the risks are reasonable and because the enterprise will have agreed that certain resources will be made available, the engineering and manufacturing groups will know what technologies are expected and in what time frame, the sales and marketing people will know that the planned product characteristics will have a high probability of being available, and the manufacturing activities will know when new technologies can be expected. Under these circumstances, all parties will have confidence that the business and technology plans have been joined together properly to provide the strategy that the enterprise deems is necessary to maintain its competitiveness.

These assurances will be forthcoming only if the management atmosphere encourages honest debate, that all parts of the organization participate in these decisions, and that everyone understands the level of the risks that the organization in undertaking when it makes its final decisions. Although it is not unreasonable for the management to expect the technologists to reach as far into the future as possible, it is essential that everyone recognize that the risk increases as the length of the reach is extended.

Willyard and McClees (1987) discuss the technology roadmap concept as used by Motorola. They provide an example using the broadcast automotive FM receiver as an example. Such a picture depicts the future technology plans for this product as would be embodied in their cycle plan. A particularly important example of technology roadmaps is given in Appendix A15.2.

This provides an example of agreement by the semiconductor industry of what technology developments will be needed to achieve a particularly aggressive goal by the year 2000. These road maps demonstrate a number of the important features that were discussed above, including:

- Specification of the goals of the program with specific time objectives.
- Identification of engineering objectives for each technology using quantifiable metrics.

- Strong interaction of manufacturing and product design objectives.
- Benchmarking against currently achievable technology.
- Identification of alternative technological opportunities during early stages of exploration.
- Identification of the technology that is needed to provide new engineering and system tools.

15.4   MODIFICATION OF THE ROADMAP

Changes should not be made in one area of the roadmap without full recognition of the possible impact on other areas. Since the environment in which a company operates is never static, it is obvious that a plan that extends across a decade of time must be subject to change. It is important, however, that frequent changes be held to a minimum, as the exploration of technology can seldom be efficiently accomplished in a mode where starts and stops are frequent. The factors that can lead to changes in the technology roadmap are:

1. Changes in the cycle plan as a result of changes that occur in the market-place. When this occurs, it will be necessary to modify matrices I and II to account for these changes.
2. Disappointments in the study of what was considered to be a promising new technology. If the exploratory work on a new technology proves that the technology is less attractive than first thought, that unforseen problems make the technology less attractive, or that the expected new products, processes, or services are not likely to be sufficiently attractive to warrant further development, the effort on this technology will need to be terminated.
3. The appearance of a new technology that was previously unknown or not considered to be attractive may make its exploration of higher priority than the exploration of some of those technologies contained in the roadmap.

It is important that a mechanism be provided for a regular review of the technology road map, at least on an annual basis. If the roadmap was carefully created and represents a consensus of the management of the corporation, the review can be one that focuses on any proposed or needed changes to the map. While the system must provide a discipline that allows occasional changes in the plan, it must also insist that a total view of the technology system be maintained and that *consistency* be maintained between timing and the resources that are available or will be provided. This emphasizes the importance of seeking to obtain a consensus of all people for all the preliminary plans.

CONCLUDING COMMENTS

As is readily apparent from the discussion above, the creation of a technology roadmap that is fully consistent with the short and long-term needs and aspirations of the enterprise can be very complicated. Although not all of the questions posed in the discussion above must be answered in all cases, most of these questions appear in any discussion of the technology needs of an enterprise with such regularity that they should be abandoned only after careful thought. As noted earlier, the objective of the TFD process is not to create a set of carefully constructed matrices that are beautifully presented, but to provide a means by which the many issues are addressed that surround the debates over what technology the company really needs. To repeat, the process will have been successful if the final product is a technology roadmap that fully supports the company's needs, is fully consistent with the company's business plan, and is adequately supported by resources that are provided by the business plan.

REFERENCES

DRUCKER, P. 1985. *Innovation and Entrepreneurship: Practice and Principles*, Harper & Row, New York: 149.

KANTROW, A. K. 1980. The Strategy-Technology Connection, *Harvard Business Review,* July-Aug.:6.

MARQUIS, D.G. 1969. The Anatomy of Successful Innovations, *Innovations* 4, Nov.

WILLYARD, C.H. and C.W. MCCLEES. 1987. Motorola's Technology Roadmap Process, *Research Management,* Sept.-Oct.:13.

READINGS FOR DISCUSSION

1. EG&G, Inc. (A), Harvard Business School Case Study 9-377-027; EG&G, Inc. (B), Harvard Business School Case Study 9-376-188.

2. Ceramics Process Systems (B), Harvard Business School Case Study 9-691-006.

3. Sun Microsystems and the UNIX Operating System: Creation of a Standard? Harvard Business School Case Study 5-690-042.

4. Honeywell Residential Division New Product Development, Harvard Business Case Study 9- 689-035.

5. Associated Instruments Corporation-Analytical Instruments Division, Harvard Business Case Study 5-690-091.

DISCUSSION

The five case studies listed above offer important contrasts in the planning of research and development activities. The EG&G case study provides details of their planning process and offers interesting information that is relevant to understanding the importance of the questions that are posed in matrices I to IV. The Ceramics Process Systems case study captures some of the problems that a small high-tech company faces. The Sun Microsystems case study presents two issues that must be addressed; the process by which a standard is created and the company's strategy in an increasingly uncertain environment. The Honeywell case and the Associated Instruments Corporation case offer insight into the various modes of accomplishing the development of new products and the business decisions that are entailed.

It is suggested that the discussion of these cases take place in two rather distinct phases. First, an analysis of each case will offer important insight into the decision processes that must be made regarding investments in technology and how these are affected by the business strategy of the company. Second, an overall comparison of the five different cases can offer interesting insights into the importance of company philosophies, company structure, and company sizes.

APPENDIX A15.1
ASSESSING THE POTENTIAL OF A TECHNOLOGICAL
OPPORTUNITY: THE TECHNOLOGY ASSESSMENT

As a manager, you will either ask for or be asked to estimate the value and risks that some new technology or process will have on an operation. This will frequently require the preparation of a critical assessment of the potential that this technology or process holds. The capability of communicating with people in the organization or of using such an assessment in making a decision can be critical to assuring that a correct decision is made. The technology assessment is an important tool in accomplishing this.

The first step in developing a useful technology assessment is identification of the audience. The second step is the creation of a clear statement of the problem that needs a solution (e.g., a new technology to improve productivity, better quality for a particular operation, reduced work-in-process inventory at some location, and so on.) or a discussion of the opportunities for new or greatly improved products that the new technology will allow. The third step is the identification of alternatives to the proposed technology and the limitations that they have in achieving the objectives that the new technology affords. The final step is a detailed analysis of the proposed technology.

Economic feasibility must be considered in the assessment. This can be accomplished by obtaining data from a variety of sources, including vendors of equipment, construction and installation costs, annual operating costs, and so on. It may be necessary to obtain assistance from a variety of people in the organization in completing this part of the study. If the study topic is an all-new technology for which there is limited production experience, estimates of changes of a current process that would be required to use the new technology may be the best that can be done. The intent here is not to offer a detailed budget and project plan, but to offer the management a financial basis on which to judge the possible attractiveness of the new technology.

A careful consideration of the risks and the barriers to implementation, as well as the benefits of this technology, is an important part of any assessment. While it is expected that the author will start the study as an advocate for the implementation of the technology, a possible conclusion is that further active efforts to develop the technology should not be undertaken, at least at this time. It may be concluded, for example, that the technology needs further development, that a pilot line is needed to prove-out the technology, or that a breakthrough is needed in a particular aspect of the technology to make it feasible for application in your company. They may even conclude that the technology is not appropriate for the problem that is being addressed. Any conclusion must follow logically from the assessment.

The topics identified on matrix III must be addressed as is appropriate. The technology assessment is a more thorough and complete analysis of the technology and its potential benefits than can be provided by matrix III.

The impact of the assessment will depend critically on its capability to communicate effectively the conclusions and/or recommendations to the people who will make a decision as to whether to proceed or not with the technology. Effective communication requires:

1. Accuracy
2. Completeness of the information conveyed
3. Ease and brevity in the presentation of information
4. Valid conclusions

As always, good communications require attention to literary style and grammar. The assessment will be a success if it presents a convincing set of arguments and conclusions in a clear and concise manner.

Since a variety of people are likely to read an assessment of this type some of whom may not be technically oriented-technical jargon should not be used in the assessment. It should be written in a style that *conveys the technical implications* of the proposal to the products and processes of the enterprise without implying or requiring that the reader know all the technical details that would be involved in implementing it.

An assessment usually contains the following sections:

1. Executive summary
2. Text
3. Conclusions and recommendations
4. Figures and tables
5. References

An executive summary is not simply an abstract. An executive summary should state the problem, the potential benefits and risks of implementing a proposed solution, and the recommendations that are being offered. A reader should be able to understand the content and the recommendations developed in the study by reading only the summary. The executive summary should not exceed two pages in length. If the reader needs to understand the rationale for the arguments and the detailed justifications, they can find this by reading the complete text of the assessment.

APPENDIX A15.2   TECHNOLOGY ROADMAP FOR THE SEMICONDUCTOR INDUSTRY*

The semiconductor industry has introduced a new generation of static random access memory (SRAM) essentially every three years, with each succeeding generation having four times the capacity of the former. With the 4-Mb SRAM available in 1991, the historical trend would suggest that the technology would evolve such that a 1-gigabit static random memory chip (1-Gb SRAM) would be available in 2003. Representatives of the U.S. semiconductor industry met in early 1991 to determine the feasibility of reducing the time to achieve this by three years, thus effectively making it available one product generation ahead of forecasts based on historical trends. To achieve this objective it would be necessary to have a manufacturable 0.12-μm design-rule process available by the year 2000. The approach that was taken was the creation of technological road maps that would provide this capability.

The following technical issues were identified as being critical to achieving this capability, with a roadmap being developed for each issue.

- Lithography: Lithography tools, masks, resists, and metrology.
- Processing: Processing equipment, facilities, defect control, yield, wafer size, sensors, minifabs, and metrology.
- Simulation: Software environment (including computer-aided design, computer-integrated manufacturing, computer-aided manufacturing, and computer-aided testing), algorithms, and computing platforms.
- Devices and Circuits: Circuit techniques, device structures, redundancy, repair, testing, and chip architecture.
- Architecture: System environment, applications, packaging, interconnections, soft errors, and data reliability.
- Economics: Economic analysis tools, fabrication costs, facility costs, economic benefits of a technology development program.

The technologies that required accelerated development to address the foregoing issues were identified according to Table A15.1.

Lithography

To meet the Micro Tech 2000 goal of 1 Gb SRAM engineering samples by the year 2000, a complete set of lithographic tools and processes needs to be developed based upon design rules with minimum features in the range of 0.1 to 0.15 μm, while meeting exacting requirements over a broad range of technical and manufacturing issues. Furthermore, an evolutionary path should be structured that addresses the incremental lithography requirements covering all semiconductor technology generations through the year 2000, while accomplishing a pull-in of the 0.12 μm generation by 3 years.

Because lithography technology is in the critical path for device and process equipment development, two different kinds of lithography requirements are defined:

Source: Adapted from the report to the National Advisory Committee on on Semiconductors entitled "Micro Tech 2000 Workshop Report: Semiconductor Technology Roadmaps", 1991.

TABLE **A15.1** **Technologies Requiring Accelerated Development**

Lithography	Processing	Simulation
Mask technology	Microcontamination	Factory models
Overlay technology	Process control and process/	Physical 3-D process models
Metrology	tool development	Tool models
Resists, resist technology	Ion implantation	0.12 µm device models
Lithography tools	Chemical vapor deposition	Design, layout, and simulation
Optical	Physical vapor deposition	tools for board, module,
X-ray proximity	Pattern transfer (etch)	and chip
E-beam direct write	Thermal treatment	Computing frameworks
E-beam projection	Large-diameter wafers	and standards
E-beam proximity	Integrated process equipment clusters	
Ion beam	Metrology	
Devices and Circuits	**Architecture**	**Economics**
Interconnection technology	High-speed interconnects	Manufacturing education
1 Gb SRAM cell technology	High-performance packaging	Manufacturing economics
Device design	Flat panel displays	Factory/product cost models
Device technology		Market simulation
Design for test		

- A development/prototyping capability to support device and process equipment development. This capability is needed by 1994.
- A production capability, with production-worthy machines available by 1997.

The lithography roadmap is shown in Figure A15.1 It shows the leading candidates for each of the technology generations through the year 2000, and the schedule and decision points required to meet the MICRO TECH 2000 goals. The philosophy behind the roadmap is to leave the options open as long as possible because of the uncertainty in attempting to choose the optimum approach at this time.

Processing

The three major topics included in the processing area are: microcontamination, which covers cleanroom facilities, purity requirements of materials, and wafer cleaning; equipment and process control, which includes generic equipment requirements, ion implantation, chemical vapor deposition, physical vapor deposition, pattern transfer, thermal treatment, wafer size, and major processing roadblocks; and metrology, which includes linear, particle, contamination, and thickness measurement, and in-process sensors. Individual roadmaps for microcontamination, process control, and metrology are given in Figs. A15.2, A15.3 and A15.4, respectively.

Simulation

As the semiconductor industry has matured, computerized modeling and simulation have assumed increasingly important roles at all levels and have enabled the design and manufacture of complex devices and systems. Micro Tech 2000 technology will be pushed near to fundamental limits, thereby placing an even greater emphasis

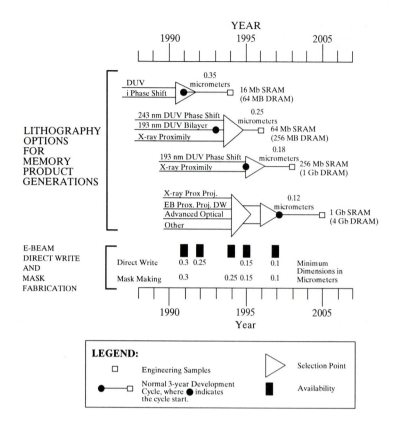

Figure A15.1   Lithography roadmap. Source: "Micro Tech 2000 Workshop Report: Semiconductor Technology Roadmaps":p.11.

Year Required	1991	1994	1997
Cleanroom	35 particles/ft³ at 0.1 µm	1 particle/ft³ at 0.08 µm	1 particle/ft³ at 0.03 µm
Liquids	<50 particles/ml at >0.5 µm	<50 particles/ml at >0.07 µm	<2 particles/ml at >0.03 µm
Gases	<10 particles/ft³ at >0.03 µm	<0.1 particle/ft³ at >0.03 µm	<0.05 particles/ft³ at >0.03 µm
Cationic Species Trace Impurities	<10 particles per billion	<100 parts per trillion	<5 parts per trillion
in Gases (N₂)	<10 parts per billion	<100 parts per trillion	1 part per trillion
% Vapor vs. Liquid Cleaning	0%	30%	60%
Contamination Level (Fe per cm², after clean)	<5 × 10¹¹	<1 × 10¹⁰	<1 × 10⁸
Participants	Materials and equipment suppliers, the National Institute of Standards and Technology, universities, national labs, and chip manufacturers.		

Figure A15.2   Microcontamination roadmap. Source: "Micro Tech 2000 workshop report":p.15.

Year Required	1991	1994	1997
Etch Equipment			
Uptime	80%	90%	97%
% Closed-Loop Control	0%	30%	80%
Ion Implantation (200 eV Development System)			
Participants	Universities, equipment suppliers, chip manufacturers, and the National Institute of Standards and Technology		
CVD			
Aspect Ratio	2:1 at 0.5 µm	4:1 at 0.25 µm	10:1 at 0.1 µm
Wafers			
Size	200 mm	300 mm	300 mm, 3.0 µm epi

Figure A15.3 Process control roadmap. Source: "Micro Tech 2000 Workshop Report: Semiconductor Technology Roadmaps":p.15.

on sophisticated models, simulation tools, and design aids. In addition, increased complexity in manufacturing will require acceleration of the trend toward factory automation and adaptive control to ensure economic viability of the enterprise.

To achieve at least one generation acceleration in the pace of semiconductor technology development by the year 2000, efficiently executable, physically based models must be developed for all phases of integrated circuit production from processing through manufacture. In addition to making the technology possible, the models will provide the predictive capability necessary for moving ideas quickly from research into manufacturing. This will depend critically on acceptance of standards at all levels and the use of modular, object-oriented software tools for flexibility and efficiency.

Year Required	1991	1994	1997
Feature Size	0.5 – 0.7 µm	0.25 µm	0.1 µm
Linear Measurement			
Accuracy	0.007 µm	0.0025 µm	0.001 µm
Particle Size Measurements			
Unpatterned Wafer	0.2 µm	0.08 µm	0.03 µm
Gas Phase	0.1 µm	0.08 µm	0.03 µm
Liquids	0.1 µm	0.08 µm	0.03 µm
Impurities			
Metals on Silicon	10^{10} per cm^2	10^9 per cm^2	10^8 per cm^2
Metals in Silicon	10^{11} per cm^3	10^{10} per cm^3	10^9 per cm^3
Liquids	1 part per billion	100 parts per trillion	5 parts per trillion
Gate Oxide			
Thickness (equivalent)	125 Å	70 Å	40 Å
Accuracy	±10 Å	±2 Å	±1 Å
In situ Measurement	a few sensors		
Coverage	available (~15%)	30%	30%
Participants	Materials and equipment suppliers, the National Institute of Standards and Technology, universities, national labs, and chip manufacturers.		

Figure A15.4   Metrology roadmap. Source: "Micro Tech 2000 workshop report":p.16.

Simulation and design can be divided into six levels, although the key to success will lie as much in interactions among the levels as within each one. In addition, several technical thrusts are required to produce standards and frameworks to integrate the models across the various levels.

Figs. A15.5, A15.6, and A15.7 provide the technology and manufacturing roadmaps for simulation and modeling.

Devices and Circuits

The devices and circuits area includes the technology issues of both the stand-alone 1 Gb SRAM chip and a microprocessor using this generation of technology (0.12 μm lithography). While most of the technology elements or building blocks are generic and needed for both the SRAM and the processor chips, each separately drives additional technology elements. For the 1 Gb SRAM chip, additional critical elements are those required to achieve a very small memory cell area of about 0.5 μm². For the processors, up to six levels of high-density and low-resistance interconnect wires will be needed to connect the huge number of transistors (greater than 100 million) and to sustain the system clock rates of greater than 400 MHz.

With respect to design issues, the main finding is that an integrated system-on-a-chip approach will need to be followed. Many previously negligible effects, such as RC delays and testing, will have to be included as an integral part of circuit and chip design. Consequently, much more sophisticated and efficient computer-aided design (CAD) tools will be required.

Two technology development vehicles, or test chips, will be needed. The SRAM test chip will drive the development of all the high-density elements, including those specific to the 1 Gb SRAM cell. The logic test chip will drive the development of the

	Year Needed
Improved diffusion/implant models	1993
Definition of frameworks for technology CAD (tools and integration)	1993
Validation of dynamic hot-carrier device degradation model based on hot-carrier distributions using 64 Mb generation devices	1994
Equipment models necessary for design of 0.12 um generation tools	1994
Validation of dynamic circuit degradation models using 64 Mb/0.25 μm circuits	1995
Lithography/deposition/etching models for fully simulating 64 Mb/0.25 μm technology	1995
3-D grid, solver, and graphics simulation models	1995
Validation of energy transport models for overshoot in 0.12 μm FETs	1995
Designation of sensor placement in 0.12 μm etch/deposition systems for use in integrated manufacturing	1997
Accurate models for all anticipated processes	1997

Figure A15.5   Critical dates for technology models. Source: "Micro Tech 2000 workshop report": p.22.

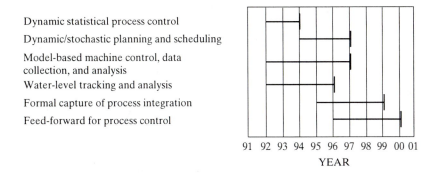

Dynamic statistical process control

Dynamic/stochastic planning and scheduling

Model-based machine control, data collection, and analysis

Water-level tracking and analysis

Formal capture of process integration

Feed-forward for process control

91 92 93 94 95 96 97 98 99 00 01

YEAR

Figure A15.6   Manufacturing model application development roadmap. Source: "Micro Tech 2000 workshop report": p.23.

multilevel planarized interconnect technology. A proposed technology development schedule and staging plan for the development of the major SRAM generations is shown in the overall Micro Tech 2000 roadmap in Figure 15.14

Architecture

The electronic systems environment of the year 2000 has been assessed to determine what will be required from end-of-the-decade semiconductor chips, and to forecast where semiconductor chips could stimulate significant, new product families. In particular, applications were examined in which a 1 Gb SRAM could stimulate enhanced or novel products, with an emphasis on products with the highest likelihood of a large market volume. Applications and system architectures, packaging and interconnect, and soft errors and data reliability are all important issues that have been addressed.

The dominant conclusion is that the availability of the technology used to build a 1 Gb SRAM and associated advanced microprocessors will provide a new era of commercial products, categorized as system-on-a-chip (SOC). For many years, high-performance silicon systems have been evolving to fewer, but larger, chips mounted

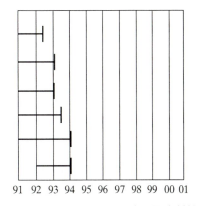

Select base technology for software environment

Selection of framework tools and adaptation for this application

Define standard semiconductor process representation

Standard, generic manufacturing model

Equipment communication in compliance with standards

Manufacturing specifications based on standard representations

91 92 93 94 95 96 97 98 99 00 01

Figure A15.7   Design and manufacturing systems integration roadmap. Source: "Micro Tech 2000 workshop report":p.23.

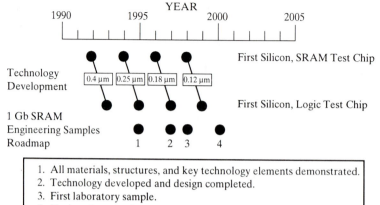

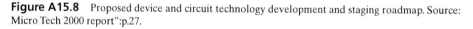

Figure A15.8 Proposed device and circuit technology development and staging roadmap. Source: Micro Tech 2000 report":p.27.

in ever-closer proximity on multichip substrates. The ability to integrate more than 6 billion transistors on a chip by the end of the decade would reduce many computer applications to a single chip.

The necessary technology adjuncts to a successful 1 Gb chip manufacturing technology include advanced display technology, low-cost packaging, adequate measures to deal with power dissipation, interface standards, and product standards.

Figure A15.09 lists a year 2000 memory chip specification in terms of pinouts, performance, and special features. Although high-density memories have traditionally been designed as stand-alone, commodity components, this memory is designed as if it were to be the on-chip memory fro a microprocessor application with a very wide, high-bit-rate interface.

The technology roadmap shown in Figure A15.10 describes the evolution of access times, burst clock rates, and corresponding microprocessor clock rates for each evolutionary generation of design rules. Stand-alone memories (the highest-density, largest-capacity memories without additional logic circuitry) are included in the Micro Tech 2000 roadmaps to provide a benchmark to measure progress against the commercial market.

Economics

The needs associated with successfully advancing the pace of U.S. semiconductor manufacturing capabilities that have not been addressed in the technology areas [are included in the economics considerations.] These include demonstrating semiconductor manufacturing competence, identifying potential markets for the Micro Tech 2000 technology, forecasting the costs of year 2000 manufacturing facilities, estimating the market share required to sustain the semiconductor industry in the United States, understanding and explaining the economic benefits of the Micro Tech 2000 technologies, and exploring mechanisms to fund the technology advances required, regardless of the source of funds. While all of the economics-related questions are important, the issues related to manufacturing require the largest accelerated effort in the development of the Micro Tech 2000 technologies.

Pinout		Performance	
72	Inputs (with 36-bit WE[†] option)	2.5 nsec	Access time
72	Outputs	1.2 GHz	Burst clock speed
24	Addresses	8 – 256	Bytes of user-controlled burst
8	Threshold adjust for input bytes	1 – 2 V/ns	Edge rates
4	Test (JTAG[*] standard)	0.5 – 1.5 V	Input/output (I/O) swings
1	Bus-burst clock		
3?	Data in/data out/address clocks		
n	Logic power supplies		
n	Output power supplies		

Features

- Processing compatibility for both memory and logic (processors)
- 6-level metal compatibility
- Pipelining
- Single port
- Built-in self-test processor
- Voltage/temperature/impedance/process compensation for I/Os
- Individual bit level timing and clocking
- Error checking
- Error correcting (in the processor?)
- Nonvolatile memory for reconfiguration and redundancy

[†]WE = Write Enable
[*]JTAG = Joint Test Action Group

Figure A15.9   System-defined 1 Gb SRAM chip specification. Source: Micro Tech 2000 report":p.30.

Year Required	1993	1996	1998	2000
SRAM				
Maximum density	16 Mb	64 Mb	256 Mb	1 Gb
Minimum design rules (μm)	0.35	0.25	0.18	0.12
Access time (ns)	10	6	4	2.5
Burst clock (MHz)	100	300	500	1200
System-on-a-Chip				
Minimum design rules (μm)		0.35	0.25	0.18
On-chip SRAM		8 MB	64 Mb	256 Mb
Common Elements				
Package — PGA	√			
— Flip chip or TAB		√	√	√
Built-in self-test		√	√	√
Error correction		?	√	√
Nonvolatile memory			√	√
Word length (bits)	72	72	72	72

Figure A15.10   Architecture roadmap. Source: Micro Tech 2000 report":p.31.

16
Managing The Technology

Continuous innovation occurs largely because a few key executives have a broad vision of what their organizations can accomplish for the world and lead their enterprises toward it. They appreciate the role of innovation in achieving their goals and consciously manage their concerns' value systems and atmospheres to support it . (Quinn,1985)

Having determined the needs of the corporation for technology, the task now is to create the environment, encourage the participants, and identify the resources that are needed to accomplish an effective management of the programs. Contrary to a common view that technology is primarily the responsibility of the R&D groups, it is essential that all parts of the organization accept some obligation for maintaining the technological viability of the company. As Marquis (1969) observed: "The management of innovation is a corporatewide task that is too important to be left to any one specialized functional department." Although some functions may not be involved in creating the technology, most will have a role to play in helping determine the appropriateness of the approaches being taken. Assuring the effective participation of all activities is a critical management objective. Although each group will approach its obligations in different ways, the objectives must be common-to ensure that technology is an effective tool in achieving a world-class competitive status.

16.1 INFLUENCE OF THE SIZE AND TYPE OF ENTERPRISE

Although the type and size of the enterprise are clearly not the sole determinants of how technology can best be managed, both of these have been found to have an important influence on the approaches that are generally followed. This is demonstrated by the contrasts that are offered in figure 16.1. While examples can be found of companies that do not fall neatly in these classifications, the normal behavior for companies is as shown.

A few comments are relevant regarding the concept of entrepreneurial. A company can be entrepreneurial whether it is large or small. The attitude of the company toward innovation, its willingness to welcome changes and to encourage a positive response to changes, and the willingness to encourage its employees to experiment and to challenge conventional ways of doing things are key characteristics of the entrepreneurial spirit. While the large company may find it somewhat harder to achieve and maintain these attitudes, size is not a good determinant as to which companies will follow this philosophy.

The way that a company includes technology in its strategy is a primary determinant of the approaches that it will take to developing and using technology. Companies that are using mature technology and are large may have a high inertia against change. Such companies are likely to limit their support of technology to developing ideas that support continuous improvement. If they are concerned about new technology they will frequently depend on sources outside their firm. Since technology will not be a key element in their competitive strategy, they are likely to be content with being technology followers. By contrast, the entrepreneurial company, many startup companies (not given a separate listing above), and many high-tech companies are so dependent on the availability of new technology that they are compelled to create and develop it actively. They invest heavily in the expectation that technology is key to their competitiveness.

The knowledge about new technologies flows throughout the technical community with remarkable speed. A product will be found in the marketplace that uses a new technology. A new discovery will be reported in a scientific journal or at a technical meeting. A new process utilizing a new concept will be offered by vendors for sale. The creation of

Some common characteristics of different size and types of companies.

Large National/Multinational Mature Companies	High-Tech Companies (large or small)
Experience difficulty in being entrpreneurial	May or may not be entrpreneurial
Multi-disciplinary teams may be difficult	Multi-disciplinary teams are frequent
Technology often focuses on development and continuous improvement	Technology derived internally or externally
Small Mature Companies	**Entrepreneurial Companies**
Being entrepreneurial is uncommon	Emphasize entrepreneurial approach
Small size fosters multidisciplinary teams	Multi-disciplinary teams are common
Technology often dirived externally	Technology often derived internally

Figure 16.1 High-tech refers to the level of technology that is commonly used in the company. It is possible for a company in any of the classes to also be considered as high-tech. Following the discussions of Drucker (1985),"an entrepreneur always searches for change, respondes to it, and exploits it as an opportunity.".

new technology can become known to an international competitor at the same time that a competitor "across town" knows of it. This has led to the notion that "technology knows no national boundaries." Some of the conceptual dimensions of the diffusion of technology are discussed in Appendix A16.1.

Since the means by which technology is managed are strongly dependent on the sources of the technology, it is important to understand the implications that arise from a focus on either internally or externally generated technology. Although it is recognized that the internal generation of technology cannot proceed without access to the external technical community, the organization and management of a system that depends heavily on external sources will be quite different from that of the system that is organized to develop much of its technology internally.

16.2   ORGANIZATION OF INTERNAL TECHNOLOGY FUNCTIONS

While the cost of staffing and funding internally supported R&D programs can be high, the benefits that the internal programs can offer can be great, particularly the capability to ensure close coupling of the programs with the needs of the company. Participants of internal programs can be expected to have an intimate understanding of the needs of the company. This can be critical in ensuring that the results have the greatest probability of being applied. Assuring that all the desirable programs are of critical size can be a problem, since it may be difficult to provide adequate resources to maintain all programs with expertise in all areas.

It is generally agreed that no specific structure is universally superior for organization of the R&D function. The philosophical orientation of the company, the company strategy regarding technology, and the personality of the principal participants must be considered in creating the most appropriate organization. A wide variety of organization can be found among companies. The following discussion summarizes the considerations that have led to the various choices.

16.2.1   Disciplinary versus Problem Focus

Many organizations use the disciplines that are a familiar part of each field as a basis for organization. This leads to the creation of departments of chemistry, electrical engineering, computer engineering, metallurgy, mechanical engineering, and so on. In contrast, the problem focus recognizes that problems are multifaceted and seldom are discipline-specific. This leads to a focus that emphasizes program areas.

Disciplinary focus.   The grouping together of people with common technical interests and backgrounds is the principal advantage of an organization that focuses on disciplines. Technological interests, often stemming from the educational system, defines the broad boundaries for each group. Professional journals, professional meetings, and professional societies are often focused on the disciplines, thus allowing the creation of groups with common purpose, uniform interests, and similar approaches to solving problems. People feel "comfortable" working in an environment in which others have similar interests.

The disadvantage of the disciplinary organization is that practical problems are seldom disciplinary and are best solved by bringing together teams of people with diverse backgrounds. The solution of most practical problems involves the use of information and methods that are not found within a single disciplinary category. Considerable effort is often necessary to couple the disciplinary groups into the problem-solving communities of the company.

Problem focus. Collecting the people that have the talents that are most likely needed to solve the problem can enhance the probability of achieving success in the shortest time. This focusing of attention onto the specific task can be effectively accomplished with a group that is problem focused. The disadvantage of a problem-focus organization can be one of lack of identity for the team members. If the problem is long term and will have many steps, the team may exist for a long time and achieve an identity of its own, as could be the case of an engine group in an automotive company-with the completion of one engine or one set of exploratory experiments, the company will need further developments or another new engine. If, however, the problem has limited life and the team is expected to disband after completion of the problem, the members must understand who will decide their next assignment, who is committed to assuring their professional health, and who is responsible for their further advancement.

Conclusion. In many organizations a combination of the these two forms of organization can be found. The longer-range research may be organized around disciplines and the shorter-range development tasks around problems. If the disciplinary focus in chosen, the challenge is to assure that the participants maintain proper awareness of the problems. If the problem focus is chosen, the challenge is to assure that the members of the problem team remain current in understanding the developments that are taking place throughout the world in the disciplines. If these two types of groupings exist within a single activity, a careful organization of seminars, program reviews, and technical discussions can assist in the transfer of information. It must be emphasized that management must make a serious effort to create the atmosphere that will stimulate these exchanges of information and provide mutual stimulation of the various groups. Accomplishing these exchanges among organizationally dispersed activities is more difficult and requires the attention of upper management to assure that the proper exchange of information takes place.

16.2.2   Centralized versus Dispersed Organization

In the centralized organization, the R&D function is localized into one, albeit perhaps large and multicomponent, activity. In contrast, the dispersed organization delegates to each activity the responsibility for the elements of R&D that directly affect its operation. Combinations of these two types of organizations are common, particularly with the centralization of research and the dispersal of development.

Centralized organization. A centralized R&D activity has the benefit of placing the primary technological responsibility with a single organization. Authority and responsibility for the activities are consolidated and the company management can look to a single organization to answer its concerns regarding technology. The synergism that is needed in the solving of technological problems is possible with the centralized organization. The management of the central organization can assure that the talent that is needed to solve a problem is available, whether that problem is long range or short range.

When it is recognized that the implementation and use of technology is a function that is shared by many people, it is often argued that the users need to control the R&D that supports their specific needs rather than having it centralized. When the "users" of the technology are separated from the "creators," an added burden is placed on the communication of ideas, results, and opportunities. In a large company, a disadvantage of a centralized activity is that the complexity of the technological tasks is often greater than can be accommodated by a single activity.

Dispersed organization The advantage of dispersing the R&D function throughout the organization is that the responsibility for exploring technology can be given to the organization that is going to use it. As such, the "creator" of the technology has an automatic ownership for what is created and is more likely to be prepared to utilize any developments that they make.

The disadvantage of the dispersed organization is the possible loss of synergism between the various elements of the R&D community. The dispersal of the R&D function often leads to a problem focus that is short term. Maintaining the proper overall view of the needs of the company can be difficult. This is particularly true for ensuring the maintenance of the proper level of long-range activities in organizations that are faced with achieving short-term profit objectives.

Organizational separation of long-range from short-range. In many instances, long-range is associated with research and short-range is associated with development. In achieving a separation between these two focuses, it is not uncommon for the long-range work to be centralized in a research activity with the short-range allowed to become problem focused in the operating groups. A combination of centralized and dispersed occurs with the centralization occurring only for the long-range.

The frequent conflict in the allocation of resources between long- and short-range programs, with the pressing needs of the short-range often leading to a loss of long-range effort, is generally offered as justification for centralizing the long-range efforts. Giving a single group the responsibility for the long-range technological needs of the company provides spokespersons that can help assure the maintenance of the proper balance between long-range and short-range. In addition, the coalescence of the long-range into a single activity enhances the probability that the research group will have the talent to attack the broad problems with multidisciplinary research teams. This allows the creation of groups that are large enough to have the necessary "critical mass."

It is also often easier to attract high-caliber research-oriented personnel to a centralized research unit than it is to dispersed units. Many examples can be found of people who joined their companies to carry on research but chose later to move into operations. The number of examples of current high-level industrial executives who followed this path testifies to the value of this as an effective means of high-level personnel recruitment.

The disadvantage of a centralized long-range activity is that the long-range needs of each operating activity must be determined and the central group must be convinced that it should work on these problems. The difficulties in creating the proper mechanisms for communication, priority setting, and the ultimate transfer of knowledge can seriously limit the successful use of those people who concentrated on long-range topics. When a central corporate research activity loses contact with the "real world" of its company, these are often some of the reasons.

Centralization of the long-range activity can make it much more difficult for the operating units to create teams that can address all elements of problems. This can deprive them of the capability to find imaginative solutions to short-term problems.

Conclusion   In large organizations with several unique operations, segregated either by product or geography, choosing to have a central research group and dispersed development groups has often been found to make it difficult to maintain good communications and, in turn, to identify the most important problems. To solve the communication problem among the groups, some companies have created liaison positions for research people. The liaison person is expected to spend time with the operations and provide a communication channel between research and development. Experience has been mixed with this approach, largely because it is easy for the research people to "let the liaison person do the work" rather than considering this to be a necessary task for all members of the technical community.

In some instances, the research management has insisted that members of the research groups dedicate a certain portion of their time to work with the development groups. Although this is almost always helpful, there is no assurance that the problems that will be chosen are the most critical to the organization. Many organizations have given the responsibility for priority setting, establishing the levels of effort, and assisting in communications to a small group, usually of staff. Such groups often fail to develop sufficient credibility and often find it difficult to develop consensus among those who have diverse fiscal and program authority, (e.g., the heads of profit centers). No easy solution has been found to these problems. If the individual managers have not embraced the system concept for the enterprise, almost all measures at correcting the problems will be met with difficulty.

16.2.3   Responsibility for the Technology Roadmap

The importance of the technology roadmap warrants special consideration. Although the creation, review, update, and critique of the roadmap must be shared among all of those who are concerned with the company's technology, it is clear that someone or some group must be responsible for seeing that the necessary review, update, and critique are accomplished in a timely and regular fashion. The challenge to the organization is to create an activity that has sufficient respect from the other elements of the organization while having sufficient authority to resolve the many conflicts that can arise during development of the roadmap.

The tendency is to create a small group isolated from the operating divisions, perhaps in a staff organization, who will be responsible for this. Although this action has many potential benefits, the result is often the creation of a group that is initially very effective but later degenerates into that of a bookkeeper with little authority for resolving important conflicts. The result is that the operations, whose concerns are dominated by short-term needs, frequently dominate the decisions.

In solving this problem, it is not uncommon to give a high-level executive the necessary authority to exercise the responsibility that will guarantee that all company interests are satisfied. This, however, may not completely eliminate the problem of the split of authority and responsibility. If the company is organized around profit centers (see Section 9.6), the heads of the centers will have the responsibility for the center's profit and loss (P&L) and will generally have final authority for allocating funds for all projects. The influence that can be exerted by an executive who does not also have ultimate responsibility for the P&L statement can be limited.

The complexity of resolving the issues of total company priorities, needs for the long-range versus, the short-range, the technology roadmap, and organizational problems suggests that the chief operating officer must take an active part in all final discussions that lead to the technology roadmap. In many large companies, it is only at the level of the COO that the authority and responsibility for expenditures, profits and the needs of the company in terms of personnel and technology coincide. As has been emphasized repeatedly, the role of the COO in a successful company must be with these broader issues and not with the detailed review of individual projects.

16.2.4   Consultants and Technology Advisory Committees

Consultants are often employed by companies to offer guidance and assist with developing solutions for particular problems. The problems that they address may be very detailed, as occur in specific technical areas, or they may be quite broad, as with the application of the techniques of total quality management (TQM) to the corporation. Consultants come from widely differing areas, including academia, retired industrialists, and consulting groups that have a recognized expertise in a specific area.

Many companies have also found it useful to create small external advisory groups that can serve as more general "sounding boards" regarding technical matters. These groups provide guidance on technology developments that are occurring outside the company, comment on trends that are occurring in the external environment and suggest how these may affect the company, and offer critiques of the directions of internal developments. While the organization and practice that these groups follow are nearly as varied as their numbers, it has generally been found that these committees are most effective when they are created as advisers to the upper management of the company. Reporting to the CEO and COO sends a strong signal of the importance that the management places on the activities of these groups.

16.2.5   Conclusions Regarding Internal Organization

As was noted earlier, no single organizational structure is uniquely best for all circumstances. All of the forms that have been discussed have their benefits and their disadvantages. The management group for a small company works together on many problems and can treat questions concerning technology in straightforward ways. Communicating needs and setting technical objectives does not require new methods. In a large company this is often not so easy. Potential conflicts among different people can present serious problems. These challenges may be made worse by the policy, justifiable in many instances, of rotating the assignment of members of the management, or because retirements or departures require the replacement of the people who helped make the earlier decisions. New people who have not participated in the creation of the roadmap may have different ideas concerning priorities that can drastically influence its composition. Resolving issues such as these requires the attention of the upper leadership of the company. As we continue to try to make technology one of the key ingredients of the competitive strategy for the organization, it is worth reiterating Marquis's (1969) admonition: "The management of innovation is a corporate wide task that is too important to be left to any one specialized functional department." To meet the needs of the company, it will be necessary for the upper leaders of the company to accept this as one of *their* principal responsibilities.

16.3   FOCUS ON EXTERNALLY GENERATED TECHNOLOGY

It should be recognized that external sources for technology have often not been given high priority by U.S. firms. Many companies are reluctant to use the results of developments that occur outside their control. This is sometimes referred to as the not-invented-here syndrome (NIH). As Rosenberg and Steinmueller (1988) observe:

> In retrospect it is obvious that there has been much that has been worth imitating [in some parts of the Japanese manufacturing system], but Americans, even when they have become aware of this, have been poor imitators. Rosenberg et al. go on to ask: "Why have the Japanese been so much better at imitation than the Americans? What has made the Japanese, if we may be permitted to use the phrase, such creative imitators?"

Rosenberg et al. proceed to offer several answers to these questions. They include the importance that the Japanese have placed on closely coupling design and manufacturing, the enhancement of manufacturing efficiencies, the creation of strong interactions with vendors, and the systematic use of engineering skills and production worker experience throughout the entire sequence of operations. We discussed these items in detail in earlier chapters and concluded that all are important in achieving world-class status, but it is still valid to ask why U.S. firms have continued to resist embracing the practices that have been demonstrated by the Japanese to be so important in their success.

Perhaps some rather different reasons are responsible for this difference in behavior. The first might be termed technological arrogance. The United States dominated the world technological landscape for at least four decades after World War II. Few of the managers who occupied responsible positions in the early 1980s had experiences in which the United States was not the leader. It was hard for a manager to accept the idea that someone could be so successful. A second reason for this reluctance to adopt the successes of others may reside with the lack of experience that many U.S. manufacturers had with competing in the world marketplace. With the development of world markets, there was a slow recognition that those with the best practices anywhere in the world will ultimately garner the market. In addition, the size of the U.S. market is so large that U.S. manufacturers have not had to commit themselves to exporting goods in order to flourish.

The competitive value of using internally or externally generated technology has been studied by Mansfield (1988) for Japanese and U.S. firms. Some of the interesting results of this study are summarized as follows:

- The Japanese seem to have cost and time advantages over U.S. firms. However, these advantages seem to be confined to innovations based on external technology.
- Among innovations based on internal technology, there seems to be no significant differences in average cost or time between Japan and the United States.
- American firms take almost as long, and spend almost as much money, to carry out an innovation based on external technology as one based on internal technology.
- The relatively higher commercialization cost for innovations based on external technology in the United States than in Japan seems to have been due in part to the fact that the Japanese, in carrying out such innovations, have been more likely that the Americans to make significant technical adaptations of the imitated product and/or to reduce its production costs substantially. The Americans have been more inclined that the Japanese to invest heavily in marketing startup costs in an effort to position such innovations optimally in the market, the emphasis being more on marketing strategies than on technical performance and production costs.

Advantage of using externally generated technology. Externally generated technology can arise from a number of sources. A company must be prepared to investigate the technical developments that are made externally and to determine those that will be useful in achieving their objectives. Having determined that a technology that is available outside the company is attractive, there are a number of ways in which it can be used. It can be developed internally, its development can be contracted with an outside source, or it can be done jointly, as with a joint venture. The advantage of contracting with external groups to accomplish the further development centers on the flexibility that this allows the company. It is possible, in principle, to enter into agreements with external groups that have the greatest expertise in any particular area rather than to be forced to depend on the expertise or to develop a particular expertise in an internal group.

Disadvantage of using externally generated technology. Principal disadvantages of using external groups for the performance of R&D is the difficulty of finding groups who have the requisite skills that also have a good understanding of the needs of the company. Just identifying an interesting external technology is not sufficient. Since it is often difficult for an external group to meet the strict time requirements that are imposed by the cycle plan on developments, it is not uncommon for the focus of external efforts to be limited to the longer term.

16.3.1   Organizations that Generate Technology

While the external sources of technology are virtually unlimited in numbers, it is useful to consider the classes of organizations that are the principal providers of technology. These can be loosely divided in the following categories:

Noncompetitor companies. As discussed in Chapter 15, technology of value to a company often arises from other industries. It is important to develop a means for understanding and evaluating the technological developments that are under way throughout industry. Within this category, it is appropriate to include the many vendors that provide technology through the products that they sell to their customers. This may include manufacturing processes or total systems, components and parts, or know-how in the use of new technology.

Competitor companies. The developments that are made by competitors should be considered as a genuine source of new technology. As has been emphasized earlier, with the sale of the first new product a company that aspires to be world-class can be expected to purchase and study it for its innovations. If some of the innovations are protected by patents, it is importnt to determine whenther rights to use the developments can or should be obtained. This will be discussed more fully later.

Consultant companies. Examples of companies that are created to serve as consultants to industry, to explore new areas, and to answer specific question, include Ricardo Inc., Arthur D. Little, Batelle Laboratories, and Arthur Andersen Consulting. These often specialize in specific areas of technology.

Not-for-profit laboratories. Examples of not-for-profit laboratories include the Batelle Laboratories, Southwest Research Laboratories, Midwest Research Laboratories, and David Sarnoff Laboratories. These laboratories and others are organized to do research under contract for enterprises that need their expertise.

Government laboratories Government laboratories are increasingly willing to share their expertise with industry and to work with industry in the solving of particular problems. Special provisions have been developed to provide protection for proprietary information and to encourage joint programs through the recently created Cooperative Research and Development Agreements (CRADAs).

Research universities. The National Science Board(1993) classification of schools of higher learning lists 105 schools in the category of research universities, meaning that they receive more than $12.5 million annually in federally sponsored research and award at least 50 PhDs each year. This group of universities do most of the basic research noted in Fig. 14.7. The long-range nature of this research generates many of the ideas that later are incorporated into the revolutionary innovations that regenerate industrial practice and products.

16.3.2   Collaborative Arrangements for Acquiring Technology

Industry-university collaboration. Collaboration between industry and the universities has been long-standing. In the early days of the development of the chemical industry this close collaboration provided many of the early ideas that resulted in successful innovations, with industry providing much of the support for the university research. Following World War II, the universities did not foster the interactions with industry, favoring instead the monies that the federal government made available for academic research. The inevitable result was a reduction in cooperation between the two. The situation is now changing with the realization that each has something unique to offer the other.

Rather than supporting basic research within their laboratories, industry has frequently chosen to participate in the basic research being carried on in research universities. Having already noted that the universities are the principal performers of basic research, the creation of partnerships with universities offers the industry a way to participate without taking the more drastic step of creating in-house basic research groups. The contribution of industry to universities takes many forms, including:

- Partial financial assistance for individual research projects
- Partial support of the research programs of individual faculty
- Partial sponsorship of major programs or research centers at the university
- Providing personnel to work in the university on programs
- Scholarships and fellowships for students
- Participation on planning and review committees of research projects
- Creation of major research programs with the provision of a major portion of funds

Mansfield (1991) has attempted to quantify the extent to which industrial technological innovation has resulted from recently completed academic research. The results can be summarized by the following quotation.

> Our findings suggest that about one-tenth of the new products and processes commercialized during 1975-85 in the information processing, electrical equipment, chemicals, instruments, drugs, metals, and oil industries could not have been developed [without substantial delay] without recent academic research. The average time lag between the conclusion of the relevant academic research and the first commercial introduction of the innovations based on this research was about 7 years [and tended to be longer for large firms than for small ones.

Advantages. For industry, participation with the research programs allows the identification of desirable future employees to be made prior to the completion of their education, thus enhancing the probability of attracting highly qualified graduates to the industry. It also offers the industry an insight into the research that is at the leading edge thus providing them a window into new or advanced technical arenas at a modest commitment of time and cost. Supporting the research of prominent faculty helps develop a cadre of people that can serve as advisers to industry when they need assistance. For the university, the involvement of industry in the research programs brings additional financial support, direct contact with practitioners who can help assure the relevance of the work to the real world, and acquaintance with potential employers of their students.

Disadvantages. The involvement of industry with the universities may lead to a stronger focus in the university on short-term applied research. The industry commitment to the universities may not be long-term, thus producing rapid fluctuations in support levels. Depending on the degree to which a program is supported by industry, the question of the ownership of any new intellectual property that arises in the course of the work must be addressed. For industry, the time and cost of the support for the university may become a burden and be hard to maintain.

Conclusions Despite these issues, there is an increasing tendency for industry to join with academic institutions to explore areas that are of mutual interest. While the financial contribution from industry is welcomed by universities, it is rare for the size of this to be large. The principal benefits to the university are more indirect. Also, since the absolute financial level of support is not large, it appears unlikely that a significant diminution of support for the universities by the federal government will be compensated for by industry. It is likely that industry will be more effective by being a strong advocate for the university research program than by attempting to be the principal source of funds for its programs.

Joint ventures. A joint venture refers to a legal structure that provides for sharing results and sharing costs among the participants. They may take a variety of forms. In one form, two companies will create a third company that is owned jointly. Ownership can be equal or unequal among the originators. Dow-Corning was created by Dow Chemical Co. and Corning Glass Works. In other cases, the joint venture may simply be a project that is jointly shared between two or more companies with the work being done at one or the other companies.

The stimulus for developing joint arrangements can be quite varied. They may be created to share technologies that both have developed independently. They may be created to generate a product for the industry of one of the partners using technology that was developed by the other, thus allowing both to share in the benefits of the development. If neither partner is willing to undertake the project alone, the joint venture may be created to allow a sharing of costs and benefits.

Although this discussion has focused on technological and financial incentives, there are also legal reasons for considering a joint venture. A joint venture involving the creation of a separate company can be constructed in such a way that the originating companies can share financial benefits without being legally responsible for the actions taken by the new company, thus limiting the liability of the originators. To accomplish this, the joint venture company must be managed as a separate independent entity without direct control being exercised by either of the originators. The benefit of this can be seen in such cases as in the breast implant verdict, where the settlement costs were largely borne by Dow-Corning and not by the originating companies.

In a sense, joint developments that are undertaken by vendors and their customers represent joint ventures. The vendor agrees to undertake a certain development using their expertise and technology with the understanding that a successful product, in terms of technical objectives and cost, will be purchased by the customer. While the vendors can be considered to be external, the use of vendors who are "members of the corporate family"-analogous to the Japanese keiretsu members-makes them nearly internal. In these circumstances the vendor will be involved in planning the new products, considering the levels of technology to be used, and developing the final product. When a long-term relationship exists between the company and specific vendors, it is often appropriate to treat them as a part of the internal development group.

The role of the vendors in joint developments can be quite extensive. In most cases the customer will require that the vendor provide some assurance of a product's quality. This may extend to requiring that the vendor agree to absorb any costs that arise from any failure of the part that they supplied. Obviously, the extent of the vendor's liability will be different for a component that is included directly in the final product from that of a vendor's material that is modified by the final manufacturer.

16.3.3   Information Sources of Technology

Having timely access to technical information is critical for any organization. Traditional sources, such as the analysis of competitive products, attendance at technical meetings and trade fairs, and publications, including technical journals, patents, and trade magazines, continue to be important. As *timely* sources for information, many of these suffer from the time delays that are involved between the creation of information and its formal dissemination. Delays between the submission of a technical paper for publication and its subsequent appearance in a journal may be several months. This places a high priority on the acquisition of information by more timely means.

This need for timely information about developments in the technical community is given as one of the principal justification for the maintenance of a research organization in a company. Active involvement in a technical area generally means that the participants will be in communication with contemporaries having similar interests, either directly or during attendance at technical meetings. These people provide the *eyes and ears* to the technical community. While association with a research organization is not a prerequisite for discussing technology with someone, it must be recognized that people will be willing to discuss their accomplishments with you only if they can also learn something about your accomplishments. This means that both must be able to make contributions to the discussion and that each has something to contribute to an understanding of the topic.

The evolution of computer networks and large computer-based data bases provides a rich source of information. As important as it is, a word of caution is needed concerning their use. Since it is a rare circumstance that the accuracy of all entries in a data base is ensured by a responsible party, it becomes critical that the user understand the limitations of the data entries or have a good knowledge of how they were obtained. Entries in nonverified data bases should be used with great care.

16.3.4   Organizing to Take Advantage of External Technology

The systematic examination of external technologies and the identification of those that have potential value for your enterprise cannot be treated as a casual assignment. If this is to be pursued successfully, the responsibility must be clearly assigned to some group or organization. The task of examining technologies that offer all new opportunities and of completing matrix III of Chapter 15 is frequently given to the research community within the enterprise, since it is this group that is usually in communication with those outside the enterprise who are engaged in long-range studies. For the examination of technologies that support the cycle plan-completion of matrices I and II of Chapter 15-the task will often be given to those who are using currently available technology. For technologies that support the development of new engineering and system tools-matrix IV-the responsibilities may be shared between the research activities and the operating activities. To accomplish the objective of being the *eyes and ears* of the enterprise to the external technical world, it is essential that the internal group, whoever they are, possess people with broad technical expertise who intimately understand the goals of the enterprise *and* the problems that are being experienced in reaching those goals. Without this dual capability and sensitivity, it is unlikely that they will be able to identify the technical opportunities that are worthy of further consideration by the company.

16.4   PERSONNEL NEEDS FOR R&D PROGRAMS

While organization and sensitivity to the sources of technology are important, an R&D activity can achieve success only if it is staffed with people having high technical competence, strong motivation, and an eagerness to search for new and imaginative solutions. As Mayo, 1993, notes (see Appendix A16.3): "Above all, a first-rate R&D community must be characterized by the excellence of people who are unmatched in their capabilities and focused on meaningful problems. Extraordinary skills must deliver extraordinary output. But individual skills are not sufficient. There also must be a high quality of teamwork."

The importance of recruiting people with the proper talents who are motivated and perform well in the type of structure that is best for meeting the objectives of the activity cannot be emphasized too strongly. Many factors enter into a person's decision to join a company, not least of which is the intellectual environment that is created by the company. Quinn (1985:266) offers the following observation by Gordon Moore, Intel's chairman, on this topic: "Young people attracted by Intel's goals and style responded, 'It's great to say you work at Intel. You know you're with the best...There's a real pride in being first, in being on the frontier. You know you are part of something very big-very important.'"

Some people perform best and are happiest in a very structured activity. Such people tend not be as comfortable in an activity that requires the exercise of a lot of individual initiative. A person who is innovative and prefers a less structured environment will tend to flourish better in a research environment. Managing these types of people requires some special sensitivity. For example, it will be very counter productive to recruit a cadre of highly innovative, highly technical people to do basic research and then attempt to set specific objectives for each research task. If people are recruited who are sufficiently good to perform basic research, they will not need a manager to tell them what to do or a manager who expects them to commit to a rigid timetable for meeting specific accomplishments. For such a group the manager needs to ensure that the areas of research that are undertaken are compatible with the goals of the organization, that the goals and objectives of the organization are clearly understood by the members of the activity, and that proper attention is given to the personal development of the individual research workers.

Similarly, a development team will not perform properly if the manager of the activity lacks the discipline of meeting schedules, sets unrealistic goals, and does not provide the resources that are needed to accomplish tasks. Since most development activities involve team participation, the ability to create an environment that is conducive to good team performance, that offers freedom of choice to the team while encouraging the meeting of specific goals, and of helping the team to interface properly to many other activities are important criteria for the successful manager of development. The reader is reminded of Question 8.6 for an earlier discussion of these issues.

16.4.1 The Needs of Various R&D Activities

Research. A research group requires people who are motivated to achieve excellence in their technical fields. They must be challenged by unanswered problems and by the opportunities to delve into the unknown. They will seek recognition by their peers in the technical fraternity, both inside and outside the company. Recognition by people outside the organization will be achieved through the publication of technical papers, attendance at technical meetings, and involvement in professional societies. As experts in their fields, these employees will expect to be given a large degree of freedom to choose the approach that is taken in solving problems. In turn, the activity must expect that these people will be sensitive to the needs of the company and be willing to address problem areas that determine the company's competitive position. As experts in their fields, many of the members of basic research groups will have advanced degrees.

Development. With requirements for meeting specific dates for completion of projects, for meeting engineering objectives, and achieving cost objectives, development groups are more structured than are research groups. People who are able to remain innovative while searching for solutions to carefully structured problems can flourish in these environments. Technical expertise is no less important for the development group than it is for the research group. The distinction is that the development groups will usually have less freedom in choosing their own problems and must be willing to work toward specific objectives that are set by management.

Operations. Operations include such areas as providing services to both external and internal customers and the operation of the facilities, including services, testing, and manufacturing. Maintaining good relations with customers, achieving quality and cost objectives, providing responsive services, and achieving high levels of productivity are essential elements of any operation. Since some of the employees in operations may be skilled tradespeople, unskilled workers, members of various ethnic groups, and many may be members of unions, technical personnel in operations must be sensitive to the needs and requirements of these groups of employees. Motivating and encouraging workers is a key aspect of operations where continuous improvement is a constant objective.

Staffs. Staffs are usually composed of highly talented people, many of whom are specialists. The staffs must maintain a sensitivity to the needs of other groups and consider all activities in the corporation as their customers. While staffs are frequently given audit responsibilities, these must not be performed in such a manner that an atmosphere of "us versus them" is created. The old description of a staff member as "someone who enters the battlefield after the war and spears the wounded" must never be allowed. Successful staff personnel require excellent technical skills as well as excellent interpersonal skills.

16.4.2   Recruitment of Personnel

Since recruitment of key personnel is of such vital importance to the success of an enterprise, the successful manager will find that it is important to devote a large amount of time to this function. Although the personnel or human resources department can offer a great deal of assistance in organizing recruitment activities, the responsibility for recruitment should never be delegated to someone who does not have a detailed knowledge of the technical needs of the organization or of the position. To paraphrase the comment by Marquis (1969), the task of recruiting highly talented people is too important to be left to any one specialized functional department. Although it may be argued that the pool of candidates is so large that special efforts are not needed, it must be remembered that the fraction of any pool that is outstanding will always be small and that the competition for these will always be large. Special efforts are needed to identify and attract the very best people. This is especially true of underrepresented groups, where the pool is almost certainly small, with the number of those who are outstanding representing only a handful.

Research.   Recruitment for research activities is difficult, not because there are few candidates, but because it is hard to find people who have the requisite combination of talents and sensitivities to the needs of the company. It is usually easy to find people who have the requisite training to solve a specific problem. Far fewer people have more general interests. To find people who are self-motivated, who have the skills and interest in attacking various problems, and who are willing to take the time to understand the needs of the company is difficult. While seminars and interviews can yield information about the technical competence of individuals, these give little information about the other attributes. Special attention must be given during the interview process to the question of the philosophic approach of the candidate to these broader matters. The expectations that management has of people in the Xerox research activity are elegantly summarized in Appendix A16.2.

As noted in Section 16.3.2, collaboration between universities and industry provides a means of helping industry make an early identification of promising candidates for recruitment. The regular visits of key industrial personnel to the campus to discuss research, offer some assistance, and express a genuine interest in the progress being made creates an atmosphere that encourages students to consider seriously any offer from these firms.

Development.   Recruitment for development activities should guard against undue emphasis on the specifics of the present problems. Future critical problems may be widely different from those of today and may demand that people take very different approaches. This suggests that people should be sought who are challenged by many different problems, are able to use a variety of tools in solving these problems, and are willing to adapt to new circumstances. Since most development activities involve teams of people, special attention must be given during the interview process to the question of the candidates' willingness and comfort with being members of a team.

The need to identify candidates who are outstanding and who will flourish in a development activity is no more easy than it is for research. As noted above, collaboration between industry and universities is an excellent way of assisting in accomplishing this. One of the most successful collaborative programs is the cooperative eduction program. Students spend alternate semesters working in industry and in completing formal course work. Although neither the industry nor the student will be making an early commitment regarding employment, both have an opportunity to learn and to make a more reasoned decision. An industry that recognizes the importance of this as a possible recruitment tool will make every effort to ensure that the work assignments that are given to students allows them to grow and mature in the company.

Operations. Recruitment for operations activities can be segregated into those areas that are largely repetitive and require a minimum of technical expertise from those that require technical expertise. For the former, the task is to find people who are competent, who are willing to learn and who are willing to work in teams. For the latter, successful technical personnel will find it challenging to identify and solve the problems that interfere with achieving a highly efficient system. The former remarks concerning the Cooperative Education Program applies to the operation activities as well as to the development activities.

Staffs. Recruitment for staff operations may occur from outside the company or from within the ranks of the corporation. If recruited from outside, efforts must be made to determine a candidates' sensitivity toward other people. This can be determined more easily for people who are already employed in the company. A key criteria for staff personnel should be that anyone that is recruited will be so talented that they will be sought by other activities in the corporation. In this way, the staff will maintain outstanding technical credibility and will be able to constantly change its membership. It is critical that a staff not become known as "the graveyard for underachievers."

16.4.3   Styles in Management

The successful manager of technology programs recognizes that the style of management must be adjusted to accommodate differences in the objectives of the programs. Much of the style that successful management requires is determined by the operating objectives that are characteristic of the activity. The more specific the objectives are, the more specific that the job assignments and reporting must be. As mentioned above, the management of a long-range basic research program demands quite a different approach than that required to introduce a near-term product change. Similarly, the more closely a task is tied to the cycle plan, the more important it is to meet specific objectives such as completion dates, costs, and engineering objectives, which, in turn, generally require more explicit assignment of tasks to the members of the activity. While the longer range tasks often have fewer specific time requirements, they demand that members of the activity exercise more individual initiative. This leads to a greater dependence on individual initiative and the need to encourage innovative solutions.

Figure 16.3 summarizes the objectives and the styles that are most commonly followed in meeting the objectives of various activities (see Kusch, 1966). The objectives of each type of technical activity is summarized in the left-hand column of the

OBJECTIVES

MANAGEMENT STYLE

Operating Activity

• Provide products, processes and services that meet the needs of the customer
• Meet cost and timing objectives for all programs
• Utilize resources in the most cost effective manner

• Insist on adherence to schedules
• Require frequent reporting of progress
• Require careful accounting of resources by all members of the activity
• Provide specific responsibilities to the whole group

Staff Activity

• Support the operation activities and management with services such as evaluations, reviews, audits,alternative plans

•Recruit members who combine technical compe tence with traits of objec tivity, tactfulness, and self-confidence
• Emphasize responsiveness to the user of services
• Encourage cooperation and non-competitiveness with the users

Development Activity

• Develop product, process and service that meet engineering and cost targets in a timely fashion

• Achieve program objectives including engineering specifications, cost and quality
• Encourage team participation
• Recruit members who have excellent technical capabilities

Applied Research Activity

• Provide eyes and ears to the technical world
• Provide in-house consulting services
• Investigate new technologies
• Transfer technologies to other activities

• Identify objectives
• Encourage innovative approaches
• Recruit those with technical excellence
• Encourage team participation

Basic Research Activity

• Provide eyes and ears to the technical world
• Provide in-house consulting services
• Perform innovative research
• Assess technologies
• Transfer technologies to other activities

• Encourage individual initiative
• Recruit those with technical excellence
• Develop innovative ways to recognize individual contributions
• Encourage team participation where appropriate

Management

• Set and maintain realistic goals and objectives
• Provide leadership in seeking imaginative solutions
• Provide timely and considered decisions
• Secure needed resources
• Provide good communications to all levels

• Encourage experimentation
• Encourage innovation
• Reward excellence
• Recognize team and individual contributions

Figure16.2   Styles in Management

table. The right-hand column gives the principal distinctions in *style* that management must follow if it is to be successful. The differences among these styles is a major reason that it is difficult for someone who has acquired much of their experience in one area to become a successful manager in one of the other areas. The talents and skills that are required of each area are quite different.

16.4.4   Training and Continuing Education

Training and continuing education are distinguished in this discussion in terms of the time objectives for these two activities. By training, it is implied that an immediate task requires that some new skills be acquired. By continuing education, it is expected that this forms a longer-term objective and is not focused entirely on a given problem. As Mayo notes in Appendix A16.3, technical personnel must accept that continuing education is a vital a part of long-term development and that people must be encouraged and rewarded for expanding their knowledge bases.

Industry provides a plethora of opportunities for training and continuing education. Eurich (1985) estimated that U.S. industry is annually investing more on continuing education than is being spent on all forms of formal higher eduction, with estimates of the former being as large as $60 billion in the early 1980s. These include courses taught by visiting college faculty, degree programs offered by universities and by a few selected in-house faculties that are accredited to confer degrees, courses offered remotely by colleges and universities using television, selected courses that focus on areas of current importance to the industry (e.g., courses on quality, TQM, or productivity), and remedial courses that are intended to provide personnel with a minimum of basic skills in reading, mathematics, and financial practice. Nancy Badore gives an example of an executive training activity at Ford Motor Co. (Appendix A6.1) that is intended to enhance a common sense of corporate mission and a common understanding of corporate goals and approaches among its worldwide managers. Practice varies widely in terms of financial support for employees who enroll in these courses. In some cases the company will pay all costs, including allowing the courses to be taken during normal working hours. In some cases the courses are offered partly on company time and partly on the employee's time. In some cases the employee underwrites the cost of the courses.

Although the emphasis here has been on courses that provide continuing education, with much of the motivation for participation being provided by the employee, there are times when a company finds it necessary to upgrade the education of a large segment of its employees. When the technology that a company is using changes abruptly, it is often necessary to undertake an extensive training/education program. Ford Motor Co. experienced this in the early 1980s when the introduction of electronic controllers for their engines required that many of their mechanical engineers needed an improved understanding of electronics. Ford and Wayne State University organized a set of courses, some taught at the company and some at the campus, to introduce and expand the knowledge of mechanical engineers with this technology. An advanced degree was given those engineers who completed these courses successfully.

Personnel

Number of People by Classification
Actual Salary of Each Person or Average by Classification
Fringe Benefits of Each Person or Average by Classification
Expected Overtime Charges

Services

Analytical Services
Test Services
Design Services
Fabrication Services
Programming Services
Legal Services
Custodial and Security Services

Capital Facilities

Computers
Analytical Facilities
Test Facilities
Modernization of Current Facilities
Depreciation on Facilities

Materials

Prototype Components and Systems
Disposable Parts and Components
Operating Materials

Figure 16.3   Elements that must be considered in creating a program budget.

16.5   BUDGETS FOR R&D PROGRAMS

As was mentioned in Section 14.5.2, there is no set percentage that can be used to determine the size of the R&D budget that should be made available by the company. Although each sector seems to approach a norm for this percentage, it is not reasonable to assume that this represents the optimal amount of funding for a particular company.

The dominant guidelines for the determination of the R&D budget must be the needs as given in the technology roadmap and the financial condition of the company. Since the technology roadmap reflects the long- and short-term needs of the company, it is critical that it be adjusted during its creation to reflect the total resources that are available. If there are insufficient funds to support the near-term needs as reflected in the cycle plan, the cycle plan must be modified. If the short-term needs of the company are so great that no funds remain for exploration of the long-term needs or for developing new engineering and system tools, it is critical that the risk of this be assessed carefully and that the roadmap be corrected as appropriate. Although it is critical that the R&D community strive to be as efficient and productive as possible, it is imperative that the company not attempt to develop technology without allocating the necessary resources. Technology is not developed by *wishful thinking* or by means sometimes described as *using smoke and mirrors*. A serious effort must be made to make the technology roadmap consistent with the availability of resources-as is sometimes said, *the words must match the music*. If not, a change in one or both must be made.

16.5.1 Base Budgets

Annual budgets for the R&D activities are determined by the programs contained in the technology roadmap. The cost of each program must be determined by careful assessment of the costs of personnel, services, capital facilities, including the costs associated with the support of the facilities, and materials. Representative elements that make up these individual costs are shown in Figure 16.3.

Using the estimate of the annual cost of each individual project the cost of all projects will usually be assembled by each activity (e.g., development, research, etc.). The capability of the organizations to absorb this level of effort must now be assessed. Questions that might be asked include: Are there sufficient people with the necessary talents available to handle all projects? If not, can someone be contracted to do some or part of the projects? Are new facilities needed in a time frame that allows them to be procured? If not, can facilities be leased or borrowed? Will test facilities be available for timely evaluation of designs?

All too often, the basis for making a decision regarding budgets focuses on the preceding year's budget. Although it is true that constraints exist on the growth or shrinkage that an activity can absorb in one year-for example, the number of high quality people that can be hired or the impact of shrinking an activity-the budget of the preceding year should not be used as a firm yardstick of the budget that is needed for the current year. The needs that are described by the roadmap should be the first basis for determining the needed resources.

Including the costs of personnel in the annual budget process emphasizes a very interesting difference in the approach taken by U.S. and Japanese firms. In Japan, with its top tier of companies following the practice of lifetime employment, personnel costs are considered as *fixed*, almost like the cost of maintaining a facility. In the United States, personnel costs are considered to be *variable*, with the number of employees being allowed to rise and fall according to economic conditions. It appears that there is movement in both countries toward a more common approach. Japanese companies are recognizing that operating costs must be reduced during a severe recession and that this may require a reduction in the number of employees. In the United States, industry is recognizing the cost of disruption that is created by frequent increases and decreases in the levels of employment. Effort are being made to stabilize the levels of employment and to make fewer changes except under severe circumstances.

As noted previously, a frequent source of disagreement is over the expenditure of funds for long-term investigations. It is essential that responsible choices be made between the immediate needs and those of the long term. It is frequently impossible to do this in a completely quantitative way. How can the value of exploring a new technology be estimated in its earliest stages? What will be the value of a new engineering tool that reduces the time to make a prototype by one-half? On the other hand, every company experiences limitations on the funds that it has available. Sometimes it is possible to enhance the effort needed to solve a short-term problem by delaying a long-term investigation. Often it is possible to reprogram a combination of programs to allow a better phasing of efforts. These examples are given as illustration of the complexity of creating a final set of budgets that recognizes the limitations of each group, realistically supports the roadmap, and reflects the overall spending limitations that the company has established. In all of these considerations, it is important to recognize that the technical community is most productive if it is able to operate in an environment that is stable and predictable. Nothing is more destructive of morale than a constant change of objectives, rearrangement of programs, and reallocation of efforts.

16.5.2   Basic Research in Industry

Industry is frequently criticized for its lack of funding of basic research. The reason for the small number of companies that support basic research in their own laboratories can be understood from the following semiquantitative argument. Basic research seldom flourishes in an environment that has a single individual researcher. There is a *critical mass* of people needed to create the intellectual environment that is stimulating to the basic research enterprise. For each area of basic research that an enterprise supports, it is reasonable to expect that the critical size will require at least three or four creative persons in the group. The annual cost of maintaining a single senior person in basic research is approximately $400,000 for salary, benefits, indirect costs, equipment, supplies, and technical support. The annual cost for a basic research group focused on a single area would therefore approach $1.5 million. Seldom is it practical to work in a single area, again because a single area does not exist in isolation and interactions among individuals is essential to ensure proper synergism. Thus the minimum annual expenditure for a basic research group would be expected to be roughly $5 million. It is generally true that a company that engages in basic research will expect to spend 10 times as much on applied research as basic research and roughly another 10 times as much on development as on applied research. Thus the R&D budget for a company willing to undertake basic research in two to three areas would probably approach $500 million. R&D as a percent of sales for the top eight sectors listed in Section 14.5.2 is 5.2 as compared to an all-industry average of 3.7. Using a value of 4 percent for R&D expenditures as a percent of sales, the sales for an enterprise supporting R&D budget of $500 million would need to be approximately $12.5 billion. Of the nearly 900 companies surveyed in Table 14.1, only 36 had sales of $12.5 billion or more. Of these, it is estimated that well under half support basic research in their laboratories. Although it may be argued that the cost of doing basic research is not large, and therefore industry should support it, the practical fact is that the number of companies that are of such a size that they can sensibly sustain a basic research effort at a reasonable level is small.

16.6   MEASURES OF EFFECTIVENESS OF TECHNOLOGY PROGRAMS

The ease with which the effectiveness of various technology programs can be assessed is related directly to the capability to assign a financial value to the programs. When a financial value can be assigned to a program, its cost-to-benefit ratio can be calculated in a straightforward fashion. The greater this ratio, the less attractive the program. In general, a cost-to-benefit analysis is done for programs that can be assessed in terms of their sales, productivity, or profitability. Even this is not as straightforward as it might appear. Using a previous example, it is extremely difficult to assess accurately the cost-to-benefit ratio of adding an additional product option to an existing product offering. Not only is the impact-and thus the cost-to the engineering, manufacturing, and sales groups difficult to calculate, the actual impact on the market of the new offering is difficult to estimate. In the final analysis, the value that is determined for the cost-to-benefit ratio is dominated by the assumptions made regarding these various factors.

Other measures of effectiveness are even less easy to translate into a well-defined financial value. While many of the metrics discussed in Chapter 4 are essential for improving efficiency, increasing customer satisfaction, and ensuring a competitive status for the enterprise, it is difficult to determine a quantitative financial value for these. It is even more difficult to calculate the value of exploring a new technology that may present uncertain opportunities to the enterprise.

Since these difficulties do not exonerate the technical community from its responsibility of making and justifying the best choices among the many interesting program possibilities, it is necessary to find alternatives to the usual financial measures that are commonly used. In attempting to find other ways of evaluating programs that generate credible results, it is useful to attempt to answer the following questions.

- *What is the likely cost of not pursuing a particular program?* Examination of this question tends to force a longer-term assessment than occurs during the usual cost-to-benefit analysis. This question appears to be asked more often by enterprises in Japan than in the United States, with the result that investments are made on longer-term programs than would be likely to result with a normal analysis.

- *Is the program directed at solving a significant bottleneck that exists in one of the operations?* Although the complexity of the operations may make it difficult to directly quantify the probable financial benefit of accomplishing this, the value of breaking a significant bottleneck can often be very large.

- *Is it premature to cancel an ongoing program? Is additional study needed to complete the study?* Although it is not productive to stop programs and start new ones simply because the alternatives look more interesting, management must be willing to stop investigations that are unsuccessful either because they have encountered serious obstacles or because they have proven to be less attractive than first believed. This question is intended to concentrate attention on the progress made with past programs and to encourage careful assessment of the need to continue them.

- *If the program is successful, is the probable value of the outcome sufficiently large to warrant the time and money spent on accomplishing it?* Although this borders on a cost-to-benefit analysis, this question anticipates that a less precise answer will result. Many examples exist of efforts being undertaken to solve problems that are not sufficiently costly to warrant the effort. In such cases it is not the cost-to-benefit ratio that should have signaled that the program not be done but the fact that the efforts of the people could be better spent on other more important problems. This question encourages attention to the issues of the best way to use the organization's talent.

- *Which programs represent the most innovative use of the available talent?* This question, as above, focuses on people rather than finances and is intended to encourage a prioritization of programs based on the talent that is available. Recognizing that people are the most important asset for a company, it is critical to use them in the most effective way.

It must be recognized, however, that these approaches are not adequate for the evaluation of several of the technology programs on the roadmap, particularly those programs that are identified through the completion of matrix III-in Chapter 15. Rather than attempting to determine first the value of each technology that is being explored in this category, which is usually almost impossible in the early stages of exploration, it is better for the corporation to determine how much it can afford to spend on this area of technology. Choosing the individual programs to be explored with this allocation of funding is better left to the decision of the technical community or a select group of technical and operation management. In the final analysis, the value of these programs can scarcely be determined except in retrospect. If the choices have been wise, the programs will have anticipated the future needs of the company. If the choices have been poor, they will not have been found to be useful. As the quotation from Quinn at the beginning of this chapter stated: "Continuous innovation occurs largely because a few key executives have a broad vision of what their organizations can accomplish for the world and lead their enterprises toward it." This emphasizes the very high premium that must be placed on the capabilities of the key technical officers of the company. In the absence of good insight and intuition, the justifications of many of the long-range programs are likely to succumb to the generation of financial estimates, the values of which will often be found to have little basis in fact.

A great deal of literature exists on a variety of approaches to the evaluation of the research activities of a company. An example is given in Appendix A16.4 of the techniques that 3M utilizes in the review of its R&D programs. It will be noted that the 3M procedures include a number of the items that were included in the creation of the matrices that lead to the technology roadmap. Although this literature merits attention, the discussion here will be limited to a discussion of three rather specific questions. How is the research activity viewed by people outside the company? How is it viewed by people inside the company? Do these views really matter?

In terms of the view by people outside, this is often created by the assessment by peers who are active in the same or similar fields. Publications, participation at technical meetings, participation in technical societies, national recognitions, and awards are all used as evidence of the status of members of the research activity. In a real sense, the view from outside will focus on the quality of the people that are in the organization. If the organization is an attractive place for people of high quality to work, it is reasoned that it is a high quality organization. There is less tendency to examine the details of what the activity is doing for the enterprise. The focus is on the people. An example of this exists with the Bell Telephone Laboratories. The Bell Laboratories have been regarded, justifiably, as one, if not *the*, premier industrial research laboratory in the world. Even though the fraction of the laboratory's effort that has been devoted to basic research has been only a very small part of its total effort, much of the laboratory's reputation has been derived from the many outstanding contributions that they made to basic science and engineering. In this case the focus on the quality of the people and their contributions generated the proper evaluation.

The internal view of the research activities is usually derived from the collective views of people not in the research activity but principally by the people in operations. What are people saying about the activities? Have they helped solve important problems? Has the research group anticipated areas that became important and, therefore been able, to have a program in place that was useful in solving problems? Or is the research group viewed as being haughty, more committed to *their* interests than to the company's problems and more interested in outside recognition than in helping solve inside problems? If management develops a view of the research laboratory that is more the latter than the former, it is unlikely that any attempts at quantifying the benefits of various programs will be of value. Such a situation can be described as the loss of support from its customers. It is as hard for a research activity to regain credibility with its internal customers as it is for a company to regain the trust of customers lost because of low-quality products.

Although some may argue that the view of a research activity by outsiders is largely irrelevant when evaluating the value of the activity, this is not completely valid. The genesis of the argument is again with the people that can be attracted to the research activity. Not only does a high-quality activity require the presence of high-quality people, the converse is also true, in that it is seldom possible for a mediocre activity to attract high quality people, and if they are successful in hiring them the chances of keeping them are low. The conclusion, then, is that favorable recognition by people both inside and outside the research activity is very important if it is to maintain the trust of its customers, the support of its management, and the interest of outstanding people in joining the organization. In many ways it is an oxymoron to consider a mediocre research activity. A mediocre research activity will not be involved in high-quality research and will be simply a problem-solving activity.

16.7 PROTECTION OF TECHNOLOGIES

The basic premise justifying the focus on technology is that it can serve as a competitive advantage even when all competitors have adopted the management practices that are advocated by the first nine foundations of world-class practice. This importance requires that companies consider seriously how they can best retain control of their technologies and limit the access that competitors have to their practices. As noted earlier, knowledge about technology diffuses rapidly once it is known. It is important, therefore, either to control access to the knowledge or to control who can use the technology when it becomes known. Various ways of accomplishing this are discussed below.

16.7.1   Patents

A patent is a contract that is entered into between the inventor(s) and the government, either U.S. or foreign (Kitner and Lahr,1975). This contract provides the inventor with an *exclusive* right to use the ideas claimed in the patent *in exchange* for *complete disclosure* of the information that forms the basis of the patent. In exchange for this right of holding a monopolistic position in the use of the patent, the inventor must provide sufficient information to allow someone who is trained in the art described in the patent to be able to learn to practice the patent. While the owner has exclusive rights to the patent and the material that it contains, full disclosure of information makes it possible for those knowledgeable in the field to determine if they have information that is relative to the patent if applicable, to file applications that are extensions of the original patent or to contest the validity of the patent.

The contract between the inventor and the government-the patent-provides the owner of the patent with exclusive rights for a specific period of time-14 years for a patent that relates to the *design* of any new, original, and ornamental article of manufacture and 17 years for a *utility patent* that relates to a new and useful process, machine, article of manufacture, or material composition. In some cases the life of a patent for a new drug can be extended beyond 17 years, in recognition of the fact that an unusual amount of time is frequently needed to test and obtain approval from the Federal Drug Administration before marketing of a newly developed drug is allowed. Some foreign countries provide lifetimes that vary from the above. In the United States, the protection period has traditionally begun with the date of issuance of the patent.

For an idea or a concept to be patentable, the following conditions must all be met:

- The idea must have utility (i.e., it must be useful and not simply be an idea that requires further research).
- The idea cannot consist exclusively of a formula, a simple rule, a plan of action, a theory, laws of nature, or scientific principles.
- The claims must be nonobvious to someone working in the field, and the content must represent a nontrivial extension of the state of the art.
- The idea must have novelty in that it has not been invented by another person and the invention has not been published, in public use, or on sale for more than one year prior to the filing of an application for the patent.
- Due diligence must have been exercised in the pursuit of the information that formed the basis of the application. This provision precludes an inventor from conceiving of an idea, allowing it to lay fallow for a long period of time, and then submitting it for a patent.

The condition that the information not have been in public use has broad implications. First, the early publication of information that is to be contained in the patent must be carefully controlled. If the information appears in print or is disclosed orally more than one year before the filing of the application for the patent, this will prevent it serving as the basis for the patent. This condition is the source of concern of company officials regarding the early publication of research results. It should be noted that the filing of a thesis constitutes publication unless the thesis is withheld from public access for a period of time. Any public disclosure prior to filing a U.S. patent application will preclude obtaining a patent in a foreign country. Second, if the information has been disclosed to anyone who is not a party to the creation of the information, this can be construed as placing the information into the public domain. Since employees of the company in which the information was created are considered as acting on behalf of the company, discussion is not restricted among fellow employees.

To allow discussions to occur between interested parties who are not employees of the same company, agreements-called confidentiality agreements-are made that preclude the receiver of the information from disclosing it to others. By making the discussion confidential, the information that is being transmitted is not considered to be placed in the public domain. By virtue of this legal agreement, the receiver of the information assumes the responsibility for maintaining its confidentiality. It is this condition that makes large companies reluctant to enter into confidentiality agreements. If one of the employees of a large company accidentally or intentionally discloses information to a noncompany employee that was not covered by the confidentiality agreement, the company may be charged with violating the agreement and may be subject to the payment of damages. In addition, if the disclosed information includes information that is already known by people within the company, legal problems can arise. Although it is common to state that all such information is excluded under the terms of the agreement, it may become necessary for a company to spend time and money proving that it knew certain things prior to the signing of the agreement.

Patents are filed in the names of the inventors. It is critical that only those people who contributed to the invention be parties to the patent. If it is proven that a name that is included as a patent inventor did not contribute significantly to the invention, the patent may be made invalid.

The patent, giving the owner exclusive rights to the use of the information for a specified length of time, represents property that can be sold, exchanged, or bartered in the same way that real property is handled. Many arrangements for accomplishing this are possible. A patent can be sold outright with all rights going to the new owner. A license can be granted that permits the licensee to use the patent, with compensation going to the owner in the form of an "up-front" payment and/or royalties based on each use of the patent. In the case of royalties, it is common for an upper limit to be set on the total royalties that will be paid. A license agreement can provide exclusivity to the licensee for a fixed time or it may be nonexclusive, allowing the owner to license others. If exclusivity is provided the licensee, the licensor will often insist on assurances that the licensee will actively pursue the application of the patent. For license agreements that permit both the owner and the licensee to practice the patent, the owner will often seek the right to use any improvements that may be made in the original patent-the *grant-back* or *license-back* provision. Certain restrictions on the use of the patent are specified by the patent act. For example, the owner may not control the territory or application to which a licensee sells *unpatented* products made with a patented process or machine. On the other hand, the patent act allows the owner of the patent to limit the application for which the patent is used, as long as this applies to the first sale of the product in which it is used. The owner can also specify the geographic area within which the patent can be practiced. Large companies often exchange the rights to use patents, thus allowing each to gain access to desirable technology. In cases where there are a large number of common interests, with both having a sizable number of patents that are of interest to the other, these agreements may take the form of cross-licensing, with both gaining access to the other's body of patents. The complexity of the agreements encourages the use of good legal counsel, both in terms of crafting the agreements and in concluding the nature of the financial terms of the agreement.

The value of a patent can range from zero to very large sums. As an example of the latter, the technology that Pilkington developed for making float glass allowed them to develop a family of patents that has generated enormous income from the licensing of this technology to companies around the world. This has become the premier technology for making plate glass. This is an example of a key patent stimulating additional patents, some of which provided improvements on the original and some of which provided extensions. All of these create a body of knowledge that is considered to be an important part of the intellectual property of the owners and licensees.

Serious conflicts have resulted from the ways in which different countries treat patents and intellectual property in general, with some countries not honoring the laws of other countries, some allowing the patenting of topics that others preclude, and some providing differing lengths of time of protection. The newly created WTO agreements attempt to bring greater consistency into the treatment of patents, with an agreement that a patent will have a life of 20 years after an inventor has filed for a patent, in distinction to the 17 years from the date of issuance, and that all 123 nations that are expected to sign the agreement will undertake to provide a more consistent treatment of patents (*Wall Street Journal*; 1994).

16.7.2   Know-How

Know-how is the term given to the collection of facts and experiences that enables operation of a system or machine. This knowledge represents a valuable asset and may be a marketable commodity. Learning to operate a new system or machine can be difficult and time consuming. Someone who has the *know-how* about the operation may be willing to transmit that information at a cost that is attractive relative to that which a new user would encounter in developing the know-how. In the acquisition of complex systems, it is not uncommon for the purchaser to require that the system be demonstrated to be fully operational- acquisition of a turn-key operation-and that the new owner be fully trained in using the system. The greater the complexity of operation, the greater is the potential value of the know-how. As an example of the value of know-how, Union Carbide has found that it is more profitable to use its know-how in making ethylene by teaching others how to use its technology than it is to be a primary producer of this chemical. The challenge in such cases is to maintain a capability to continue to remain at the cutting edge of the technology, thus maintaining a marketable commodity.

Other forms in which know-how is purchased are quite common. The information that is supplied on troubleshooting of product faults represents a transfer of know-how from the manufacturer to the purchaser of the equipment. Service manuals represent a form of know-how. Consultants are employed to provide their know-how to their customers. Teachers are expected to transmit their know-how to their students.

16.7.3   Trade Secrets

Trade secrets represent information that is critical to the creation of a product that is maintained as a secret by a specific group of employees. They may include a formula, process, computer program, or business or product plan. A trade secret can prove to be a useful way of protecting the technology if that technology is expected to have a life that exceeds the life of any patent that might have offered initial protection. The required full disclosure of the technology in a patent makes this information available to anyone upon expiration of the patent. Of course, a trade secret is effective only as long as it remains a secret. A classic example of the effective use of a trade secret is the detailed formula for the mixture that gives the unique flavor to Coca-Cola.

Employees who move from a company to one of its competitors must be very careful about the information they convey to the new employer. Disclosure of the secrets of the former company to the new employer has often led to legal action being taken against both the former employee and the new employer. Prosecution of one or both have not been uncommon.

16.7.4   Copyrights

A copyright is an enforceable right that is used to prevent unauthorized copying by others of intellectual materials, including literary material, musical scores, and artistic works. It applies to "particular expressions of ideas, but not the ideas themselves." Computer programs can be protected by copyrights. The copyright grants protection for 50 years beyond the life of the creator and 75 years if the holder of the copyright is an institution.

16.7.5   Trademarks

A trademark provides the identify of the source and quality of goods and services. Although this does not protect the technology directly, the value of recognizable trademarks is an important assistance in marketing products. Trademarks must be registered in each country in which exclusive rights are desired. Instances are known of the registration of common trademarks in obscure locations by other than the original owners. When the original owners later sought to introduce their product into that country, they found that they had to pay substantial amounts to those in whose name the trademark was registered originally.

16.7.6   Other Ways of Gaining Protection

The protection gained by the procedures listed in Sections 16.9.1 to 16.9.5 are all dependent, in one or another form, on the protection that the law provides for creators and users of technology. Since it is not always possible to predict the outcome of legal actions, it is worth asking whether there are actions that can be taken that will effectively protect the technology without resorting to the use of the legal procedures outlined above.

The answer to this query is yes. Effectively reducing the ease with which a competitor can use your technology is possible under a variety of circumstances, most of which are under the control of the company.

First, if technology is viewed as a principal element in achieving a competitive position in the marketplace, the company will be active in creating the new technology. If, as Schumpeter would have described the situation, the company is continuously replacing [destroying] the technology that it is using with new technology, a competitor will find it very difficult to remain competitive by trying to copy what is in the marketplace. By the time that the competitor has copied a technology, a new technology will appear. This is clearly the philosophy that is taken by the PC companies with regard to the makers of the PC clones-as soon as a clone appears in the marketplace the original manufacturer announces a major enhancement in technology or an all-new technology. The protection against competitors is created by remaining ahead of the competitor in terms of the technology that is being introduced into the marketplace. Although this philosophy can be very effective, it requires the maintenance of a very talented cadre of technologists, it can be expensive, and it is a philosophy that is unforgiving of mistakes. An incorrect choice in the direction to be pursued or the choice of a technological advancement that proves to be unusually difficult can effectively destroy the lead that the company had over its competitors.

Second, if the technology requires a large investment, this will effectively limit the number of people who can compete. With a pilot line for a new semiconductor process costing roughly $500 million and the production line for a new computer CPU costing approximately $2 billion, the number of companies that can afford to develop and introduce an all new computer is very small. Although financial barriers do not eliminate competition, it is usually easier to compete with a few companies of comparable size than it is to compete with a myriad of small companies, each of which may have a unique approach to a problem that could be difficult for a company to offset.

Third, if neither of the foregoing circumstances prevail, a company may find that it is attractive not to compete in a general market for the products but to identify and become the leader in a niche market. The size and uniqueness of the niche may be such as to be very profitable for the leader but it may be sufficiently limited that others will find it unattractive to attempt to dislodge the leader. For such cases, the risk is that a technology will arise that completely obsoletes the niche or makes it so easy to enter the niche that the original leader is readily displaced.

REFERENCES

DRUCKER, P. 1985. *Innovation and Entrepreneurship: Practice and Principles*, Harper & Row, New York: 28.

EURICH, N. 1985. *Corporate Classrooms: The Learning Business*, Carnegie Foundation for the Advancement of Teaching, Princeton, N.J.

KITNER, E. W. AND J. L. LAHR. 1975. *An Intellectual Property Law Primer*, Macmillan Publishing Company, New York.

KUSCH, P. 1966. *Style and Styles in Research*, Robert A. Welch Foundation Conference on Chemical Research, Nov. 21-23. Houston, Texas.

MANSFIELD, E. 1988. The Speed and Cost of Industrial Innovation Japan and the United States: External versus Internal Technology, *Management Science*, 34 (10), October:1157.

MANSFIELD, E. 1991. Academic Research and Industrial Innovation, *Research Policy*, 20, Feb.1.

MARQUIS, D. G. 1969. The Anatomy of Successful Innovations, *Innovations*, Nov.

MAYO, J. S. 1993. Learning to Stay Ahead, *ChemTech*, July:8.

NATIONAL SCIENCE BOARD. 1993. *Science and Engineering Indicators*, U.S. Government Printing Office, Washington, DC.

ROSENBERG, N. AND W. E. STEINMUELLER. 1988. Why Are Americans Such Poor Imitators? *AEA Papers and Proceedings,* May:229.

QUINN, J. B. 1985. Innovation and Corporate Strategy: Managed Chaos, *Technology in Society*, 7:263.

WALL STREET JOURNAL. 1994. GATT Proviso Stirs Ire of Inventors and Their Backers, Oct,5:B2.

QUESTIONS

16.1 Drucker's book on innovation and entrepreneurship is now a decade old. Would you expect his message to be different if he were writing it today? How can the message concerning the importance of innovation and entrepreneurship be reconciled with the importance of continuous improvement? Can one have more innovation and entrepreneurship in a firm than is optimal? Why? How would you organize to achieve what you believe to be the optimal?

16.2 Analyze the PARC story as told by Smith and Alexander. Summarize the critical decisions that were made in the organization of PARC.

What were the principal reasons for the failure of Xerox to capitalize on the inventions at PARC? Why did they exist? What could have been done to reduce the impact of these flaws? Based on the Xerox experience, how would you organize a research activity that is to be at the cutting edge of a rapidly changing technological field? Would the plan be different for a research activity for a company that is experiencing a lower level of technical activity?

16.3 It was noted in Section16.5 that less than half of the companies with annual sales of over $12.5 billion support significant basic research. What are some of the reasons for this? If you were CEO of one of these companies, would you advocate a greater focus on basic research?

16.4 You have been hired as vice-president of research for an organization that has lost the confidence of its customers. What actions would you take to recover credibility?

READINGS

DRUCKER, P. 1986. *Innovation and Entrepreneurship: Practice and Principles*, Harper & Row, New York.

SMITH, D. K. AND ROBERT C. ALEXANDER. 1988. *Fumbling the Future: How Xerox Invented, Then Ignored, the First Personal Computer*, William Morrow, New York.

APPENDIX A16.1 CONCEPTUAL DIMENSIONS OF TECHNOLOGY DIFFUSION*

The notion of technology diffusion must be taken today to "include adoption by other users as well as more extensive use by the original innovator. More generally it encompasses all those actions at the level of the firm or organization taken to exploit the economic benefits of the innovation" (OECD, 1988a:49).

Thus, diffusion cannot be reduced to the introduction of new machinery onto the factory floor or to the adoption by firms of new intermediate goods. It must include the other active and vital steps taken by firms to adapt technology to their needs and thus increase the economic efficiency with which new technology is utilized. These steps include the reorganization of factory work and materials flows (such as just-in-time production programming) and improved management practices on the shop floor, in production development and in marketing. More generally, the notion of technology diffusion must also include the process whereby knowledge and technical expertise spread through the economy.

Conceptual work on technology diffusion has advanced hand in hand with the changes in the way policy makers regard the process. As a result, a number of useful related notions are emerging. These include the need to distinguish between the cost of adoption and the cost of purchasing and the importance of analyzing diffusion as a competitive selection process between old and new technologies. The recognition that *every act of adoption involves certain transformations and is thus an act of incremental innovation in itself* is central to these new views.

Underlying these issues is a fundamental conceptual distinction between two types of technology diffusion. It is the distinction between *disembodied* and *equipment-embodied* technology diffusion. The former is the process whereby technology and know-how spread through channels other than embodiment in machinery. It originates in the *externalities* that characterize the innovation process and the *research spillovers* that occur when the firm developing a new idea or process cannot fully appropriate the results of its innovation. Equipment-embodied diffusion on the other hand is the process whereby innovations spread in the economy through the purchase of technologically-intense machinery, components and other equipment."

APPENDIX A16.2 LETTER TO A YOUNG RESEARCHER**

When we hire someone at PARC, there is one qualification we consider more important than technical expertise or intellectual brilliance: intuition. A well-honed intuition and the ability to trust it are essential tools for doing the kind of research we do here.

Our approach to research is 'radical' in the sense conveyed by the word's original Greek meaning: 'to the root.' At PARC, we attempt to pose and answer basic questions that can lead to fundamental breakthroughs. Our competitive edge depends on our ability to invent radically new approaches to computing and its uses and then bring these rapidly to market.

This is different from what goes on at most corporate research centers, where the focus is on improving current technology and advancing the status quo. If you take a job somewhere else, when you embark on a project you will probably have a pretty good idea of how and when your work will pay off. The problems you address will be well defined. You will help to improve computer technology state of the art by going one step farther along a well-plotted path.

If you come to work here, there will be no plotted path. The problems you work on will be ones you help to invent. When you embark on a project, you will have to be prepared to go in directions you couldn't have predicted at the outset. You will be challenged to take risks and to give up cherished methods or beliefs in order to find new approaches. You will encounter periods of deep uncertainty and frustration when it will seem that your efforts are leading nowhere.

That's why following your instinct is so important. Only by having deep intuitions, begin able to trust them, and knowing how to run with them will you be able to keep your bearings and guide yourself through uncharted territory. The ability to do research that gets to the root is what separates merely good researchers from world-class ones. The former are reacting to a predictable future; the latter are enacting a qualitatively new one.

Another characteristic we look for in our research staff is a commitment to solving real problems in the real world. Our focus is on technology in use, and people here are passionate about seeing their ideas embedded in products that shape the way people work, think, interact, and create.

At Xerox, both corporate executives and research scientists are strongly committed to making research pay off. Over the last few years, new channels of dialogue have opened between research and other parts of the company. In particular, corporate strategy and research shape and inform each other. PARC's strategic role will undoubtedly be further strengthened by the emergence of digital copying and the company's new focus on documents of all kinds, whether in digital or paper form. The fusion of two previously separate Xerox businesses-information systems and copying-means that the company will be able to capitalize on PARC's expertise in ways it has been unable to do in the past.

This is an exciting time to be embarking on a career in systems research. New tools and technologies make it possible to deliver large amounts of computing power to users, and this increase in power opens up possibilities for using computation in new ways.

If you come to work here, you will sacrifice the security of the safe approach in which you can count on arriving at a predictable goal. But you will have an opportunity to express your personal research 'voice' and to help create a future that would not have existed without you.

Sincerely,

John Seely Brown   Frank Squires
Corporate Vice President   Vice President, Research Operations

---

[*] *Source* OECD. 1992. Technology and the Economy: The Key Relationships , OECD, Paris, France:48.

[**] *Source* Brown, J. S. 1991. "Research That Reinvents the Corporation," Harvard Business Review, January-February:105. [*]

APPENDIX A16.3 LEARNING TO STAY AHEAD*

Challenges for R&D in the third millennium are driven by the richness and power of technology. But the push of technology is not enough. The expanding needs of domestic and global customers provide the marketplace pull that must complement that push, because technical feasibility alone can't bring about a technology-based vision of the future. The requirements and challenges are derived from the combined forces of technology and the marketplace.

What do these requirements and challenges demand? They demand that an R&D community constantly enhance its effectiveness in generating new technology. They also demand that it accelerate the translation of that technology into new products and services and pursue ever-more effective research, development, and technology conversion.

"The first element that's needed is a sense of mission, a stability of direction, and a constancy of purpose. Research people have always needed a sense of mission, but today's dynamic environment makes it difficult to maintain. Leaders have to work to see that the mission is clear at all times.

At the heart of competitiveness is closeness to customers, both internal and external. Because of its vital link to customer satisfaction, quality must ultimately be defined by the customer. Moreover, in my industry, alignment of R&D with business requires that we focus on customers.

But the researchers and developers must retain their sense of professional community. That sense is characterized by an ability to affect one's own destiny, by meaningful mechanisms for coupling within the community, and by managing force imbalances across the community.

There must also be an enhanced synergy between personal goals and company needs. This should be easy, for these people are generating the technology for exciting new products and services that enrich our society. Most of us delight in doing such useful things.

Quality of technical supervision is another factor that has contributed heavily to R&D excellence. A combination of excellent technical skills and people skills, as well as some business common sense, is required.

The quality of the reward system, especially merit, compensation, and promotion, is yet another factor. This includes the opportunity to rise either by excelling in one's function or by broadening into other functions. Opportunities should be available for self-development as well.

Whatever the environment, the prowess of an R&D organization is closely tied to its mechanisms for intergroup learning or organizational learning, and especially for on-the-job learning in groups of more-than-critical mass that contain many stimulating co-workers.

Finally, we must ensure that all employees have challenging work, that they have access to needed information, that they have the tools they need, and that they are utilized to the limit of their personal capabilities.

Above all, a first-rate R&D community must be characterized by the excellence of people who are unmatched in their capabilities and focused on meaningful problems. Extraordinary skills must deliver extraordinary output. But individual skills are not sufficient. There also must be a high quality of teamwork.

Organizational learning is the collective learning among the disciplines within an organization. Peter Senge of MIT's Sloan School of Management defines a learning organization as 'a group of people continually enhancing their capacity to create what they want to create.' And this, of course, is central to competitive innovativeness. In the global arena, the organization that learns the fastest will eventually win. If you can't apply it competitively, you haven't learned it.

There are both informal and formal approaches to organizational learning. Organizational learning aimed at establishing a basic infrastructure capability is usually done informally either from the top down or from the bottom up. Bell Labs' thrust in software research and engineering, for example, started top-down with a broad commitment initiated by company management. That thrust led to deployment of a large and growing group of researchers.

Once a group is deployed, the informal organization learning is usually bottom up....

Another, somewhat different example of formal organizational learning is the increasingly pervasive use of what we call Best Current Practices. These attempt to promote tried and proven approaches-for example, for software development-that are deemed to be the best in class by people inside and outside AT&T. Each best practice represents a systematic or structured method for approaching one aspect of the development process. The idea of using them is based on the notion that great benefits come from analyzing successful projects in order to understand and apply whatever is being done right. I must stress that we do not view these practices as a substitute for the creative and innovative processes so central to an R&D organization. Instead, we see them as a supplement to these processes to provide proven guidance without damaging creativity and innovation. Most important, Best Current Practices provide broad, flexible guidelines, not rigid commandments.

Organizational learning should not be legislated, but rather enabled. It's enabled, for instance, by making successful examples available to a critical mass of technical talent, along with opportunities for self-development. Researchers and developers must be encouraged to expand their personal knowledge bases. They need not be forced to do it in a structured manner. It might be as simple as setting minimum requirements on hours per year spent in formal education and training courses.

Whatever curriculum technical graduates have absorbed, rapid changes in technology will quickly discount the more applied portions. In fact, the vast majority of engineering jobs deal with subjects that inevitably change and disappear. And the engineer who does not continuously renew his or her learning will eventually become a non-engineer. It is impossible to overemphasize how important lifelong education is in the engineering profession. And effective R&D management must help ensure the availability of such education...."

*Source: Mayo, J.S. July, 1993 Learning to Stay Ahead: Organizational Learning is the Key to R&D Excellence in the Third Millennium. *ChemTech* July:8.

APPENDIX A16.4 HOW 3M EVALUATES ITS R&D PROGRAMS*

Nearly 25 years after its initiation, 3M's technical audit system continues to provide regular internal peer reviews of laboratory activities. By this mechanism, major research programs are assessed, the overall health of laboratories is evaluated and recommendations on product programs, technology areas and laboratory issues are provided to management.

A director of Corporate Technical Planning and Coordination (CTP&C) administers 3M's technical audits with the aid of a full-time staff which includes managerial, technical analyst and secretarial members. The atmosphere and attitudes surrounding the audits are recognized as key to their acceptance and success. CTP&C strives for a constructive and supportive approach that will truly benefit the units being reviewed. Throughout the process, an emphasis is placed on honest and open exchange. In addition, CTP&C tries to remain flexible, allowing modifications in the audit procedure when this will make it more meaningful to an individual laboratory.

From the beginning, ground rules were established to ensure that the technical audit process would not stifle 3M's valuable innovative climate. These emphasized the preservation of individual autonomy. For example, implementation of audit recommendations is not mandatory (i.e., a low rating does not result in the automatic termination of a program). Rather, the audit is positioned as a tool to help the operating unit plan and implement programs and allocate resources. Also, to shelter embryonic efforts, small programs and ideas which are still in the early stages are exempted from the audit. In addition, auditors are not forced to arrive at a consensus opinion, but are free to make independent program ratings, which are averaged later. Thus, the audit system is compatible with a 3M environment which nurtures entrepreneurs. However, at the same time, the considerable attention 3M's upper management gives to audit results provides a positive endorsement that is critically important to the system's acceptance.

[S]ix to seven weeks before the audit is to be held, a CTP&C manager meets with the laboratory head to select the programs to be evaluated. These include: 1) major new product programs (i.e., those on which significant investments are being made and/or those projected to have the largest sales impact); 2) technology building efforts; 3) process development and cost savings programs; 4) product maintenance work (e.g., incremental product improvements). Since the results of sales service and production service efforts are readily measured by more traditional means, these types of laboratory programs are excluded from this review.

The actual audit is conducted as an all-day meeting at which members of the laboratory being reviewed make presentations describing their programs. These oral descriptions are supplemented by detailed written materials which the laboratory has provided prior to the audit. A typical team of auditors includes invited members from 3M's laboratory management and senior scientist ranks (external auditors), plus management and technical personnel from within the unit being audited (internal auditors)....With regard to product programs, it has proven valuable for the following areas to be examined:

*Source: Krogh, L. C., J. H. Prager, D. P. Sorensen, and J. D. Tomlinson. 1988. How 3M Evaluates Its R&D Programs, *Research Technology Review*, 31 (6):10.

Technical factors

- Overall technology strength (breadth, patentability, competitiveness)
- Personnel (numbers, skills)
- Competitive factors (knowledge of competition, 3M product performance)
- Remaining R&D Investment (vs. time to complete the program)
- Manufacturing implementation (feasibility, cost, protectability)
- Probability of technical success

Business Factors

- Financial potential (sales, profits)
- 3M Competitive position (marketing channels, product value)
- Probability of marketing success

In terms of the laboratory as a whole, the factors that are assessed include:

Overall factors

- Organization/planning (strategy, focus, clarity of goals)
- Staffing (numbers, skills)
- Program balance (product maintenance vs. related/unrelated new product efforts)
- Coordination/interaction (with marketing, manufacturing, other 3M labs)

The auditors are asked to provide both numerical ratings and essay comments in each of the above areas. Further evaluative information is gathered in a breakfast meeting held a day or two following the audit. This get-together, which is attended only by the external auditors, the lab head being reviewed and the CTP&C staff, is designed as a highly informal, free-wheeling discussion where participants can openly share concerns and ideas.

Working with a rich data base containing nearly 25 years of information, the CTP&C Department has spent considerable time studying auditors' ratings vs. the actual results of 3M's product programs. It has been found that the overall probability of program success, obtained by multiplying the probability of technical success by the probability of marketing success, correctly predicts a program's outcome a high percentage of the time....A low audit probability of success rating correlates with failed (i.e., abandoned) programs while a significantly higher audit rating is typical of those that have actually succeeded in the marketplace (i.e., met certain 3M financial criteria).

3M's program success rate has also been analyzed by type of laboratory effort (i.e., maintenance of existing business, related new business, and unrelated new business). Although unrelated programs can have a very high payoff, they are also more difficult, having an actual success rate that is less than half that of related programs and only one-fifth of what maintenance programs achieve....A correlation exists between the success rate of a related new product program and its estimated sales potential at maturity. For related new product programs rated as likely to succeed, 3M operating units have actually succeeded more often on those aimed at larger targets. One possible reason for this is that the smaller targets are perceived as less attractive and, therefore, lack strong commitment on the part of both researchers and management.

Ongoing studies of related new product programs have also shown that, in addition to probability of success and sales, the following factors have a statistically significant impact on R&D success or failure.

- Competitive position of the 3M business unit developing the product
- 3M product performance vs. competition
- Degree to which the technology is related to 3M's existing technical base
- Degree to which the market is related to 3M 's existing business base

[A]s far as unrelated new product programs are concerned, two factors in addition to probability of success have been found to be most significant in affecting success or failure

- Uniqueness or "newness to the world" of the product being developed;
- Competitive position [in the closest industry] of the 3M business unit developing the product....

Index